Nadleh Whut'enne 'ink'ez Stella Whut'enne Hubughunek

NADLEH AND STELLA DAKELH DICTIONARY

Nadleh Whut'enne 'ink'ez Stella Whut'enne Hubughunek
Nadleh and Stella Dakelh Dictionary

February 2023

©2023 Nadleh Whut'en First Nation and Stella First Nation

ISBN: 978-1-989940-88-4

Nadleh and Stella
DAKELH DICTIONARY

NADLEH WHUT'EN INDIAN BAND
Box 36, Fort Fraser, BC V0J 1N0
Tel: (250) 690-7211
Fax: (250) 690-7316
www.nadleh.ca

STELLA FIRST NATION
Box 760, Fraser Lake, BC V0J 1S0
Tel: (250) 699-8747
Fax: (250) 699-6430
www.stellaten.ca

NADLEH WHUT'EN DEDICATION

On behalf of Nadleh Council, we would like to give recognition to the Nadleh Stella Dakelh Language Working Group Committee who work tirelessly in developing our Dakelh Dictionary.

Our language is our identity, it connects our communities and our Dakelh language is tied to our traditional land. We know who we are and where we come from.

We are preserving our cultural heritage, traditional values and the valuable knowledge for generations yet to be born.

Larry Nooski, Chief Nadleh Whut'en – May 2019

Stella and Nadleh First Nations continue to revitalize and maintain our language and culture. Our communities have invested and supported language revitalization planning, advocacy and immersion for almost a decade. We have been able to accomplish the mapping and developing of a dictionary. Access to the language supports and tools is critically important in reaching our goal of building the fluency of our language to our children and adult learners.

The positive development of these important tools was accomplished with the assistance of our Knowledge Keepers and valuable Elders input and guidance. We acknowledge both of our Communities' Elders who also contributed their time, efforts and energy. We also thank both the Stellat'en and Nadleh Language Committees for their assistance and coordination.

We dedicate these new investments, progress and achievement to our relations and ancestors that dreamed of this day. There is still more work to do to get to a place where there will be a smooth inter-generational transfer of our language and culture.

Chief Robert Michell, Stella Whut'en First Nation – February 2023

Lhulh yats'oolhduk, Dakelh k'una.

Recently, the BC First Nations Cultural Council announced an increase in language learners and immersion opportunities according to a long term study completed in mid-February 2023. Apparently, there are 3000 new learners in all of Indigenous BC of the 34 distinct languages and that this was completed by many champions and Elders.

The growth of the positive development is because of the availability of new tools from on line systems to immersion program with mentors. Although, there is still much more work to be done to undo the damage from the discriminatory past and not so long ago history, our communities are making progress and celebrate our achievements.

Let us converse with one another and honour our ancestors. Since time immemorial we have survived and endured the hardship of colonialism, contact and assimilation. The hard work for those who sat on the committee from the beginning and current members we owe a debt of gratitude. Our identity, language and culture are tied to the land, this we shall not forget. This is part of saving our knowledge based on culture and heritage values.

Chief Archie Patrick, Stella Whut'en First Nation – July 2019

The Nadleh Whut'en Council expresses profound appreciation for the unwavering dedication of Nadleh and Stellat'en fluent speakers, whose remarkable efforts have culminated in the creation of the Dakelh Dictionary. This monumental endeavor is a testament to the commitment of these individuals to preserving and revitalizing the Dakelh language, a linguistic embodiment of our identity, inherent rights, and profound relationships that govern Yinka Dene communities.

The Dakelh language serves as a conduit to our cultural heritage, connecting us intrinsically with nature and embodying the wisdom by which our ancestors responsibly managed the environment throughout time immemorial. Our oral ancestral knowledge, beliefs, historical experiences, and teachings are the keystones that distinguish us and provide invaluable insights into our existence on the land from which we originated. This reservoir of wisdom holds immeasurable value for the continual progress of future generations yet to grace our communities.

This expression of gratitude encapsulates the importance of celebrating and acknowledging the tireless endeavors of individuals and communities dedicated to the preservation and revitalization of indigenous languages and cultures. The Nadleh Whut'en Council extends its deepest appreciation to those whose efforts ensure the enduring legacy of the Dakelh language.

Chief Martin Louie,

Councillor Roy Nooski, Councillor Mark Lacerte, Councillor Eleanor Nooski, Councillor Ashley Heathcliff, Councillor Damien Ketlo.

– February 2024

ACKNOWLEDGEMENTS

The Nadleh Whut'en and Stella Whut'en would like to thank and acknowledge the following for their contributions to the development and publication of this dictionary.

Stella Whut'en First Nation Chief and Council

Nadleh Whut'en First Nation Chief and Council

Nadleh/Stella Dakelh Language Committee Members past and present

Stella Whut'en Language Writer:
Dennis Patrick

Nadleh Whut'en Language Coordinator:
Eleanor Lowe

Nadleh Stella Dakelh Dictionary Past Contributors:

Emma Baker
Robert Luggi Sr.
Andrew Casimel
Mary George
George George Sr.
Roy Nooski
Adeline Nadon
Andrew Louie †
Azeline Patrick †
Archie Patrick
Bernadette Ketlo †
Fred George †
Josephine Carter
Mary Casimel
Virginia Reynolds
Gary Michell
John Casimel
Zaa Louie
Mary Twist
Bruce Allen
Violet Kennedy
Lillian Louis
Amy Williams
Rose Luggi
Kenny Luggi
Martina Stehr
Theresa Luggi
Peter John
Bernadette McQuary †
Teddy George †
Peter George †
Tommy George †
Matthew Ketlo †
Norma Ketlo †
Edgar Ketlo †
Leah Patrick †
Robert Louie †
Margaret Nooski †
Ida George †
John Ketlo Sr. †
Ernie Nooski †
Kenny Nooski
Joseph Louie
Bev Ketlo (typist)
Marlene "Rita" Morin
Martin Louie
Nellie Nooskie
Sally Adolph
Leonie Spurr
Allen "Shaboom" Louie

Compiler:

William J. Poser

Over the years, dictionary work in our communities has received financial support from Lonnie Hindle and PGNAETA and from the First People's Cultural Foundation and from the First Nations Health Authority.

CONTENTS

ABOUT THIS DICTIONARY

This draft lexicon contains material from Dakelh as it is spoken by the Nadleh and Stella Whut'en people. It began with about 1200 words collected over the years by Bill Poser, to which were added several hundred words, mostly body parts and kinship terms, compiled by George George, Sr. in 2003 and hundreds of words compiled by Emma Baker in 2015. More recently, beginning in 2017, a group of elders assisted by Dennis Patrick and Eleanor Lowe made further additions and revisions.

There are a total of 4050 entries in the Dakelh-to-English section and 4509 entries in the English-to-Dakelh section. 135 roots are listed in the Root List. 269 stems are listed in the Stem List. The dictionary contains 106 place names. Although the place names listed here are in general believed to be accurate (a few are marked as uncertain), the inclusion of these place names should not be taken as the official position of Nadleh Whut'en, Stella Whut'en, or the Carrier Sekani Tribal Council for legal or diplomatic purposes.

It is hoped that the lists of stems and roots will prove useful in identifying verb forms not to be found in the main portion of the dictionary. However, as the aspectual system is not well understood, the aspectual categorizations of stems should be regarded as particularly tentative and likely to contain errors.

EXPLANATION OF THE DAKELH WRITING SYSTEM USED IN THIS DICTIONARY

’	Glottal stop	ʔ
a	Low back unrounded vowel	a
b	Unaspirated bilabial stop	b
ch	Aspirated palato-alveolar affricate	tʃ
ch’	Ejective palato-alveolar affricate	tʃ’
d	Unaspirated apico-alveolar stop	d
dl	Unaspirated lateral affricate	dl
dz	Unaspirated apico-alveolar affricate	dz
dz	Unaspirated lamino-dental affricate	dz
e	Mid front unrounded vowel	e
f	Voiceless labio-dental fricative	f
g	Unaspirated velar stop	g
gh	Voiced velar fricative	ɣ
gw	Unaspirated labio-velar stop	g^w
h	Voiceless laryngeal glide	h
i	High front unrounded vowel	i
j	Unaspirated palato-alveolar affricate	dʒ
k	Aspirated velar stop	k
k’	Ejective velar stop	k’
kh	Voiceless velar fricative	x
kw	Unaspirated labio-velar stop	k^w
kw’	Ejective labio-velar stop	k'^w
l	Voiced lateral approximant	l
lh	Voiceless lateral fricative	ɬ

m	Bilabial nasal stop	m
n	Alveolar nasal stop	n
ng	Velar nasal stop	ŋ
o	Mid back rounded vowel	o
oo	High back rounded vowel	u
p	Aspirated bilabial stop	p
r	Voiced rhotic approximant	r
s	Voiceless apico-alveolar fricative	s
<u>s</u>	Voiceless lamino-dental fricative	<u>s</u>
sh	Voiceless palatal fricative	ʃ
t	Aspirated apico-alveolar stop	t
t'	Ejective apico-alveolar stop	t'
tl	Aspirated lateral affricate	tl
tl'	Ejective lateral affricate	tl'
ts	Aspirated apico-alveolar affricate	ts
ts'	Ejective apico-alveolar affricate	ts'
<u>ts</u>	Aspirated lamino-dental affricate	<u>ts</u>
<u>ts</u>'	Ejective lamino-dental affricate	<u>ts</u>'
u	Mid central unrounded vowel	ʌ
w	Labio-velar glide	w
wh	Voiceless labio-velar fricative	x^w
y	Voiced palatal glide	j
z	Voiced apico-alveolar fricative	z
<u>z</u>	Voiced lamino-dental fricative	<u>z</u>

A colon (:) indicates that the preceding vowel is long.

GRAMMATICAL INFORMATION

Nouns

This lexicon provides three types of grammatical information about nouns. First, if a noun has different forms for singular, dual, and plural, the non-singular forms are listed. In most cases the non-singular form is a *duo-plural*, meaning that it refers to two or more. In a few cases the *dual*, referring to two, is distinguished from the *plural,* referring to three or more. In general, only nouns referring to people or dogs have distinct non-singular forms.

Second, it indicates whether a noun is *inalienably possessed.* A noun is said to be inalienably possessed if it must have a possessive prefix attached, indicating to whom it belongs. For example, the stem meaning "arm" is *-gan. gan,* however, is not a word that can be used by itself. It is necessary to indicate whose arm it is, e.g. *sgan* "my arm". To refer to an arm without indicating whose it is, it is necessary to say *'ugan* "someone's arm". In general, body parts and kinship terms are inalienably possessed. When a noun is inalienably possessed, it is entered with a preceding hyphen.

Finally, the possessed stem of the noun may be listed in the *poss* field. For many nouns the stem to which the possessive prefixes are attached is not the same as the free form. For example, "fish" is *lhukw* when free, but has the stem *-lukw* when possessed, as in *nelukw* "our fish".

Verbs

The Carrier verb is based upon a *stem*, which is invariably the last syllable of the verb. This is the part that contributes the basic meaning of the verb. For example, the stem of *nuske* "I am going around by boat" is *ke*. The same stem is seen in related forms, such as *nuts'uke* "we are going around by boat". We may associate the meaning "go by boat" with the stem *ke*.

The stem may change according to the tense and mode of the verb, whether it is affirmative or negative, and the aspect. For example, "I am going to go around by boat" is *nuteskelh*. Here the stem is *kelh*. The corresponding negative, meaning "I am not going to go around by boat", is *nuzteskel*. Here the stem is *kel*. To say "we customarily go around by boat" we say *nuts'ukih*. Here the stem is *kih*.

The various stems may be associated with an abstract *root*. For example, we may say that the root meaning "to go around by boat" is *ke*. At least historically, the various stems were derived from a root by adding suffixes. For example, the future affirmative stem *kelh* is derived from the root *ke* by adding the suffix *lh*. The system for deriving stems from roots is complicated, and it is not entirely clear whether it is still functioning today as it once did. At the very least, the root is a useful name for a set of related stems.

This dictionary contains a *Root List* and a *Stem List*. The Root List lists the roots known for the language and explains their meanings. Since there may be more than one root with the same form, roots have subscript numbers. For example, the root meaning "to go around by boat" is listed as ke_2 since there is another root, listed as ke_1, meaning "for two people to sit or be located", as in the form *huzke* "they are sitting".

The Stem List tells you what roots a particular stem is associated with, in which tense/mode, negative, and aspect forms. For example, the stem *kelh* is listed as associated with the root ke_2 in the FA tense/mode/negation combination and the continuous aspect.

If a verb is not listed in the dictionary, you may be able to figure out what it means by using the Stem List and the Root List, together with your knowledge of Carrier grammar and of the meanings of the various prefixes, many of which have their own listings in the dictionary. First, find the stem and look it up in the stem list. Then look up in the root list the root or roots with which your stem is associated. This will give you possibilities for the basic meaning of the verb.

Glosses of verbs use certain conventions in order to specify information that is not straightforwardly translated into English. English does not distinguish the dual (referring to two people) from the plural (referring to more than two people), and does not distinguish singular, dual, and plural in the second person, glosses use numbers to specify this information. Thus, "you (1)" means that the form refers to a single person, "we (2)", means that the form refers to the two of us, and "we (3+)" means that the form refers to the three or more of us. A gloss that uses the term "non-plural" means that the form may refer to either one or two but not to three or more.

Valence prefix-stem combinations shown represent the underlying form, with valence prefix and stem separated by a hyphen. Thus, *najeh* "he is getting better" is represented as *d-yeh,* the underlying form that is converted into the surface *jeh* by the D-Effect rule. Note that Carrier permits the valence prefix sequences *d-lh* and *lh-d.* Listings of non-zero valence prefixes are believed to be accurate; however, some stems listed as having the 0-valence prefix probably have non-overt d-valence prefixes. These are gradually being identified.

Similarly, the verb stem shown may not be precisely as it surfaces in the actual form due to the D-Effect in first person dual subject forms. Thus, the verb stem shown for *'oonidujeh* "we (2) are picking berries" is *-yeh,* which becomes *-jeh* in the first person dual under the influence of the underlying d of the subject prefix *idud-.*

The terms *d-class, n-class,* and *wh-class* refer to categories in the system of noun classification. These refer roughly to things that are stick-like, round, and saliently two- or three-dimensional respectively. For more detailed explanation consult the grammar sketch.

The tenses and modes of verbs are indicated by the following abbreviations in square brackets following the gloss.

[FA]	Future Affirmative
[FN]	Future Negative
[IA]	Imperfective Affirmative
[IN]	Imperfective Negative
[OA]	Optative Affirmative
[ON]	Optative Negative
[PA]	Perfective Affirmative
[PN]	Perfective Negative

HOW TO USE THIS DICTIONARY

1. Finding Words

1.1 Parts of the Dictionary

This dictionary has three main parts. The first two are organized alphabetically, one from Carrier to English, the other from English to Carrier. The third is organized by topic. This is useful for reviewing vocabulary for a particular purpose but also for finding out how to say something in Carrier. If you cannot find what you want in the English to Carrier section, you may be able to find it in the topical index.

1.2 Alphabetical Order

The Carrier entries are organized alphabetically. In most respects the order is the same as that used in English, but there are some differences. Where a cluster of two or three letters is used to write a single Carrier sound, that group of letters is treated as a unit and alphabetized separately. For example, since *lh* represents a single sound, it is ordered after *l*. Thus, *lhukw* comes after *lumudak* even though *h* comes before *u*. The complete alphabetical order used is listed on page 29.

The first entry on a page is shown at the upper left of the page; the last entry on the page at the upper right. In the following example, the first entry is *lhuki*, the last *Melijayah*. Note that the end of the last entry from the previous page spills over on to the page shown. The word listed at upper left is the first word whose entry starts on that page.

mai *N* berries.

lhuki *V* [l-ki] it (generic) is sweet. *[IA]*

lhuk'i *V* [0-k'i] it tastes good, it is sweet. *[IA]*

lhulcho *ADV* halfway, half.

lhulcho dezbun *V* [0-bun < bun_1] it is half full. *[IA]*

lhulh *PPC* together with each other. (1) **Lhulh** 'uhut'en. 'They are working together.'

lhulhcho *N* triplets.

lhulhtuske *N* sisters, female first cousins. See also: -ulhtus.

lhum *N* a piece of ice.

lhuna *N* captive.

lhuschuz *N* knot.

lhustsis *N* ash.

lhustsis bekuk *N* ash tray.

lhut *N* smoke.

lhut be'hanajut *N* chimney.

lhuyul *V* [l-yul < yul_1] it (generic) is white. *[IA]*

lhuz *N* urine, pee. Poss: -luz.

-lhuzkah *N* urinary tract.

lhuzk'et *N* urinary tract.

-lhuzzus *N* bladder (urinary).

madus *N* mattress.

madli *V* [0-dli] he/she is foolish. *[IA]*

mahudli *V* [0-dli] they are foolish. *[IA]*

mai *N* berries.

mai be'ulha *N* jam.

mai nesdliz *N* jam.

mai nezgi *N* dried berries.

mai nudutleh *N* jam.

mai nuhudleh *N* the distribution of fruit at a balhats.

maitoo *N* fruit juice.

maitoo dot'en *V* [d-'en < $'en_1$] it is purple. Literally, "it is the colour of berry juice". *[IA]*

maitlus *N* fruit leather.

maiyun *N* a song sung during the distribution of fruit at a balhats. Never sung at a funeral balhats.

mandah *N* canvas. Etymology: Loan from Spanish manta "coarse cotton cloth" introduced by Mexican pack-train men.

masdli *V* [0-dli] I am foolish. *[IA]*

mba *N* your (1) father.

mbalh'i *V* [lh-'i < $'i_1$] I am waiting for you (1). *[IA]*

mbat *N* your (1) mitts.

mbe *QWH* who?. (1) **Mbe** nyuts'ulhni? 'What is your name?'

mbez *N* your (1) mother-in-law.

mboo' *N* your (1) rectum.

mboos *N* your (1) cat.

mbun *N* your (1) roof.

mbunt'ah *N* your (1) attic.

mbut *N* your (1) stomach.

mbuttl'ah *N* the bottom of your (1) stomach.

me *N* **omen.**

meljah *N* white person. Etymology: Said to be a loan from Chinook Jargon, but Chinook Jargon source is unknown.

Meljayah *N* Stellakoh village. An old

1.3 Spelling Variation

One difficulty in looking up words is variation in spelling. Here are a number of things to look for:

- **a.** The difference between the "fronted" consonants *s, z, ts, dz, ts'* and their plain counterparts *s, z, ts, dz, ts'* is difficult for many people to hear. Even fluent native speakers of the language often get this wrong. Therefore, you should consider the possibility that a word written with a plain consonant will be found in the dictionary with the fronted equivalent, as well as the possibility that a word written with a fronted consonant will be found in the dictionary with the plain equivalent. If you hear a word and have difficulty telling plain and fronted consonants apart, you should look under both.
- **b.** The plain consonants *ts* and *ts* and their glottalized counterparts *ts'* and *ts'* are often confused. If you do not find a word with one, it may be listed under the other.
- **c.** The diphthong *ui* that occurs in words like *k'ui* "birch tree" is often written *i.*
- **d.** The apostrophe that marks glottalization is sometimes written before the consonant that it goes with rather than after. For example, you will see *Saik'uz* spelled *Sai'kuz.*
- **e.** The apostrophe that marks glottalization is sometimes written after the first of the two consonants of a digraph rather than after the second as is standard. For example, you may see *ts'* written *t's* and *kw'* written *k'w.*
- **f.** The glottal stop is frequently omitted, especially by non-native speakers, for whom it is difficult to hear. Thus, if you see a word with no consonant preceding a vowel, you should consider the possibility that a glottal stop is missing. For example, a word spelled *a* may really be *'a.*
- **g.** Many people have difficulty hearing syllable-final h, so it may be left out when it should be written or written when it should not be.
- **h.** When next to a glottal stop or *h,* the *u* sound may become similar enough to an *a* that some people will hear it and write it as *a.* Words with an *a* in this context may sometimes be found in the dictionary with a *u.*

i. English speakers unfamiliar with the sounds of Carrier tend to hear *g* for *k'*, *d* for *t'*, and *kl* or *gl* for *tl'* and syllable-initial *lh*. If you are trying to look up a word such as a placename written down by someone who does not know Carrier, you should be on the lookout for such mistakes. For example, the *Takla* of *Takla Lake* reflects Carrier *tatl'ah.*

j. English speakers unfamiliar with the sounds of Carrier are likely to write syllable-final *lh* as *thl* or *th.*

You should also be aware that written material may be in another writing system. The Carrier Linguistic Committee writing system used in this dictionary is the writing system in common use today and the only one in which much material in this dialect has been written, but you may encounter another writing system, the Morice system, especially if you read scholarly work about Carrier language and culture.

This is the system used by Father Adrien-Gabriel Morice in his many scholarly publications. Since his work is widely read, his spellings of Carrier words have also found their way into other people's writing. The 1938 edition of the Roman Catholic prayerbook also uses this writing system. If you need to read material in this writing system, you can obtain an explanation of it from the Yinka Dene Language Institute website.

1.4 Forms Not Listed as Such

The greatest difficulty in using this dictionary (and other dictionaries of Athabascan languages) is that many words are not listed in the dictionary as such. This situation arises because word-formation in Carrier is so productive: many words can occur in many forms. In the case of verbs, a single verb root may underlie hundreds of thousands of forms. Of course, it is easier to find a word if it is listed as such, but it is often possible to find a word even if it is not listed in exactly the form that you are trying to look up.

1.4.1 Nouns

Nouns take on a number of different grammatical forms. In general, these forms are not listed. Only the basic form is listed. If you encounter the inflected form of a noun, you must be prepared to figure out the basic form. In possessed nouns a prefix precedes the noun stem. The prefix must be removed to obtain the stem. For example, to look up the word *hubuludab,* the prefix *hubu* must be removed to obtain the stem *ludab. ludab* is listed in the dictionary, so you will find out that it means "table". To work out that *hubuludab* means "their table", you must also know that *hubu* is the third person duo-plural possessive prefix. The section on grammatical information provides an outline of the most important aspects of Carrier grammar. Many affixes are listed in the dictionary. If you suspect that something might be an affix but are not sure, you can also try looking for it in the list of affixes. Some possessed nouns differ from their unpossessed counterparts not only in having a possessive prefix but in changing their stem. Since such stem changes are no longer completely systematic, many possessed stems are listed separately with a cross-reference to the main entry. You should, however, familiarize yourself with the common stem changes.

Nouns referring to people also have plural forms. Irregular plurals are listed, but regular plurals are not. Regular plurals ending in *ne* and *ke* should be recognized and the plural suffix removed in order to look up the noun.

The suffixes *yaz* "small" and *cho* "large" are very common and may be added to just about any noun. Most such forms are not listed separately. If you encounter a noun ending in *yaz* "small" or *cho* "large", it may be listed under the form obtained by removing the suffix.

1.4.2 Postposition Prefix Complexes

Many postpositions cannot follow pronouns but instead take prefixes that mark their object. An example is the postposition *ba* "for". It is not possible to say **si ba* "for me"; instead, one says *sba.* Although their meaning is different, the inflected forms of postpositions look very much like possessed nouns. These combinations of a postposition and a pronominal prefix are assigned the part of speech *Postposition Prefix Complex,* abbreviated *PPC.* Only irregular PPCs are listed as such. Otherwise, only the postposition itself is listed. The reader is expected to be able to recognize that a form like *sba* might consist of the postposition *ba* with first person singular object and to look it up under *ba.*

1.4.3 Verbs

If a verb is not listed in the dictionary, you may be able to figure out what it means by using the Stem List and the Root List, together with your knowledge of Carrier grammar and of the meanings of the various prefixes, many of which have their own listings in the dictionary. First, find the stem and look it up in the stem list. Then look up in the root list the root or roots with which your stem is associated. This will give you possibilities for the basic meaning of the verb. You can then work out the full meaning of the verb by combining the basic meaning with the meanings of the various prefixes.

For example, suppose that you encounter the verb *nuhikel.* The Stem List identifies *kel* as the the Perfective Negative and Future Negative stem of the root ke_2 "to go by boat". This form is obviously not a Future Negative form as it does not contain the *t* prefix that appears in all future forms. It is, however, a possible Perfective Negative form, with the *h* the third person plural subject prefix. The initial *n* might be the prefix meaning "in a loop, around", with the following *u* inserted by rule to separate it from the consonant *h.* Thus, we arrive at the correct interpretation: "they did not go around by boat".

One difficulty in analyzing a verb form is determining its stem. To a first approximation, the stem is the last syllable, which will consist of a consonant, a vowel, and possibly another consonant. For example, the last syllable of *nuteskelh* is *kelh,* which is also the stem. Unfortunately, the stem is not always the same as the last syllable.

One situation in which the stem is different from the last syllable is in D-Effect contexts. When the D-Effect applies, the final *d* of a prefix combines with the initial consonant of the verb stem. The first consonant of the last syllable, therefore, is not the same as the first consonant of the stem. To find the stem, we must undo the D-Effect. For example, if the verb form is *'ut'en* "he is working", there are two possibilities. This could be a verb with stem *t'en* and no valence prefix, or it could be a verb with stem *'en* and valence prefix *d.* In this case, the latter is correct.

Several other phonological rules complicate the relationship between the last syllable and the stem. If the stem appears to begin with *s*, it probably really begins with *ts*, where the *t* has dropped out due to a preceding *s*. In *ussi* "I am dirty", for example, the stem is actually *tsi*. Another such case is the rule that takes *tl* to *lh* following *lh*. For example, the true stem of "to knead" is *tlus*, but in *nelhlhus* "you kneaded" it shows up as *lhus* because of the preceding valence prefix *lh*.

2. Understanding Entries

Some entries are very simple, while others contain a great deal of information. The simplest entries consist of the headword, an abbreviation for the part of speech, and a gloss, that is, a short definition, as in the following example.

Many entries contain one or more examples. Examples are numbered and followed by an English translation. The headword is highlighted in the example.

nayulh'ool *V* [lh-'ool] he/she is braiding it. *[IA]* (1) Dutsigha **nayulh'ool**. 'He or she is braiding her hair.'

Entries beginning with a hyphen indicate that the word cannot appear by itself but must be preceded by a prefix, if the word is a stem, or by a stem, if it is a suffix. Entries whose headword ends in a hyphen are prefixes that must be followed by a stem.

Entries for verbs are more complex. Between the part of speech and the gloss is information about the valence prefix, verb stem, and verb root. The valence prefix is separated from the stem by a hyphen. The stem is separated from the root by a "less than" sign, indicating that the stem is a form of the root. Roots may have subscript numbers. These are not pronounced but serve to distinguish stems and roots with the same pronunciation.

The following example indicates that the valence prefix is *lh*, the stem *tuk*, and that this stem is derived from the root tuk_1. It is not a mistake that the valence prefix is given *lh* while the consonant preceding the stem is *l:* the valence prefix is given in its underlying form. In this case, the underlying form *lh* is realized *l* because it follows the first person dual subject marker, whose underlying form is *idud.* (See the discussion of the D-Effect in the section on Grammatical Information for further information.)

> **yaidulduk** *V* [lh-duk < duk_1] we (2) speak. *[IA]*

Entries for verbs also include information about Tense/Mode/Negation, in square brackets following the gloss. In the example, the Tense/Mode/Negation information is *IA,* indicating that this verb form is in the Imperfective Affirmative. The full set of categories and abbreviations is given in the section on Grammatical Information.

The Tense/Mode/Negation information may be followed by a verb's aspect. In the following example, the verb is in the Customary aspect of the Imperfective Affirmative.

> **hawhulhjuk** *V* [l-lhuk] it rots out. *[IA–customary]* (1) Hubutsul **hawhulhjuk.** 'Their anuses rot out.'

Entries for classificatory verbs indicate that the verb is a classificatory verb and specify the class. The classificatory information may also include information about absolutive classification. In the following example, the verb is a classificatory verb for handling long rigid objects that is also marked for a *d*-class absolutive argument.

> **sghadutetilh** *V* [0-tilh < tan_1] he/she will give me (st). [abs: d] [class: lro] *[FA]* (1) Tuz **sghadutetilh.** 'He is going to give me a cane.'

Some entries contain miscellaneous other pieces of information that should be readily recognizable. The entries for some placenames contain the latitude and longitude of the place. Quite a few entries give the etymology of the word, that is, an explanation of the word's origin. Some entries contain a more detailed discussion of the meaning of the word in addition to the gloss, a note on the cultural significance of the word, or an explanation of the grammar needed to use the word.

Glosses of verbs use certain conventions in order to specify information that is not straightforwardly translated into English. English does not distinguish the dual (referring to two people) from the plural (referring to more than two people), and does not distinguish singular, dual, and plural in the second person, glosses use numbers to specify this information. Thus, "you (1)" means that the form refers to a single person, "we (2)", means that the form refers to the two of us, and "we (3+)" means that the form refers to the three or more of us. A gloss that uses the term "non-plural" means that the form may refer to either one or two but not to three or more.

The abbreviation *uo* stands for "unspecified object" and is used to mark transitive verbs in the form used without an overt noun phrase object. For example, *'usyi* is an "unspecified object" form meaning "I am eating" and is used without a noun phrase indicating what is being eaten. It can form a sentence all by itself. The corresponding specified object form is *usyi.* This cannot form a sentence by itself, but must be used together with a noun phrase indicating what is being eaten, as in the sentence *lhes usyi* "I am eating bread."

Glossing the optative form of verbs poses a particular problem since the optative does not have any single English translation. Optatives are therefore glossed as if they were not optatives, with the information that the form is optative supplied by the *[OA]* or *[ON]* indicating the verb's Tense/Mode/Negation category. Thus, even though the most likely meaning of *'uts'ooyi* is "let us (three or more) eat (something)", it is glossed "we (3+) may eat (u.o.)", because it might be used in a variety of other contexts where it would not have this meaning, such as *'aw gak 'uts'ooyi ghait'ah* "we can't eat anything at all".

The abbreviation *q.v.* stands for the Latin phrase *quod vide* "which (you should) see" and is used for cross references. In the example below, we are told that *chilhukah* is an irregular plural of the noun whose singular is *chilh* and that for further information, we should look at the entry for *chilh.*

chilhukah *N* Irregular duo-plural of **chilh**, q.v.

Alphabetical Order

a	n
b	o
ch	oo
ch’	p
d	r
dl	s
dz	s̪
d̪z̪	sh
e	t
f	t’
g	tl
gh	tl’
gw	ts
h	ts’
i	t̪s̪
j	t̪s̪’
k	u
k’	v
kh	w
kw	wh
kw’	y
l̲	z
lh	z̪
m̲	

Grammatical Category Abbreviations

ADJ	Adjective
ADV	Adverb
CONJ	Conjunction
INT	Interjection
N	Noun
NP	Noun Phrase
NUM	Number
PART	Particle
PP	Postposition
PPC	Pronoun Postposition Complex
PPP	Postpositional Phrase
Q	Quantifier
RC	Relative Cause
V	Verb
VP	Verb Phrase

DAKELH TO ENGLISH

'a *ADV* fast, quickly.

'a *N* fog. (1) ‘’

'a be'ut'es *N* microwave oven.

'a buzkai tujulh *V* [0-julh] he/she is quick to anger. *[IA]* Etymology: “his/her blood gushes out quickly”.

'a naints'ut *V* [0-ts'ut] it is foggy. *[IA]*

'a'alh *V* [0-'alh < 'alh$_1$] he/she chews (u.o.). *[IA*–customary] (1) Yuba **'a'alh.** ‘He chews it for her.’

'a'antal *V* [0-tal < tulh$_1$] he/she kicked into a hole. *[PA]* (1) Yutsul **'a'antal.** ‘He kicked him up the buttcrack.’

'acho *ADV* quickly.

'adelhts'ine *N* the people sitting at a balhats, that is, the members of the invited clans as opposed to the host clan.

'adilhti *V* [lh-ti < ti$_1$] he/she is buried. *[PA]*

'ahulih *V* [0-lih] they are gambling. *[IA]*

'aidudlih *V* [0-lih] we (2) are gambling. *[IA]*

'ailhghis *V* [lh-ghis < ghis$_1$] you (1) screw in. *[IA]*

'ak'i *N* someone's mother's sister.

'alih *V* [0-lih < lih$_1$] he/she is gambling. *[IA]*

'an *N* cave, den.

'ana'dutsih *N* electrical outlet.

'ana'dutsih *N* electrical outlet or plug.

'andih *N* ant.

'andihghez *N* ant eggs.

'andihken *N* ant hill.

'awundooh *INT* no.

'ane *INT* come here! Addressed to a young child.

'angwul *N* pine cone, acorn.

'anih *INT* come here!

'ankw'us *ADV* almost.

'aslih *V* [0-lih] I am gambling. *[IA]*

'at *N* wife. Duoplural: **'atke**.

'at ut'i *V* [0-t'i < t'i$_2$] he/she is married. Literally,“he/she has a wife”. *[IA]*

'atanlih *V* [0-lih < lih$_1$] you (1) will gamble. *[FA]*

'atehlih *V* [0-lih < lih$_1$] you (2+) will gamble. *[FA]*

'atke *N* wives.

'ats'alhti *V* [lh-ti < ti$_1$] we (3+) buried. [class: body] *[PA]*

'ats'ulih V [0-lih] we (3+) are gambling. *[IA]*

'aw *PART* not. (1) **'Aw** ombuz ghait'ah. 'You can't stretch it.' (2) Dune **'aw** 'oh nuchaiyal. 'The man has not been walking around.'

'ilhtsul *N* low-bush blueberry.

'awetze *INT* that's all, the end.

'awhenadintl'as *V* [0-tl'as] he/she clings to. *[IA]*

'awhenadustl'as *V* [0-tl'as] I cling to. *[IA]*

'awhulyiz *ADV* always.

'awhuz *ADV* still. (1) **'Awhuz usjiz.** 'I'm still breathing.'

'ayah *INT* ouch!

'ayalhti *V* [lh-ti < ti_1] he/she buried him/her. *[PA]*

'ayoh ho *INT* ouch!

'az *ADV* outside.

'az natanda *V* [0-da] you (1) will go to the washroom. *[FA]*

'azdeh *ADV* from outside, from away. (1) **'Azdeh** uyalh. 'He is coming.'

'aztelih *V* [0-lih < lih_1] we (3+) will gamble. *[FA]*

'buzkai nawhudleh *N* beets.

'eghilh *V* [0-ghilh < ghe_1] he/she is packing (u.o.). *[IA–prog]*

'en *PRO* he/she, him/her.

'enne *PRO* they/them.

'esghilh *V* [0-ghilh < ghe_1] I am packing (u.o.). *[IA–prog]*

'et *ADV* there.

'et ze *CONJ* only if.

'ezdlai *N* purse.

'iduyi *V* [0-yi < yi_1] we (2) are eating (u.o.). *[IA]*

'ighun *PPC* by reason of that.

'ilyez *V* [l-yez < yez_1] he/she is comparatively tall. *[IA]* (1) Nyanus **'ilyez.** 'She is taller than you.' (2) ''

'ilha hoolat *NUM* nine *[multiplicative].*

'ilhah *NUM* once, one *[multiplicative].*

'ilhah hoolah *NUM* nine *[generic].*

'ilhah hoolah lanezi *NUM* ninety *[generic].*

'ilhah hoolanun *NUM* nine *[human].*

'ilho hooladun *NUM* nine *[locative].*

'ilhohoolawh *NUM* nine *[abstract].*

'ilhowh *NUM* one *[abstract].*

'ilhtsul *N* low-bush blueberry.

'ilhtsultoo dohot'en *V* [d-'en < $'en_1$] it is purple. [abs: wh] *[IA]* Etymology: "it is the colour of blueberry juice".

'ilhtsultoo dot'en *V* [d-'en < $'en_1$] it is purple. *[IA]* Etymology: "it is the colour of blueberry juice".

'ilhtsultoo doonat'en *V* [d-'en < $'en_1$] it is purple. [abs: n] *[IA]* Etymology: "it is the colour of blueberry juice".

'ilhuk'i *NUM* one *[generic].*

'ilhuk'i *NUM* one *[locative].*

'ilhunun *NUM* one *[human].*

'inchooh isdak dohot'en *V* [d-'en < 'en$_1$] it is pink. [abs: wh] *[IA]* Etymology: "it is rose-coloured".

'inchooh isdak dot'en *V* [d-'en < 'en$_1$] it is pink. *[IA]* Etymology: "it is rose-coloured".

'inchooh isdak doonat'en *V* [d-'en < 'en$_1$] it is pink. [abs: n] *[IA]* Etymology: "it is rose-coloured".

'indak *N* flower.

'indakya *N* flowers.

'indamanuk *N* appeal to spouses for funds at a balhats. In general, at a balhats, the funds distributed have been raised previously by the clan or family making the distribution. In some circumstances, however, an appeal is made to the spouses of those making the distribution (who will belong to other clans, since one is not supposed to marry a member of one's own clan). This is said to be **'indamanuk.**

'indawuz *N* Black Gooseberry.

'indumanuk *N* A dance with one's partner in balhats.

'indumanukshun *N* sweetheart song. A song sung when couples dance into a balhats. Never sung at a funeral balhats.

'indzi *N* strawberry.

'Indzik'et *N* Yensischuck. A berry-picking area and hay-cutting meadow at I.R. 3. Etymology: "strawberry place".

'inja *V* [0-ja] you (1) did something, something happened to you (1). [PA] (1) Sla **'inja.** 'You have helped me.'

'ink'ez *CONJ* and.

'inlootsan *N* hailstone.

'inlootsan nanukat *V* [0-kat < kat$_1$] it is hailing. *[IA]*

'int'ah *V* [0-t'ah] it is. *[IA]*

'int'ah *V* [0-t'ah < t'oh$_1$] you (1) are. *[IA]* (1) Soo **'int'ah**? 'Are you well?'

'int'ah *V* [0-t'ah] you (1) are. *[IA]*

'int'lus tsih *N* left side.

'intl'us *ADV* left.

'ints'ilh *N* Varied Thrush.

'inyi *V* [0-yi < yi$_1$] you (1) eat (u.o.). *[IA]*

-'lunatdultsiyan *N* great-grandfather.

'oh *ADV* around. (1) '' (2) **'Oh** nut'ah. 'He is flying around.' (3) **'Oh** nuduldlut. 'He is going around talking loudly.'

'ok'et *N* whirlpool, eddy.

'olulh *N* spring.

'on'at *ADV* 'on'at.

'on'at *CONJ* plus.

'on'at yusk'ut *N* next year.

'on'un hanit'ai *V* [0-t'ai] it is over-ripe. *[IA]*

'onduk *ADV* above.

'ont'e *V* [0-t'e < t'oh$_1$] you (1) may be. [OA] (1) Soo **'ont'e**! 'Be well!'

'onyaz ts'i *ADV* on the other side.

'ooduket *V* [0-ket < ket$_1$] we (2) are buying (u.o.). *[IA]*

'oojoonih *N* porcupine.

'ook'et'as *N* cook stove.

'oolnih *V* [l-nih] he/she is jealous. *[IA]*

'oolhjas *V* [lh-jas < jas_2] he/she is fishing with a hook. *[IA]*

'oolhkes *V* [lh-kes] he/she exults. *[IA]*

'oosa' *N* pot, kettle, pail.

'oosyi *V* [0-yi < yi_1] I may eat (u.o.). *[OA]* (1) **'Oosyi** hukwa'nuszun. 'I want to eat.'

'oot'e ghainil *V* [0-nil] he/she is overwhelmed. *[IA]*

'oozi *N* name.

'u'an *N* cave.

'uba *N* father. (1) **'Uba** lhti sghanintan. 'Father gave me a rifle.'

'ubalt'i *N* twine used for stringing nets, making dipnets, and so forth. This term is used in reference to the twine when it is not part of the net.

'oojoonih *N* porcupine.

'ubez *N* someone's mother-in-law.

'ubezan *N* father's sister.

'uchanyoo *N* peritoneum. The fatty lining of the abdominal cavity.

'uda *ADV* long ago.

'udada *N* story, history, legend.

'udai-a *N* dining room.

'ude *N* antlers, horns. Indefinitely possessed form.

'udidughas *V* [0-ghas < $ghaz_2$] we (2) are planing (u.o.). *[IA]*

'udulhtsiyun *N* a song sung to embarass someone.

'udutl'us *V* [0-tl'us] he/she is earning income. *[IA]*

'udutsut *V* [d-tsut] it (Ruffed Grouse) is drumming. *[IA]*

'udzich'ooz *N* aorta.

'ugha *N* fur.

'ugha *N* hair, fur.

'ughak *N* fish bones. Indefinitely possessed form.

'ughez *N* testicles, eggs.

'ughuts *N* cartilage.

'uheghilh *V* [0-ghilh < ghe_1] they are packing (u.o.). *[IA–prog]*

'uhint'ah *V* [0-t'ah < $t'oh_1$] they are. *[IA]*

'uhooja *V* [0-ja] it happened. *[PA]*

'uhooket *V* [0-ket < ket_1] they are buying (u.o.). *[IA]*

'uhoolhkes *V* [lh-kes] they exult. *[IA]*

'uhuldzus *V* [l-dzus < $dzus_1$] they dance in the Indian way. *[IA]*

'uhut'en *V* [d-'en < $'en_2$] they are working, doing. *[IA]* (1) Lhulh **'uhut'en.** 'They are working together.'

'uhuyi *V* [0-yi < yi_1] they are eating

(u.o.). *[IA]*

'uja *V* [0-ja] he/she did something, something happened to him/her . *[PA]* (1) Sbut soo **'uja.** 'My stomach is nice and full.'

'ujooh *N* drying poles. The poles from which fish or meat is hung to dry.

'ujoohnih *N* porcupine.

'ujoohtsun *N* brisket.

'ukechun *N* legs.

'ukoh *N* river.

'ukoh whuniz *PPP* midstream.

'uk'a *N* fat.

'uk'ah *N* tracks, footprints.

'ujooh *N* drying poles. The poles from which fish or meat is hung to dry.

'uk'enus *ADV* most.

'uk'o' *N* hunchback.

'uk'oon *N* fish egg.

'uk'oondzai *N* fish trap.

'uk'une'huguz *V* [0-guz] they are writing. *[IA]*

'uk'une'hut'en *V* [d-'en < 'en$_2$] they are obedient. *[IA]*

'uk'une'inguz *V* [0-guz] you (1) write. *[IA]*

'uk'une'uguz *N* secretary.

'uk'une'ulhtsi *V* [lh-tsi] I draw, paint. *[IA]*

'uk'unets'uguz *V* [0-guz] we (3+) are writing. *[IA]*

'ul *N* branch of conifer.

'ul'en bayah *N* brothel.

'ula *N* bark.

'ulagi *N* fingernail.

'ulats'i *N* spruce bark canoe.

'uldzus *V* [l-dzus < dzus$_1$] he/she dances in the Indian way. *[IA]*

'uloo *N* mother.

'ulh *N* dam.

'ulhti *N* rifle.

'ulhticho *N* big rifle. E.g. .306 or .303.

'ulhudzus *V* [l-dzus < dzus$_1$] I dance in the Indian way. *[IA]*

'unaghoo *N* eyetooth.

'undunulhmulh *V* [lh-mulh < mulh$_1$] he/she is boiling (u.o.). *[IA]*

'undunulhmulh *V* [lh-mulh] I am boiling (u.o.). *[IA]*

'uni *N* bait.

'unibek *N* pin. This includes both straight pins and sewing pins.

'unizus *N* back pack.

'untelhah *V* [0-lhah] he/she choked on something solid. *[PA]*

'untelhuyiz *V* [l-yiz] he/she choked on somthing liquid. *[PA]*

'unt'oh *N* avalanche.

'unubek *N* straight pin.

'unulh'uz *V* [lh-'uz < tl'es$_1$] he/she is hammering (u.o.). *[IA]*

'uscheh *N* shady spot.

'Usdas *N* name of trickster culture hero.

'Usdas Tezdluz *N* A rock formation along the shore of Fraser Lake west of I.R. 1 in front of the present day Bill Evans ranch. It shows the impression of **'Usdas's** body where he fell from the sky, as recounted in the legend of **'Usdas** and the swans. Etymology: "**'Usdas** urinated".

'usdloos *N* toboggan.

'usgoo *N* earthworm, maggot.

'uskai suli *V* [0-li < li_1] he/she is bleeding. *[IA]*

'uskai yuzelhghi *V* [lh-ghi < ghi_1] he/she bled to death. Literally, "blood killed him/her". *[PA]*

'uskehne *N* children.

'uski benunulbas *N* stroller.

'uski ts'ai *N* bassinet.

'uskidune *N* boy.

'uskiyaz *N* small child.

'ustes *N* mattress.

'ust'ah *V* [0-t'ah] I am. *[IA]*

'ust'ot *V* [0-t'ot < $t'ot_1$] I smoke. *[IA]*

'ustlen *N* subject (of ruler), servant.

'ustl'oo *V* [0-tl'oo < $ts'it_1$] I am knitting (u.o.). *[IA]*

'usyi *V* [0-yi < yi_1] I eat (u.o.). *[IA]*

'utadujun *V* [d-yun] we (2) will sing. *[FA]*

'utahkw'ak *V* [0-kw'ak] you (2+) burp. *[IA]*

'utai *N* uncle. father's brother.

'utancho *N* cabbage. Etymology: Literally, "big leaves".

'utanjun *V* [d-yun] you (1) will sing. *[FA]*

'utant'ilh *V* [d-'ilh] you (1) will work. *[FA]*

'utanyilh *V* [0-yilh < yi_1] you (1) will eat (u.o.). [FA]

'utehjun *V* [d-yun] you (2+) will sing. *[FA]*

'utejun *V* [d-yun] he/she will sing. *[FA]*

'utes *N* knife.

'utesjun *V* [d-yun] I will sing. *[FA]*

'utesyilh *V* [0-yilh < yi_1] I will eat (u.o.). *[FA]*

'utet'ilh *V* [d-'ilh] he/she will work. *[FA]*

'uteyilh *V* [0-yilh] he/she will eat (u.o.). *[FA]*

'utinkw'ak *V* [0-kw'ak] you (1) burp. *[IA]*

'utna *N* non-Athabaskan Indian. Especially, but not exclusively, those of the West Coast.

'ut'an *N* leaf.

'ut'antsi *N* Lambs Quarters, Pigweed, White Goosefoot. Etymology: Literally, "bad leaves".

'ut'en *V* [d-'en < $'en_2$] he/she is working. [IA] (1) Sulh **'ut'en.** 'He is working with me.' (2) ''

'ut'ooz *N* bark.

'utl'ah *N* a bottom. Indefinitely possessed.

'utl'oo *V* [0-tl'oo] he/she is knitting (u.o.). *[IA]*

'utselh *V* [0-tselh] he/she is chopping (u.o.). *[IA]*

'utsigha *N* head hair.

'utsitah *N* crown of head. Indefinitely possessed form.

'utsiyan *N* grandfather.

'utsiyancho *N* great-grandfather.

'utsoh *N* awl.

'utsoo *N* breast. Indefinitely possessed form.

'utsoo *N* maternal grandmother.

'utsukw *N* penis.

'utsukw *N* someone's penis.

'utsun *N* meat.

'utsun duneldus *N* ground meat.

'utsun sugi *N* dried meat.

'utsun sut'e *N* fried meat.

'utsun tazul *N* hamburger soup.

'utsun tazul *N* meat soup.

'Utsun Ts'usyih *N* Friday. Etymology: "one does not eat meat".

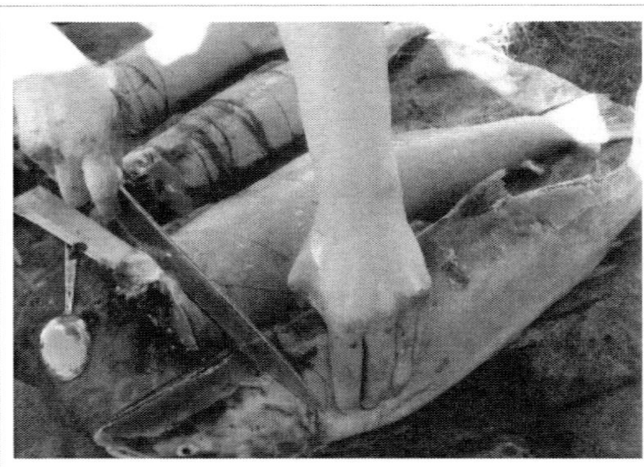

'utselh *V* [0-tselh] he/she is chopping (u.o.). *[IA]*

'utsungui *N* dried meat.

'utsut *V* [0-tsut < $tsut_1$] it is drumming. This describes the way a Ruffed Grouse drums with its wings. *[IA]* (1) **'Utsut 'utsut.** 'The grouse is drumming.'

'utsut *N* Ruffed Grouse.

'utsutle *N* colon, tripe.

'uts'anyi *V* [0-yi < yi_1] we (3+) ate (u.o.). *[PA]*

'uts'eh *N* sinew.

'uts'ooket *V* [0-ket < ket_1] we (3+) are buying (u.o.). *[IA]*

'uts'un *N* paternal grandmother.

'uts'uncho *N* paternal great-grandmother.

'uts'uyi *V* [0-yi < yi_1] we (3+) eat (u.o.). *[IA]*

'uts'uyi ba whuz'ai *N* dining room.

'uye' *N* a person who makes a contribution to his or her father's clan at a balhats . Although this literally means "man's son", it applies to both men and women.

'uyemba *N* green moss (on ground or tree trunks).

'uyi *V* [0-yi < yi_1] he/she is eating (u.o.). [IA] (1) Ligok **'uyi.** 'The chicken is eating.'

'uyoocha *N* some more.

'uyooschum *N* branch of deciduous tree.

'uyust'e *N* body. Indefinitely possessed.

'uz'e *N* someone's uncle.

'uza' *N* noble in the clan system, of either sex. Duoplural: **'uza**'ne

'uzdughut *V* [0-ghut < $ghut_1$] we (3+) are sawing (u.o.). *[IA]*

'uzkai *N* blood. Indefinitely possessed form.

'uztejun *V* [d-yun] we (3+) will sing.

[FA]

'uztet'ilh *V* [d-'ilh] we (3+) will work. *[FA]υ*

'uzteyilh *V* [0-yilh < yi_1] we (3+) will eat (u.o.). *[FA]*

'uztukw'ak *V* [0-kw'ak] we (3+) burp. *[IA]*

'uzus *N* bag.

'uzut *N* liver.

'uzuz *N* cured hide. (1) **'Uzuz** ulhlhah. 'She is greasing a hide.'

'uzuz *N* hide, skin.

'uzuz dzoot *N* skin jacket.

a *INT* yes.

a'ah *INT* okay, alright.

adih *INT* well. (1) **Adih**, dahooja? 'Well, what happened?'

ahyi *V* [0-yi < yi_1] you (2+) eat. *[IA]*

-ak'i *N* mother's sister.

-anus *PP* than.

anyi *V* [0-yi < yi_1] he/she ate. *[PA]*

anyi *V* [0-yi < yi_1] you (1) ate. *[PA]*

Azilenchonoo *N* The larger of the two small rock islands in the mouth of the Nadleh River where it flows out of Fraser Lake at Nadleh. Latitude: 54° 5' 3" N. Longitude: 124° 36' 55" W. Etymology: "big Angeline island", after Angeline Ketlo, who fished there and used these islands for preparing her catch. A char and whitefish netting site.

Azilenyaznoo *N* The smaller of the two small rock islands in the mouth of the Nadleh River where it flows out of Fraser Lake at Nadleh. A char and whitefish netting site. Latitude: 54° 5' 4" N. Longitude: 124° 36' 53" W. Etymology: "little Angeline island", after Angeline Ketlo, who fished there and used these islands for preparing her catch.

balt'i *N* twine used to string net. This form is used in reference to the twine when on the net. Compare: **'ubalt'i.**

ba *N* edge.

ba *PP* for, on behalf of.

-ba *N* father.

ba 'alha'hoont'ah *V* [C-t'ah < $t'oh_1$] he/she believes it to be true. *[IA]*

ba we'hooja *V* [0-ja] he/she is bewildered. *[IA]*

ba't'en lhai *V* [0-lai] he/she is busy. *[IA]*

Babine whut'en *N* Babine person.

bahne *N* soldiers.

bak'i *N* his/her mother's sister.

bal'ah *N* rim of a basket. Usually made of a piece of k'en dulk'un "Red Willow", usually but not always peeled.

balt'i *N* twine used to string net. This form is used in reference to the twine when on the net. Compare: **'ubalt'i.**

balhats *N* potlatch. Etymology: A loan from Chinook Jargon, ultimately from Nuuchanulth pała·č; 'make a ceremonial gift.'

Banghan Nuts'uke *N* November. Etymology: "We go by boat half the time", referring to the fact that in the second half of November the ice typically prevents travel by canoe.

Basghelh *N* Mary Mountain. A sacred rock bluff that rises on the North side of Nadleh village. Ceremonial cremations were carried out on the highest point. It was also a spot where men went to acquire spiritual powers. It was also the purification site where a pair of moccasins disinterred from a grave were hung for spiritual/ceremonial purposes. Referred to as Mary Mountain after Mary Pius, who lived just West of the bluff.

Basghelh Bunk'ut *N* Mary Lake. The small lake behind (North) of Basghelh, the rock bluff right behind Nadleh village. Etymology: "**Basghelh** lake".

bat *N* mittens.

be *PP* by means of. (1) Duyeztli yoo **be** yulhlhah. 'He is smearing his horse with salve.' (2) Duda **be** lhukw ookaih. 'He (a pelican) scoops up fish in his beak.' (3) Yeztli yoo **be** idullhah. 'We are smearing the horse with liniment.'

be'didil *V* [0-dil] it faded. *[PA]*

be'dudughut *N* saw.

be'dutedulh *V* [0-dulh] it will fade. *[FA]*

be'indak *N* his/her flower.

be'lunatdultsiyan *N* his/her great-grandfather.

be'nezk'uz *N* refrigerator.

be'nezul *N* thermos.

be'ooget *N* fork.

be'ook'et'as *N* his/her cook stove.

be'ook'utudai *N* his/her table.

be'tunadugus *N* clothes washer.

be'ts'ut'en *N* things we (3+) use.

be'ul'uz *N* hammer.

be'ulje *N* bone skin scraper.

be'ulht'uwk *V* [lh-t'ukw < t'ukw$_1$] I breast feed him/her. *[IA]*

be'usdudzak *N* canister set.

be'ustes *N* his/her mattress.

be'ustlen *N* his/her subject, servant.

be'utun *N* freezer.

be'ut'es *N* frying pan, oven. Etymology: "that by means of which it is baked".

be'utsun *N* his/her meat.

be'uzdla *N* dresser, cupboard.

bech'anuszun *V* [0-zun < zun$_1$] I think he's full of it. *[IA]*

bedahudut'as *N* can opener.

bedutleh *N* pie. Etymology: "it has mushy stuff inside".

beghoosuski *V* [0-ki < ke$_2$] I passed him in a boat. *[PA]*

behananudughas *N* hair clippers.

behoonjai *N* a person who must be shown respect.

behooyeznuleh *N* seeder. A machine for planting seeds.

bela'ts'uldeh *N* wash basin.

belhk'a *V* [l-k'a] it is fruitful. Means that an area not only produces a lot but that whatever comes from it is fat and of good quality. *[PA]*

Belhk'achek *N* Belgatse. A former village site on Cheslatta Lake. Ety-mology: Literally, "the mouth of Belhk'akoh "Knapp Creek".

Belhk'akoh *N* Knapp Creek.

belhyal *N* war club. Made from long bone of moose.

bena'dukuk *N* chest of drawers, clothes chest.

bena'ulkuk *N* waste basket.

benaznuldooz *N* hair curlers.

bene'ts'ut'en *N* things we (3+) use.

benek'edugus *N* camera.

benek'ehugis *N* camera.

beni *N* his/her mind. (1) **Beni** lhe'nintananesja. 'He is weird.'

beni hoolah 'ut'ot *N* marijuana. Etymology: Literally, "a crazy person smokes it".

beni hooni *V* [0-ni] he/she is smart, brainy. Literally, "his mind exists". *[IA]*

beniduljut *V* [l-jut < $joot_1$] we (2) are afraid of it. *[IA]*

beningoo-i *N* your (1) vehicle.

benlhujut *V* [l-jut < $joot_1$] I am afraid of it. *[IA]*

benugoo-i *N* vehicle.

benusgoo-i *N* my vehicle.

benuts'ugoo *N* car, vehicle. Etymology: "that by means of which one drives around".

benuts'ugoo tl'it *N* carbon dioxide. Etymology: "car fart".

benuts'ulgaih *N* bicycle. Etymology: "that by means of which one runs".

benuts'ut'ah *N* airplane. Etymology: "that by means of which one flies around".

beooduget *N* fork.

beoonezul *N* heater.

beoonujut *V* [0-jut < $joot_1$] it is dangerous, powerful. *[IA]*

besk'i *N* seagull.

bet *N* char.

Bet Ooza *N* October. Etymology: "month of the char".

betadzih *N* dipper.

betandulh *V* [0-dulh] you (1) are going to fight. *[FA]*

betanizilh *N* kettle.

betaszih *V* [0-zih] I am bailing out (st:boat). *[IA]*

betats'utnai *N* drinking glass.

betehdulh *V* [0-dulh] you (2+) are going to fight. *[FA]*

betinawhultsas *N broom.*

betoonuts'uya *N* bath tub.

bets'ujun *N* stereo.

beyatuk *N* telephone, grammar.

beyulhdzeh *V* [lh-dzeh < $dzeh_1$] I am pouring it into it. *[IA]*

beyusk'ut *N* age. Literally, "his winters".

-bez *N* mother-in-law, paternal aunt.

-bezan *N* father's sister, father's female first cousin.

beznuljut *V* [l-jut < $joot_1$] we (3+) are afraid of it. *[IA]*

beznulha *N* cold cream.

beztedulh *V* [0-dulh] we are going to fight. *[FA]*

bilh *N* snare.

bindak *N* its flowers.

bindak khastl'ah *N* raspberries.

bosdun *N* white person. Etymology: Loan, probably via Chinook Jargon, from English Boston, after the home port of many of the coastal trading ships.

-boo' *N* anus, rectum. *[babytalk]*

boona *N* his/her older brother.

boos *N* cat.

booscho *N* cougar. Etymology: "big cat".

boosyaz *N* little cat.

boozi *N* his/her name.

bu- *PREFIX* third person singular possessor. Attaches to class 1 nouns.

bu'at *N* his/her wife.

bu'at hoolah *V* [0-lah] he/she is a bachelor, he/she is single. Literally, "he/she has no wife". *[IA]*

bu'et *PPC* without it.

buba *N* his/her father.

bubat *N* his/her mitts.

bube'uzdla *N* his/her cupboard, dresser.

bubez *N* his/her mother-in-law.

bubezan *N* his/her paternal aunt.

buboo' *N* his/her rectum.

buboos *N* his/her cat.

bubun *N* his/her roof.

bubunt'ah *N* his/her attic.

bubut *N* his/her stomach.

bubuttl'ah *N* the bottom of his/her stomach.

buchahooncha *V* [0-cha < cha_1] he/she is a glutton. *[IA]*

buchai *N* his/her grandchild.

buchaicho *N* his/her great-grandchild.

buchaike *N* his/her grandchildren.

buchak *N* his/her ribs.

buchets'un *N* his/her tailbone.

buch'i' *PRO* his, hers, its.

buda *N* his/her lips.

budadeltel *N* his/her porch.

budadentanghunez'ai *N* his/her doorknob.

booscho *N* cougar. Etymology: "big cat".

budadentanghunezdla *N* his/her doorknobs.

budadent'az *N* his/her window.

budadent'azts'asdla *N* his/her curtains.

budagha *N* his/her moustache.

budaningi *V* [0-gi] he/she is skinny. *[IA]*

budati *N* his/her doorway.

budes *N* his/her lungs.

budesyaz *N* his/her younger sister.

budune delts'i *N* his/her living room.

buduneti kw'uts'uda *N* his/her reclining chair.

budusneke *N* his/her ancestors.

budzek hah'elts'ul *V* [l-ts'ul] he/she has an earache. *[IA]*

budzi *N* his/her heart.

budzi hoolah *V* [0-lah] he/she is merciless. Literally, "his heart is absent". *[IA]*

budzi nzoo *V* [0-zoo < zoo_1] he/she is good-natured. Literally, "his/her heart is good". *[IA]*

budzidool *N* his/her artery.

budzik'ut *N* his/her chest.

budzoozt'an *N* his/her shirt.

bugan *N* his/her arm.

bughah *PPC* next to it.

bughe *N* his/her brother-in-law, sister-in-law.

bughez *N* his/her testicles.

bughezzus *N* his/her scrotum.

bughoo *N* his/her teeth.

bughootsun *N* his/her gums.

bughundan *N* his/her son-in-law.

bugwaz *N* his/her nephew/niece.

bugwut *N* his/her knee.

bugwutduts'un *N* his/her kneecap.

bugwutt'uk *N* the back of his/her knee.

bugwutyuschun *N* his/her shins.

bujook *N* his/her toque.

bujootsun *V* [0-tsun] he/she has body odour. *[IA]*

bukawetezut *V* [0-zut] he/she fainted. *[PA]*

bukechun *N* his/her leg.

bukechunnah *N* his/her ankle.

bukechunt'ak *N* the back of his/her legs.

bukelagi *N* his/her toenail.

bukelamai *N* his/her toes.

bukelamaicho *N* his/her big toe.

bukelamaiyaz *N* his/her little toe.

bukelamaiyazts'unts'e *N* his/her fourth toe.

bukelamaiyez *N* his/her long toe.

bukelamaiyezts'unts'e *N* his/her middle toe.

bukelascho *N* his/her big toe.

bukelatsul *N* his/her heel.

bukengi *N* his/her toenail.

bukent'ak *N* the top of his/her foot.

bukentsul *N* his/her heel.

buketl'ah *N* his/her sole.

buketl'i *N* his/her toe spaces.

buketsul *N* his/her heels.

bukoo *N* his/her house.

bukui *N* his/her husband.

bukuninta *V* [0-ta < ta_1] you (1) are looking for him/her . *[IA]*

buk'elh'az *PPC* after him/her .

buk'enahoolhah *V* [0-lhah] he/she is buried in an avalanche or landslide. *[IA]*

buk'i *N* his/her pelvic region.

buk'its'un *N* his/her hipbone.

buk'une'idut'en *V* [d-'en < $'en_2$] we (2) obey him/her. *[IA]*

buk'une'ts'ut'en *V* [d-'en < 'en_2] we (3+) obey him/her. *[IA]*

buk'une'ust'en *V* [d-'en < 'en_2] I obey him/her. *[IA]*

buk'ut *PPC* on it.

buk'uz *N* half of it, one side of it.

bukw'ulh *N* his (boy's) penis.

bukw'uts'uda *N* his/her chair.

bukw'uts'udacho *N* his/her easy chair.

bukw'uts'udayez *N* his/her chesterfield.

bukw'uz *N* his/her kidneys.

bula ncha *V* [0-cha < cha_1] he/she is generous. Literally, "his hand is large". *[IA]*

bulachunah *N* his/her wrist.

bulagi *N* his/her fingernail.

bulak'et *N* his/her palm.

bun *N* lake.

bulasge N his/her finger. (1) Tsalhts'ul **bulasge** lhe'ultsolyaz nulh'en. 'I am looking at a baby whose fingers are oh so small.'

bulasts'ah *N* his/her pinkie.

bulat'ah *N* his/her wrist.

bulat'ak *N* the back of his/her hand.

bulatl'i *N* his/her finger spaces.

bulik *N* his/her dog.

bulikcho *N* his/her big dog.

bulili *N* his/her bed.

bulilik'usulhchooz *N* his/her sheet.

buloo *N* his/her mother.

buludab *N* his/her table.

buluz *N* his/her urine.

buluzkah *N* his/her urinary tract.

buluzzus *N* his/her bladder.

bulh *PPC* with him/her . (1) **Bulh** hunlhuch'e. 'He got mad at him.'

bulh 'et'ah *V* [0-t'ah < $t'oh_2$] he/she is going by plane. *[IA–prog]*

bulh 'utet'ah *V* [0-t'ah < $t'oh_2$] he/she will go by plane. *[FA]*

bulh nahnideltsut *V* [l-tsut] you (2+) are sleepy. *[IA]*

bulh ne'uzt'a *V* [0-t'a < $t'oh_2$] he/she has gone in a loop by plane. *[PA]*

bulh nenideltsut *V* [l-tsut] we are sleepy. *[IA]*

bulh nideltsut *V* [l-tsut] you (1) are sleepy. *[IA]*

bulh sdutez *N* pajamas.

bulh'ults'ulh *V* [l-ts'ulh] it aches, has pain. [abs: 0] *[IA]* (1) Sghoo **bulh'ults'ulh.** 'My tooth aches.'

bulh'unults'ulh *V* [l-ts'ulh] it aches. [abs: n] *[IA]* (1) Stsi **bulh'unults'ulh**. 'My head aches.'

bulhtus *N* his/her sister.

bulhutsin *N* his/her brother.

bumai *N* its (plant's) seeds.

bumusdus *N* his/her cow.

bun *N* lake.

bun *N* roof.

buna *N* his/her eye.

buna duguz *V* [0-guz] he/she is cross-eyed. *[IA]*

bunabagha *N* his/her eyelashes.

bunalhti *N* his/her blanket.

bunasdlal *V* [0-dlal < $dlal_1$] I dreamt about him/her. *[PA]*

bunat *PPC* around it.

bunats'osdooz *N* his/her eyebrows.

bunda *N* this morning.

bundada *N* morning.

bundada totsuk *N* every morning.

bunde *N* tomorrow.

bunde bundada *N* tomorrow morning.

bunde lhgha *N* tomorrow evening.

bunde ombun *N* day after tomorrow.

Bundzi Bunk'ut *N* Angly Lake. Latitude: 54° 13' 57" N. Longitude: 124° 39' 32" W. Etymology: "lake with a heart". This lake has a warm spring bubbling to the surface in a spot that never freezes over.

Bundzikoh *N* Angly Creek. The creek flowing from Angly Lake to Oona Lake. Latitude: 54° 11' 6" N. Longitude: 124° 43' 27" W. Etymology: "heart lake river".

bunen *N* his/her face. (1) **Bunen** nulyul. 'Her face is white.' (2) **Bunen** dunughai. 'His face is hairy.'

buniltal *V* [l-tal] it (tire) is flat. [PA]

buninchustah *N* his/her sinus cavity.

buninchuz *N* his/her thumb.

bunints'uzti *N* his/her elbow.

bunniz *N* centre of lake.

bunt'ah *N* ceiling, eaves, attic.

buntsilh *N* porch.

bunul'en-i *N* his/her television.

bunut'uk *N* his/her muscle.

busdi *N* stretching frame.

busih *N* his/her wall.

busih *N* their wall.

bustocho *N* his/her big stove, heater.

-but *N* stomach, belly.

butah *PPC* among them.

butai *N* his/her uncle other than mother's brother , step-father.

butesk'et *N* his/her bed, bedroom.

buti *N* its handle. (1) Lubo: **buti.** 'The cup's handle.'

butoo *N* his/her water.

buttl'ah *N* its bottom.

-buttl'ah *N* bottom of the stomach.

butus *PPC* over it.

butuz *N* his/her cane.

but'ak *N* his/her back.

but'akts'un *N* his/her backbone.

but'eke *N* his/her friend.

but'ukts'un *N* his/her collarbone.

butl'ah *N* his/her bum.

butl'asus *N* his/her pants, dress.

butl'egha *N* his/her pubic hair.

butl'et *N* his/her groin.

butl'it *N* his/her crotch, its exhaust

pipe.

butl'uz *N* his/her gall bladder.

butsah *PPC* before him/her.

butsakelhgwus *N* his/her navel.

butsan *N* his/her feces.

butsanbayah *N* his/her outhouse.

butse' *N* his/her daughter.

butsi *N* his/her head.

butsi'alh *N* his/her pillow.

butsidah *N* the top of his/her head.

butsigha *N* his/her head hair.

butsinghai *N* his/her brain.

butsint'ak *N* the back of his/her head.

butsints'un *N* his/her skull.

butsit'ak *N* the back of his/her head.

butsitl'ah *N* the back of his/her head.

butsiyan *N* his/her grandfather.

butsiyancho *N* his/her great-grandfather.

butsiyanyaz *N* his/her younger grandfather.

butsoo *N* his/her maternal grandmother.

butsoocho *N* his/her maternal great-grandmother.

butsoola *N* his/her tongue.

butsooyaz *N* his/her younger grandmother.

butsuk hooni *N* gloves.

butsukw *N* his penis.

butsul *N* his/her rectum.

butsul kwun dilhk'ai *N* tail lights of vehicle. Etymology: Literally, "lights that burn in its anus".

buts'aha'elts'utne *N* one's father's clan.

buts'ahainalhyi *V* [l-yi] it grows out of it, branches from it. E.g. a limb from the trunk of a tree. *[IA]*

buts'ahbul *N* his/her cowboy hat.

buts'e *N* her daughter.

buts'ik N his/her intestines.

buts'ikcho *N* his/her large intestine.

buts'ilchun *N* his/her neck.

buts'ilchunt'ak *N* the nape of his/her neck.

buts'oo *N* his/her breast.

buts'oola *N* his/her nipples.

buts'udusnik *V* [0-nik] I dislike him/her. *[IA]*

buts'utelhk'az *V* [l-k'az] he/she has chills. *[IA]*

buts'uyi ghutna-a *N* his/her kitchen.

buts'uyibesula *N* his/her food cupboard.

buwuda *N* his/her calf.

buwus *N* his/her shoulders.

buwuz *N* his/her thighs.

buwuzts'un *N* his/her femur.

buyah *N* his/her house.

buyas'at *N* his/her daughter-in-law.

buyat *N* his/her older sister.

buyaz *N* her son.

buye' *N* his son.

buyeztli *N* his/her horse.

buyun *N* his/her floor.

buyunlatel *N* his/her flooring.

buyunsubul *N* his/her floor covering.

buyunsulchooz *N* his/her floor covering.

buyunts'un *N* his/her backbone.

buyust'e *N* his/her body.

buz'e *N* his/her mother's brother.

buzaz *N* his/her father-in-law.

Buzdaiyus *N* Pitka Mountain. Latitude: 54° 13' 31" N. Longitude: 124° 35' 21" W. Etymology: "stop famine mountain" because it is said always to produce game in time of need.

buzek *N* his/her mouth.

buzesdak *N* his/her throat.

buzesdak'al *N* his/her uvula.

buzesdalos *N* his/her uvula.

buzesdamai *N* his/her uvula.

buzesdazum *N* his/her tonsils.

buzets'un *N* his/her chin.

buzkai *N* his/her blood.

buzkaich'ooz *N* his/her veins.

buzkaich'oozcho *N* his/her artery.

buzkeh *N* his/her children.

buzool *N* his/her wind pipe.

buzoolts'un *N* his/her larynx.

buzumcho *N* his/her pancreas.

buzut *N* his/her liver.

buzuz *N* his/her hide.

cha *PART* also, too.

cha'dudilti *V* [l-ti < ti_2] he/she is not proud, he/she is meek. *[IN]*

chaditi *V* [0-ti < ti_2] it is not expensive, it is cheap. *[IN]*

chadizoo *V* [0-zoo < zoo_1] it is not good. [abs: d] *[IN]* (1) Duchun **chadizoo**. 'The tree is not nice.' (2) Shun **chadizoo.** 'The song is not nice.'

chahesgooh *V* [0-gooh < goo_1] they are not driving. *[IN–prog]* (1) Burns Lake ts'i **chahesgooh**. 'They are not driving to Burns Lake.' (2) Lheidli ts'i **chahesgooh**. 'They are not driving to Lheidli.'

chaheskelh *V* [0-kelh < ke_2] they are not going by boat. *[IN–prog]*

chahizoo *V* [0-zoo < zoo_1] they are not good. *[IN]*

chahoodudzum *V* [0-zum] we (2) are hungry. *[IA]*

chahoonzum *V* [0-zum] you (1) are hungry. *[IA]*

chahoozoo *V* [0-zoo < zoo_1] it is not good. [abs: wh] *[IN]*

chahuhoonzum *V* [0-zum] they are hungry. *[IA]*

chahusyi *V* [0-yi < yi_1] they do not eat. *[IN]*

chahutesdulh *V* [0-dulh < dil_1] they (3+) will not walk. *[FN]* (1) 'Aw Fraser Lake ts'i **chahutesdulh?** 'Won't they go to Fraser Lake?' (2) 'Aw Fraser Lake ts'i **chahutesdulh.** 'They will not go to Fraser Lake.'

chahutesgooh *V* [0-gooh < goo_1] they will not drive. *[FN]* (1) Burns Lake ts'i **chahutesgooh.** 'They will not drive to Burns Lake.'

chahuteskelh *V* [0-kelh < ke_2] they will not go by boat. *[FN]*

chahutesyi *V* [0-yi < yi$_1$] they will not eat. *[FN]*

-chai *N* grandchild, sibling's grandchild, first cousin's grandchild. Duoplural: -chaike

-chaicho N great-grandchild.

chailah *V* [0-lah] it is not. *[IN]* (1) Syah whuduldzan **chailah.** 'My house is not blue.'

chaizoo *V* [0-zoo < zoo$_1$] it is not good. *[IN]*

-chak *N* rib.

-chak'ests'ah *N* arm pit.

chalhyal *N* large basket. A large bark basket suitable for carrying on back, into which smaller baskets are emptied.

chalhyalsto *N* air-tight stove.

chalhzus *N* tripe.

chan *N* rain.

chehhaindzool *N* duff, the pine needles, pine cones, and other matter that form the carpet of a conifer forest.

-chan *N* belly.

chan nainli *V* [0-li < li$_1$] rain is pouring down. *[IA]*

chanildzan *V* [l-dzan < dzan$_1$] it is not blue. [abs: n] *[IN]* (1) 'Indzi **chanildzan.** 'Strawberries are not blue.'

chanilk'ah *V* [l-k'ah] he/she is not obese, fat. *[IN]*

chanilhdza *V* [lh-dza < dza$_1$] it is not far. *[IN]*

chanizoo *V* [0-zoo < zoo$_1$] it is not good. [abs: n] *[IN]* (1) Nukuk **chanizoo.** 'The ball is not nice.'

chaoosyi *V* [0-yi < yi$_1$] I may not eat. *[ON]*

chasangooh *V* [0-gooh < goo$_1$] you (1) are not driving. *[IN–prog]*

chaseskelh *V* [0-kelh < ke$_2$] I am not going by boat. *[IN–prog]* (1) Nadleh ts'i **chaseskelh.** 'I am not going to Nautley.'

chasgooh *V* [0-gooh < goo$_1$] he/she is not driving. *[IN–prog]*

chasidugooh *V* [0-gooh < goo$_1$] we (2) are not driving. *[IN–prog]*

chasinyi *V* [0-yi < yi$_1$] you (2) do not eat. *[IN]*

chasit'us *V* [0-'us < 'as$_1$] we (2) are not walking. *[IN–prog]*

chasjun *V* [d-yun < yun$_1$] I did not sing. *[PN]*

chaskelh *V* [0-kelh < ke$_2$] he/she is not going by boat. *[IN–prog]*

chasusjun *V* [d-yun < yun$_1$] I do not sing. *[IN]*

chasusyi *V* [0-yi < yi$_1$] I do not eat. *[IN]*

chasyalh *V* [0-yalh] he/she is not walking. *[IN–prog]* (1) Dune Fraser Lake ts'i **chasyalh.** 'The man is not going to Fraser Lake.'

chasyi *V* [0-yi < yi$_1$] he/she does not eat. *[IN]*

chateskel *V* [0-kel < ke$_2$] he/she will not go by boat. *[FN]*

chatesyalh *V* [0-yalh < ya$_1$] he/she will not walk. *[FN]*

chatesyi *V* [0-yi < yi$_1$] he/she will not eat. *[FN]*

chatuszangooh *V* [0-gooh < goo$_1$] you (1) will not drive. *[FN]*

chatuzadugoo *V* [0-goo < goo$_1$] we (2) will not drive. *[FN]*

chatuzanyi *V* [0-yi < yi$_1$] you (1) will not eat. *[FN]*

chatuzat'us *V* [0-'us < 'as$_1$] we (2) will not walk. *[FN]* (1) 'Aw Fraser Lake ts'i **chatuzat'us?** 'Will we not go to Fraser Lake?'

chatuzehyi *V* [0-yi < yi$_1$] you (2+) will not eat. *[FN]*

chatuzesgooh *V* [0-gooh < goo$_1$] I will not drive. *[FN]*

chatuzeskel *V* [0-kel < ke2] I will not go by boat. *[FN]* (1) Nadleh ts'i **chatuzeskel.** 'I will not go to Nautley by boat.'

chatuzesyi *V* [0-yi < yi$_1$] I will not eat. *[FN]*

chats'eskelh *V* [0-kelh < ke$_2$] we (3+) are not going by boat. *[IN–prog]*

chats'uhoonzum *V* [0-zum] we (3+) are hungry. *[IA]*

chats'usyi *V* [0-yi < yi$_1$] we (3+) do not eat. *[IN]*

chawes'en *V* [0-'en < 'en] he/she is blind, cannot see. *[IN]*

Chawts'unk'ut *N* Chowsunkut Lake. Latitude: 53° 59' 19" N. Longitude: 124° 44' 1" W. Etymology: "Chao's grave", after a Chinese man named Chao who lived there and is buried there.

chawzes'en *V* [0-'en < en$_1$] I am blind, cannot see. *[IN]*

chawhuszum *V* [0-zum] I am hungry. *[IA]*

chazahyi *V* [0-yi < yi$_1$] you (2+) do not eat. *[IN]*

chaztesdulh *V* [0-dulh < dil1] we (3+) will not walk. *[FN]* (1) 'Aw Fraser Lake ts'i **chaztesdulh.** 'We will not go to Fraser Lake.'

chazteskel *V* [0-kel < ke$_2$] we (3+) will not go by boat. *[FN]*

chaztesyi *V* [0-yi < yi$_1$] we (3+) will not eat. *[FN]*

cheh *PP* at the base of.

chehhaindzool *N* duff, the pine needles, pine cones, and other matter that form the carpet of a conifer forest.

Chestl'ada *N* Cheslatta village.

-chets'un *N* tailbone, coccyx.

chigamin *N* money. Etymology: Loan ultimately from Nuuchanulth, presumably via Chinook Jargon.

cikamin *N* "metal, steel, money".

chilchunk'oon *N* herring eggs.

chilh *N* young man. Duoplurals: chilhuka, chilhke

chilhukah *N* Irregular duo-plural of chilh, q.v.

chilhuke *N* young men.

-chist'loh *N* younger brother.

-cho *SUFFIX* big.

chos *N* canoe paddle, oar.

Choostl'o *N* Ormond Lake. Latitude: 54° 10' 39" N. Longitude: 124° 41' 24" W. Etymology: After the grass that grows in a marshy area at one end. Site for catching whitefish and trout.

Choostl'okoh *N* Ormond Lake River.

-chul *N* younger brother or male cousin.

chunalhduz *N* Creamy Peavine.

Chunbaz Bunk'ut *N* Alf lake. Latitude: 54° 4' 47" N. Longitude: 124° 55' 33" W. Etymology: "wooden raft lake".

chundoo *N* lodgepole pine. Locally known as "jack pine", which standardly refers to Pinus banksiana.

Chundooyus *N* unnamed hill. The hill on the sooutheest side of Ormond Lake. Latitude: 54° 0' 0" . Longitude: 0° 124' 0". Etymology: "lodgepole pine hill".

chundulht'a *N* woodpecker.

chunih *N* marten.

chunihcho *N* fisher.

chunkhelh *N* box.

chunlai *N* salamander. The only species of salamander found in the region is the Long-toed Salamander Ambystoma macrodactylum. No lizards are found in the region. However, this term is applied to other varieties of salamander and to lizards, such as the geckos sold as pets.

chuntulhi *N* coyote.

chuntunuye *N* coyote.

chunzool *N* cedar.

chus *N* down (of birds).

-ch'e' *N* property. With possessive prefixes forms the possessive pronouns mine, yours, etc.

ch'ok *N* Sitka Mountain Ash.

ch'uk *N* Red-winged Blackbird.

ch'usghak *N* peaked roof.

-da *N* lips, beak. (1) Du**da** be lhukw ookaih. 'He (a pelican) scoops up fish in his beak.'

dada *N* ailment, illness.

dadacho *N* cancer. Etymology: "big illness".

dadel'ez *V* [l-'ez < 'es$_1$] he/she put his foot inside. *[PA]*

dadeltel *N* porch.

dadembun *V* [0-bun < bun$_1$] it overflowed. *[PA]*

dadentan *N* door.

dadentanghunez'ai *N* doorknob. Duoplural: dadentanghunezdla

dadentanghunezdla *N* doorknobs.

dadentank'et *N* doorway.

dadent'az *N* window.

dadent'az ts'oh sula *N* curtain.

dadent'az ts'osdla *N* curtains.

dadent'azts'asdla *N* window curtain.

dadzi *N* Common Loon.

-dagha *N* moustache, facial hair.

dagwut *N* fish spear.

dahalhts'i *V* [lh-ts'i < ts'i$_2$] they caused a draft. This is said of people who allow the wind to enter by leaving a door or window open. *[PA]*

dahgha *N* Brittle Horsehair Lichen (tree moss).

dahidental *V* [0-tal < tulh$_1$] they kicked the door in. *[PA]*

dahint'ah *V* [0-t'ah < t'oh$_1$] how are they? *[IA]*

dahjolh *N* black fly.

dahot'e *V* [0-t'e < t'oh1] how is it? *[OA]*

dahooja *V* [0-ja] what happened? *[PA]*

dahooldzoh *V* [l-dzoh] how far?, how long a time? *[IA]*

dahoolk'uz *V* [l-k'uz] how cold is it? *[IA]*

dahoont'ah *V* [0-t'ah < t'oh$_1$] how is it? *[IA]*

dahuneltsuk *V* [l-tsuk] how many of them? *[IA]*

dahunindil *V* [0-dil < dil$_1$] they (3+) entered. *[PA]*

dahyus *N* snow up high.

dahzo *N* frost up high.

dai *N* hunger.

daidental *V* [0-tal < tulh$_1$] he/she kicked it (the door) in. *[PA]*

Daideyus *N* A mountain between Nadleh and Nak'azdli. Etymology: "the peak that blocks hunger". So-called because the area has lots of game and so helps people to avoid starvation.

daint'ah *V* [0-t'ah < t'oh$_1$] how are you (1)? *[IA]*

dakelh *N* Carrier person.

dak'et *N* fall, autumn.

dalcho *V* [l-cho < cha$_1$] how big is it? *[IA]*

daldzoo *V* [l-d-zoo < zoo$_1$] how pretty is it? *[IA]*

dalghen *N* cream.

daltsi' *N* Peamouth Chub. Also known as Northwest Dace (1)."

daltsuk *V* [l-tsuk] how much? *[IA]*

dalyiz *V* [l-yiz] how long is it? *[IA]*

danelhutsuk *V* [l-tsuk] how many of you? *[IA]*

danidutal *V* [0-tal < tulh$_1$] we (2) kicked the door in. *[PA]*

danidut'az *V* [0-'az < 'as$_1$] we (2) entered. *[PA]*

danilhk'az *V* [lh-k'az] a draft blew in. *[PA]*

danilhts'i *V* [lh-ts'i < ts'i$_2$] it (wind) is blowing into. This refers to a substantial wind blowing in, as through an open window. *[IA]* Compare: dats'i.

daninya *V* [0-ya < ya$_1$] he/she entered. *[PA]*

dant'ah *V* [0-t'ah < t'oh$_1$] how is he? *[IA]*

dant'i *QWH* what? (1) **Dant'i** ka'ninzun? 'What do you want?'

danustal *V* [0-tal < tulh$_1$] I kicked the door in. *[PA]*

danusya *V* [0-ya < ya$_1$] I entered. *[PA]*

danzoo *V* [0-zoo < zoo$_1$] he/she is generous. *[IA]*

daskwun *N* skewer, fire poker.

dati *N* entrance, doorway.

datuk *N* doorway.

datsa *V* [0-tsa] he/she is begging. *[IA]*

datsan *N* American Crow.

Datsan T'acho *N* The open, grassy hills on the north side of Fraser Lake, visible from Lejac. Etymology: "raven's wing".

datsan'algut *N* juniper.

datsancho *N* Common Raven. Etymology: "big crow".

Datsanchonun *N* The open, grassy hills on the north side of Fraser Lake, visible from Lejac. Etymology: "raven hillside"

dats'i *V* [lh-ts'i < $ts'i_2$] it (wind) is blowing into. This refers to a draft, where a little bit of wind blows in, as through a crack. *[IA]* Compare: danilhts'i.

datsan *N* American Crow.

dats'ooz *N* mouse. Etymology: Literally, "squeaker".

dawtenilh *V* [0-nilh] what will happen? *[FA]*

dayi *N* chief, generally with reference to band chief. Etymology: Loan from Chinook Jargon taye, itself from Nuuchanulth *ta:yi:* "elder brother, senior". According to Tommy George, this term was first used in reference to Joz **Dayi** "Chief George", in 1862.

dayun *N* midsummer.

-dayun *N* the area surrounding the backbone. (1) S**dayun** whunduda. 'The area surrounding my backbone is sore.'

dazdental *V* [0-tal < $tulh_1$] we (3+) kicked the door in. *[PA]*

dazneltsuk *V* [l-tsuk] how many of us? *[IA]*

daznindil *V* [0-dil < dil_1] we (3+) entered. *[PA]*

dazsai *V* [0-tsai] he/she died. *[PA]*

dazul *N* echo.

de' *INT* give me. In spite of the curtness suggested by the English translations, this expression is perfectly polite. It is perfectly polite by itself; indeed, Carrier has no word for "please".

-de' *N* horn, antler.

de'indak *N* his/her own flower.

dech'ulh *N* cloth.

dedah *V* [0-dah < dah_1] it is of (st) kinds. *[IA]* (1) Nawh **dedah.** 'There are two kinds.'

dedohneyun *N* clan song.

dedohshun *N* clan song.

Deduk Usk'en *N* Mount Greer. Latitude: 53° 59' 47" N. Longitude: 124° 30' 58" W.

dedulti *V* [lh-ti < ti_2] we (2) honour (s.o.). *[IA]*

dek'a *N* tobacco.

dek'atse *N* tobacco pipe.

datsancho *N* Common Raven. Etymology: "big crow".

Dek'e Yusk'ut *N* A small hill on the northeast side of Nadleh Village, where the water tower is now located. It is an important feature in the legend of Ketlangai, and was formed by the villagers who fell to the earth after being carried into the air by the eagle or fish hawk that captured Ketlangai.

Dek'ezyusk'ut *N* The mountain East of Nadleh. Etymology: "Last snow".

del'us *V* [l-'us] he/she is walking in a slow, stately manner. *[IA–prog]*

delhti *V* [lh-ti < ti$_2$] I honour (s.o.). *[IA]*

delhuts'i *V* [l-ts'i < ts'i$_1$] you (3+) are sitting. *[IA]*

dena'whudults'it *V* [l-ts'it] he/she is bragging about himself. *[IA]*

denawhudinges *V* [0-ges] you (1) tie yourself down, put on your seatbelt. *[IA]*

denayulhdzeh *V* [lh-dzeh < dzeh$_1$] he/she is pouring it back into (st). *[IA]*

-des *N* lungs.

desgwut *V* [0-gwut] it is sharp. *[IA]*

-destl'oh *N* younger sister.

-desyaz *N* younger sister.

detnik 'utni *V* [d-ni] it thunders. *[IA]*

detnik na'nanguz *V* [0-guz] there is a rainbow. *[IA]*

detnik oolhgis *V* [lh-gis] lightning strikes. *[IA]*

deyoh naih *N* underwear.

deyoh tl'asus *N* slip.

deyulhdzeh *V* [lh-dzeh < dzeh$_1$] he/she is pouring it into (st). *[IA]*

-dezcho *N* female second cousin older than ego.

dezghel *V* [0-ghel < ghel$_2$] it is calm, becalmed. *[PA]*

dezti *V* [0-ti < ti$_2$] it is expensive. *[IA]*

deztulh *V* [0-tulh] he/she kicked (d-class object). *[PA]* (1) Skechun **deztulh.** 'He kicked me in the leg.'

-dezyaz *N* female second cousin younger than ego.

di *QWH* what? (1) **Di** ts'utni? 'What is it called?'

dich'ah *PRO* himself, by himself.

didughai *V* [0-ghai < ghai$_1$] we (2) are hairy. *[IA]*

didughut *V* [0-ghut < ghut$_1$] we (2) are sawing (st). *[IA]*

didun *NUM* four *[locative]*

didut *PRO* he, him/her, her, she.

dihcho *N* Blue Grouse.

dika *QWH* why?

dillhuk *V* [l-lhuk] it is flammable. *[IA]*

dilh *N* Sandhill Crane.

dilhghis *V* [lh-ghis < $ghis_1$] you (1) turn (st). *[IA]*

dilhtsul *N* Whooping Crane.

Dimos k'elh'az *N* Monday.

Dimosdzen *N* Sunday.

dincha *V* [0-cha < cha_1] it is big, it (snow) is deep. [abs: d] *[IA]* (1) Yus **dincha.** 'The snow is deep.'

dindukw *V* [-dukw < dukw1] it is short. [abs: d] *[IA]*

dink'i *NUM* four *[generic].*

dintel *V* [0-tel] it is wide. [abs: d] *[IA]*

dintsi *V* [0-tsi < tsi'_1] it is bad. *[IA]* (1) Shun **dintsi.** 'The song is bad.' (2) Duchun **dintsi.** 'The stick is bad.'

dinun *NUM* four *[human].*

dinyiz *V* [0-yiz] it is long. [abs: d] *[IA]* (1) Duchun **dinyiz.** 'The stick is long.'

dinzoo *V* [0-zoo < zoo1] it is good, beautiful. [abs: d] *[IA]* (1) Shun **dinzoo.** 'The song is nice.'

dit *NUM* four [multiplicative], fourth.

dit lanezi *NUM* forty *[generic].*

ditnikwun *N* dandelion.

ditnikwun *N* Indian Paint Brush.

diwh *NUM* four *[abstract].*

diyooh *V* [0-yooh < $yooh_1$] it smoulders. *[IA]*

diyooh *N* birch conk. Etymology: Literally, "it smoulders", after its use as mosquito repellant. **diyooh** is set on fire and used as mosquito repellant.

Dohcho *N* a mountain near the Endako mine.

Dohyaz *N* a mountain near the Endako mine.

dolulh *N* floats for fish net.

duchuntah *N* forest, bush.

dondeti *N* dinosaur.

doso *N* burlap, denim, canvas.

doso tl'asus *N* blue jeans.

doocha *ADV* again.

dooda *N* butter.

du- *PREFIX* third person reflexive possessor for class 1 nouns.

du'ek *V* [0-'ek < $'ek_1$] it is breakable. *[IA]*

duba nadesuti *VP* he/she has delusions of grandeur. *[IA]*

dubanawhuldzit *V* [l-dzit] he/she makes it look good for himself. This could be said, for example, of someone who looks at himself in a mirror, sees that he looks bad, but tells himself he looks good. It could also be said of someone who covers up for a mistake. *[IA]*

dubat'eoonaoonudzun *V* [d-zun] he/she is opiniated. *[IA]*

dube *N* goat.

duchun *N* stick, tree.

duchun be'ts'uyi *N* chopsticks. Etymology: "Sticks by means of which we eat".

duchun khelh *N* trunk.

duchuntah *N* forest, bush.

duchunyoo *N* herbal medicine.

duch'ukw *N* porcupine.

duda *N* his/her own lips. (1) **Duda** be lhukw ookaih. 'He (a pelican) scoops up fish in his beak.'

dudi'a *V* [0-'a < 'ai$_1$] you (1) pick it up. [class: sdo-gen] *[IA]* (1) Sba **dudi'a.** 'Pick it up for me.'

dugoos *N* sucker species.

Dugoos Ooza *N* May. Etymology: "month of the sucker fish".

dugha *N* his/her own hair, fur. (1) **Dugha** hallhuk. 'He jumps out of his fur.'

dughai *V* [0-ghai < ghai$_1$] he/she is hairy, furry. [abs: 0] *[IA]*

dughas *V* [0-ghas < ghaz$_2$] he/she is planing. *[IA]* (1) Dzihtel **dughas.** 'He is planing a board.'

dughoo *N* his/her own teeth. (1) **Dughoo** ulhgoos. 'He grinds his teeth.'

dughul *N* safety pin.

dughuntezghal *V* [0-ghal < ghal$_1$] he/she tripped and fell. *[PA]*

dughut *V* [0-ghut < ghut$_1$] he/she is sawing (st). *[IA]* (1) Tsuz **dughut.** 'He is sawing wood.'

dujooz *V* [0-jooz < jooz$_1$] it is sharp, pointed. *[IA]*

duk duntuldus *V* [l-dus] it is bouncing up and down. *[IA]*

dukdunenkai *N* thimbleberry.

duke *N* his/her own feet.

duke be 'ulhghalh *V* [lh-ghalh < ghalh$_1$] he/she is tap-dancing. *[IA]*

duk'ai *N* rainbow trout.

Duk'ai Hooni Bunk'ut *N* Barlow Lake. Latitude: 54° 11' 13" N. Longitude: 124° 27' 2" W. Etymology: "lake where there are trout". Seymour Thomas and others used to cut hay here.

Duk'ai Ooza *N* June. Etymology: "month of the trout".

dul suli *V* [0-li] it (blood) has clotted. *[PA]*

dulagi *N* his/her own fingernail. (1) **Dulagi** ulhgoos. 'He bites his fingernails.'

dulai *V* [0-lai < lai$_1$] they are numerous. [abs: d] *[IA]*

dulba *V* [l-ba < ba$_1$] it is off-white. This includes shades such as cream, tan, and blond. *[IA]*

duldlut *V* [l-dlut] he/she speaks very loudly, orates. *[IA]*

duldzan *V* [l-dzan < dzan$_1$] it is blue. *[IA]*

dulgi *V* [l-gi < gi$_1$] it is grey. *[IA]*

dulgiyaz *N* sucker species.

dulgoosmai *N* Bitter Cherry, Chokecherry.

dulghi *V* [l-ghi] he/she is hollering. *[IA]*

dulk'un *V* [l-k'un < k'un$_1$] it is red. *[IA]*

dulk'us *V* [l-k'us] it is crackling. Describes the sound made by a fire or by cracking a joint. *[IA]*

dulkwus *V* [l-kwus] he/she is

coughing. *[IA]*

dulkw'ah *N* frog.

dulkw'ah besk'um *N* clam.

dultl'uz *V* [l-tl'uz < tl'uz$_1$] it is yellow. *[IA]*

dultso *N* bile.

dulyul *V* [l-yul < yul$_1$] it (d-class) is white. [abs: d] *[IA]* (1) Stuz **dulyul.** *'My cane is white.'*

dulhbai *N* dentalium shell.

-dulhchai *N* great-grandchild.

dulhgus *V* [lh-gus < gus$_1$] it is black. *[IA]*

dulhnat *V* [lh-nat] he/she is splitting (st). *[IA]* (1) Tsuz **dulhnat.** 'He is splitting wood.'

dulhtus *N* his/her own sister.

Dulhts'ehyoo *N* The name of a clan. It's crests are: frog, crane, raven.

dulhukwus *V* [l-kwus] I am coughing. *[IA]*

duna'ulh *V* [0-'ulh < 'alh$_1$] he/she is chewing. *[IA]* (1) Dunitsun **duna'ulh.** 'He is chewing moose meat.'

dune *N* man, person. Duoplurals: dunene, duneke.

dune 'adilhti *N* grave.

dune ba yalhduk *N* lawyer.

dune besulhti *N* coffin.

dune delts'i *N* living room.

dune dunuyuz *VP* he/she is bossy. *[IA]*

dune hodulh'eh *N* teacher.

dune oolhdzai *N* extinct giant bird.

dune tl'asus *N* pants. Etymology: "man's bum covering".

dune tsalhts'ul *N* baby boy.

dune uyi *N* cannibal.

dunecho *N* tall person, older man. Duoplural: dunechone

dunene *N* people, men.

duneti *N* foot path.

duneti *N* old man. Duoplural: dunetike

duneti kw'uts'uda *N* reclining chair. Etymology: Literally, "male elder's chair".

dunetsi' *V* [0-tsi' < tsi'$_1$] he/she is ugly. Applicable only to men. *[IA]*

duneyaz *N* boy.

duneyez *N* tall man.

duneyun *N* personal song.

duneza' *N* male noble in the clan sytsem. Duoplural: duneza'ne

duneza' *N* porcupine.

dunezoo' *V* [0-zoo'] he is handsome. Applicable only to men. *[IA]*

dulkw'ah *N* frog.

dunezoolh *N* ghost.

duni *N* moose.

duni'at *N* cow moose. Etymology: Literally, "moose wife".

duni’ulh *V* [0-’ulh < ’alh$_1$] you (1) are chewing. *[IA]* (1) Dunitsun **duni’ulh.** ‘You are chewing moosemeat.’ (2) Soocho **duni’ulh.** ‘Chew properly.’

dunih *N* kinnikinnick.

duniht’an *N* kinnickinnick leaves.

duninchuz *N* his/her own thumb.

dunink’ooz *V* [0-k’ooz] it is sour. *[IA]*

dunitsun *N* moose meat.

dunizuz *N* moose hide. (1) **Dunizuz** yubuz. ‘He is stretching the moose-hide.’

Dunt’emyoo *N* The name of a clan. Its crests are black bear, grizzly bear, wolf, full moon.

dunughai *V* [0-ghai < ghai$_1$] he/she has facial hair, it is hairy. [abs: n] *[IA]* (1) Bunen **dunughai.** ‘His face is hairy.’

dunuldzan *V* [l-dzan < dzan$_1$] it is blue. [abs: n] *[IA]* (1) ’Ilhtsul **dunuldzan.** ‘Low-bush blueberries are blue.’

dunulk’un *V* [l-k’un < k’un$_1$] it is red. [abs: n] *[IA]*

dunultl’uz *V* [l-tl’uz < tl’uz$_1$] it is yellow. [abs: n] *[IA]*

dunulhdus *V* [lh-dus] he/she is grinding (st). *[IA]* (1) ’Utsun **dunulhdus.** ‘He is grinding meat.’

dunulhgus *V* [lh-gus < gus$_1$] it is black. [abs: n] *[IA]*

dunusghai V [0-ghai < ghai$_1$] I have facial hair. *[IA]*

dunusyus *V* [0-yus] I crumple (st). *[IA]*

dunutsun *V* [0-tsun] it is dirty. [abs: n] *[IA]*

dunuts’un *V* [0-ts’un < ts’un$_1$] it is hard, unyielding. [abs: n] *[IA]*

dunuyo *V* [0-yo] it is shaggy, he/she has peach-fuzz on his face. [abs: n] *[IA]*

dusneke *N* ancestors.

dustl’us *V* [0-tl’us] I paint (st). [abs: 0] *[IA]*

dustl’us *N* paper.

dustl’us chunkelh *N* cardboard box.

dustl’us dakat *N* post office.

dustl’us dutsun *N* card board.

dustl’uszuz *N* paper bag.

dusts’i *ADV* upper.

Dushin *N* Cree person.

dutni *N* marmot.

dutoo *N* his/her own water.

dut’ai *N* bird, duck.

dut’ai ooltas *N* hawk.

dut’aitsun *N* bird meat.

dutleh *V* [0-tleh < tle$_2$] it is soft. *[IA]*

dutsiz *V* [0-tsiz] it is rough. *[IA]*

dutsun *V* [0-tsun] it (generic) is dirty. *[IA]* (1) Nyints’iz **dutsun.** ‘Your elbow is dirty.’

duts’un *V* [0-ts’un < ts’un$_1$] it is hard. [abs: 0] *[IA]*

duts’un suli *V* [0-li] it hardened, became hard. [abs: 0] *[PA]*

duyeztli *N* his/her own horse. (1) **Duyeztli** yoo be yulhlhah. ‘He is smearing his horse with salve.’

duyo *V* [0-yo] he/she is shaggy. [abs: 0] *[IA]*

Duyuk *N* The canyon on the Nechako River downstream from its junction with the Nautley River. Etymology: "canyon". A fishing station for gaffing, spearing, and netting Sockeye Salmon, and for angling for Dolly Varden, Char, and Rainbow Trout.

duyuk *N* canyon, waterfall.

duyun *N* shaman, medicine man.

duyun ts'eke *N* female shaman.

duyun-ishun *N* ritual song. Ritual songs are sung only by 'uza'.

duzah *V* [0-zah] he/she is dishonest. *[IA]*

dlat *N* weeds in water.

dli yuzelhghi *V* [lh-ghi < ghi_1] he/she died of exposure. Literally, "feeling cold killed him/her". *[PA]*

dlim dats'olh *N* Masked Shrew.

dlimcho *N* packrat, Bushy-tailed Wood Rat.

dlohuninzun *V* [0-zun] they are smiling. *[IA]*

dlonus'i *V* [0-'i] I am amused. *[IA]*

dlonuszun *V* [0-zun] I am smiling. *[IA]*

dlots'uninzun *V* [0-zun] we (3+) are smiling. *[IA]*

-dza *N* ear.

-dzabal *N* earlobe.

dzah 'uhooja *V* [0-ja] something bad happened. *[PA]*

dzah nusuzut *V* [0-zut] he/she suffered greatly, endured an ordeal. *[PA]*

dzah nuszut *V* [0-zut] I am suffering. *[IA]*

dzeh *N* pitch, chewing gum.

-dzek *N* ear canal.

-dzek whudadilhchooz *N* eardrum.

dzekw'ul *N* earrings.

dzen *N* day.

Dzen Dilhdukw *N* December. Etymology: "short days".

dzenes *N* daytime.

Dzenwhuyaz *N* Saturday. Etymology: "little day".

dzetniz *N* noon.

dzetniz hukw'elh'az *N* afternoon.

dzezoh *N* kerchief.

-dzi *N* heart.

-dzidool *N* artery.

dzihtel *N* board. (1) Dzihtel dughas. 'He is planing a board.'

-dzik'ut *N* chest.

dzohtsulyaz tink'us unli *V* [0-li] they (teeth) are offset. *[IA]* (1) Bughoo **dzohtsulyaz tink'us unli.** 'She has offset teeth.'

dzoot *N* coat.

dzootdukw *N* vest. Etymology: "short coat".

dzoozt'an *N* shirt. Etymology: "thin coat".

dzulh *N* mountain.

dzulhcheh *N* base of a mountain.

dzulhcho *N* big mountain.

dzulhyeguz *ADV* between mountains.

edughilh *V* [0-ghilh < ghe_1] we (2) are packing. [IA–progressive]

eghilh *V* [0-ghilh < ghe] he/she is packing. *[IA–progressive]*

Endakoh *N* Endakoh River. Latitude: 54° 3' 41" N. Longitude: 124° 55' 41" W.

esjun *V* [d-yun < yun_1] I sang. *[PA]*

esyi *V* [0-yi < yi_1] I ate. *[PA]*

gah *N* rabbit. The only species native to the territory is the Snowshoe Hare, also known as Varying Hare, Lepus americanus. The term is, however, applied to other varieties of rabbit and hare.

gahdzo *N* rabbit ears. Eaten skewered and scorched over a fire.

gahgi N dried rabbit meat.

gahk'ah *N* rabbit tracks.

gahti *N* rabbit trail.

gahtsun *N* rabbit meat.

gal be 'ut'en *V* [d-'en < $'en_2$] he/she is walking quickly. *[IA]*

gal be nuye *VP* he/she is walking quickly. *[IA]*

gambilh *N* rabbit snare.

-gan *N* arm.

-gant'uk *N* collar bone.

ges *N* Spring Salmon.

Ges Hutaghe *N* The trail from the old Nechako River bridge crossing to the canyon along the ridge of the hill. Etymology: "they pack Spring Salmon".

gescho *N* A variety of salmonid fish.

Gestlah Ooza *N* September.

gestl'ah *N* kokanee.

gesya *N* a variety of slug. Etymology: Literally, "Spring Salmon louse".

gooh *N* deadfall trap.

goolh *INT* dear, sweetie. Endearment addressed children and younger brothers and sisters.

goos *N* Cow Parsnip.

goosibai *N* sucker species.

goozeh *N* Whiskey Jack. Also known as Gray Jay and Canadian Jay.

gugoos *N* pig. Etymology: Ultimately from French *coche*, probably via Cree.

gugoos k'atsun *N* bacon.

gugoostsun *N* pork.

gha *PP* because of.

-gha *N* fur, hair of body.

ghada *N* last year.

ghait'ah *PART* cannot. This particle is used to express inability. It follows the optative affirmative form of the verb, which is in turn preceded by the negative particle 'aw. (1) 'Aw noosbuz **ghait'ah.** 'I can't stretch it.' (2) 'Aw wusbuz **ghait'ah.** 'I can't stretch it.'

ghanususguz *V* [0-guz] I cased. This refers to skinning by pulling the carcass out of the hide rather than cutting the hide away with a blade. *[PA–distributive]* (1) Gah lhai **ghanususguz.** 'I cased a bunch of rabbits.'

-ghe *N* spouse's sibling, sibling's spouse.

-ghez *N* testicles, eggs of birds.

-ghezzus *N* scrotum.

ghidudli *V* [0-li] we (2) are guarding (s.o.). *[IA]*

-ghoo *N* teeth.

-ghootsun *N* gums.

ghuhinli *V* [0-li] they take care of

(st), baby-sit (s.o.). *[IA]*

ghunadulhu'oo' *V* [l-'oo' < tl'oo$_1$] he/she put on (st). Something like a belt or sash that one "ties" on. *[PA]* (1) Se **ghunadulhu'oo'.** 'He put on a belt.'

ghunadulhu'oo' *V* [l-'oo' < tl'oo$_1$] I put on (st). Something like a belt or sash that one "ties" on. *[PA]* (1) Se **ghunadulhu'oo'.** 'I put on a belt.'

ghunadutsa *V* [0-tsa] they are crying. *[IA]*

ghunasdutsa *V* [0-tsa] we are crying. *[IA]*

-ghundan *N* son-in-law.

ghuninedutsa *V* [d-tsa] you are crying. *[IA]*

ghusli *V* [0-li] I am guarding (s.o.). *[IA]*

-ghuts *N* cartilage.

ghuts'inli *V* [0-li] we (3+) are guarding (s.o.). *[IA]*

gwada' *N* quarter. That is, a twenty-five cent coin. Etymology: Loan from English quarter.

-gwastl'oh *N* niece (younger than ego).

-gwaz *N* nephew, niece.

-gwut *N* knee.

-gwutduts'un *N* kneecap.

-gwutt'uk *N* back of the knee.

-gwuttsi *N* kneecap. Etymology: Literally, "knee head".

-gwutyuschun *N* shin.

gwuzih *N* Whiskey Jack, Grey Jay, Canadian Jay.

ha'dunet'ah *V* [0-t'ah] he/she cheats. *[IA]*

ha'hunule *V* [0-le] they harvest (u.o.). *[IA]*

ha'nidulye *V* [0-le] we (2) harvest (u.o.). *[IA]*

ha'nule *V* [0-le] he/she harvests (u.o.). *[IA]*

ha'nusle *V* [0-le] I harvest (u.o.). *[IA]*

ha'ts'unule *V* [0-le] we (3+) harvest (u.o.). *[IA]*

ha'ulih *V* [0-lih] he/she ferrets things out. *[IA]* (1) Tube **ha'ulih.** 'She really ferrets things out.'

ha'ulih *V* [0-lih] he/she ferrets things out. [class: mdo-c-gen] *[IA–customary]* (1) Tube **ha'ulih.** 'She really ferrets things out.'

hadanwul *V* [0-wul] it got chipped. *[PA]*

hadi' *INT* hello!, oops!, what! An expression of surprise.

hadih *INT* hello.

hadilhghis *V* [lh-ghis < ghis$_1$] you (1) unscrew. *[IA]*

hahonkai *V* [0-kai] he/she dug (with a shovel). *[PA]* (1) Nadun **hahonkai.** 'He dug two holes.'

hainya *V* [0-ya] you (1) came from. *[PA]* (1) Nts'ez **hainya?** 'Where did you come from?'

halghok *V* [l-ghok < ghok$_1$] it (fish) is jumping. *[IA]*

hallhat *V* [l-lhat] he/she jumps out of. *[IA]*

hallhuk *V* [l-lhuk] he/she jumps out of. *[IA–customary]* (1) Dugha **hallhuk.** 'He jumps out of his fur.'

haltoo' *N* trigger for rabbit snare.

hananulyeh *V* [l-yeh] it (plant) is

growing back. *[IA]*

handunelmulh *V* [l-mulh] it came to a boil. *[PA]*

handunelhmulh *V* [lh-mulh] he/she is bringing (st) to a boil. *[IA]*

hanulhyeh *V* [lh-yeh] he/she is growing (st:plant). *[IA]*

hanusle *V* [0-le] I harvest. *[IA]*

hanuyeh *V* [0-yeh] it (plant) is growing. *[IA]*

hanuyeh buts'uyi *N* fertilizer.

hanyi *V* [0-yi < yi_1] they ate. *[PA]*

hasadanat *V* [0-nat] it is sunny. *[IA]*

Hasilhos *N* 'uza' name in Dulhts'ehyoo clan. Formerly held by Bernadette McQuary.

hasya *V* [0-ya] I came from. *[PA]* (1) Southbank ts'i **hasya.** 'I came from Southbank.'

hasyis *V* [0-yis] I am plucking (st). *[IA]*

hat'al *N* cedar bark. Etymology: Loan from Witsuwit'en.

hats'unule *V* [0-le] we (3+) harvest. *[IA]*

hawukaih *N* cache.

hawus *N* foam.

hawhulhjuk *V* [lh-juk < jut_1] it rots out. *[IA–customary]* (1) Hubutsul **hawhulhjuk.** 'Their anuses rot out.'

hayanyuz *V* [0-yuz] he/she plucked it. *[PA]*

hayuyis *V* [0-yis] he/she is plucking it. *[IA]*

hedulh *V* [0-dulh < dil_1] they (3+) are walking. *[IA–prog]* (1) Fraser Lake ts'i **hedulh.** 'They are going to Fraser Lake.' (2) Fraser Lake ts'i lak **hedulh?** 'Are they going to Fraser Lake?'

hegooh *V* [0-gooh < goo_1] they are driving. *[IA–prog]* (1) Lheidli ts'i **hegooh.** 'They are driving to Lheidli.' (2) Lheidli ts'i lak **hegooh?** 'Are they driving to Lheidli?'

hekelh *V* [0-kelh < ke_2] they are going by boat. *[IA–progressive]*

hidelhti *V* [lh-ti < ti_2] they honour him/her. *[IA]*

hoolht'o *N* bee.

hidilhti *V* [lh-ti] they buried him/her. *[PA]*

higha'ooli *V* [0-li] they are careful with it. *[IA]*

highunli *V* [0-li] they are guarding him/her. *[IA]*

hika'ninzun *V* [0-zun] they want it. *[IA]*

hikatelhughaz *V* [l-ghaz] they (3+) are running after it. *[PA]*

hikatezdil *V* [0-dil < dil_1] they (3+) are going after it. *[PA]*

hik'untal *V* [0-tal < $tulh_1$] they kicked a hole in it. *[PA]*

hincha *V* [0-cha < cha_1] they are big. *[IA]*

hindunulhmulh *V* [lh-mulh < mulh$_1$] they are boiling it (water). *[IA]*

hinilh'en *V* [lh-'en] they look at it. *[IA]*

hintsi V [0-tsi < tsi'$_1$] they are bad. *[IA]* (1) Ts'ekoo **hintsi.** 'The women are bad.'

hinyez *V* [0-yez] they are tall. *[IA]*

hinzoo *V* [0-zoo < zoo$_1$] they are good. *[IA]*

hiyeneyulhgoo *N* vehicle.

hiyenuljut *V* [l-jut < joot$_1$] they are afraid of it. *[IA]*

hobudulh'eh *V* [lh-'eh < 'eh$_1$] he/she is teaching them. *[IA]*

hodul'eh *V* [l-'eh < 'eh$_1$] he/she is learning. *[IA]*

hodulh'eh *V* [lh-'eh < 'eh$_1$] he/she is teaching. *[IA]*

hoh *ADV* around.

hon'en *V* [0-'en < 'en$_1$] he/she sees (wh-class object). *[IA]*

hoodutal'eh *V* [l-'eh] you (1) will learn. *[FA]*

hoodutelhu'eh *V* [l-'eh] you (2+) will learn. *[FA]*

hoolah *V* [0-lah] there is not, he/she is absent. *[IA]*

hoolil *V* [0-lil] he/she disappeared. *[PA]*

hooloh *V* [0-loh] he/she is absent. *[IA]*

hoolhghulh *N* Devil's Club.

hoolhkw'ul *N* beetle.

Hoolhtan Bunk'ut *N* Top Lake. A small lake west of Peta Lake, at the head of Duncan Creek. Latitude: 54° 13' 24" N. Longitude: 124° 49' 8" W. Etymology: "stops raining lake". Camping and trapping area. The original wagon road runs along the west side of the lake.

hoolht'as *N* Peregrine Falcon.

hoolht'o *N* bee.

hoolht'ocho *N* bumblebee.

hoolht'oghe *N* honey. Etymology: "bee grease".

hoolht'ot'o *N* beehive.

hoolhts'ik *N* Stinging Nettle.

hooncha *V* [0-cha < cha$_1$] it is large. [abs: wh] *[IA]*

hoondukw *V* [0-dukw < dukw$_1$] it is short. [abs: wh] *[IA]*

hoongwun *N* cedar chest.

hoonk'az V [0-k'az] it got cold. [abs: wh] *[PA]*

hoonliz *N* skunk.

hoontel *V* [0-tel < tel$_1$] it is wide. [abs: wh] *[IA]*

hoont'i *V* [0-t'i] he/she is happy. *[IA]*

hoont'i *V* [0-t'i < t'i$_1$] he/she likes (st). [abs: wh] *[IA]* (1) Nudaih **hoont'i.** 'He likes to dance.'

hoont'i *V* [0-t'i] you (1) reside. *[IA]* (1) Nts'e **hoont'i?** 'Where do you live?'

hoontsi' *V* [0-tsi' < tsi'$_1$] it is bad. [abs: wh] *[IA]* (1) Buzek **hoontsi'.** 'His mouth is dirty.'

hoonust'i *V* [0-t'i] I am happy. *[IA]*

hoonust'i *V* [0-t'i < t'i$_1$] I like (st). [abs: wh] *[IA]* (1) Nusdaih **hoonust'i.** 'I like to dance.' (2) Usjun **hoonust'i.** 'I like to sing.'

hoonyan *N* elder.

hoonyiz *V* [0-yiz] it is long, tall. [abs: wh] *[IA]* (1) Yah **hoonyiz.** 'The house is tall.'

hoonzoo *V* [0-zoo < zoo_1] it is good. [abs: wh] *[IA]*

hoot'as *N* horsefly.

hoot'ukw *N* leech. Also known as "blood sucker". (1) **Hoot'ukw** suzt'ukw. 'A leech has sucked on me.'

hooyi *V* [0-yi < yi_1] they may eat (st). *[OA]*

hoozdutel'eh *V* [l-'eh] we (3+) will learn. *[FA]*

huba 'alha'hoont'ah *V* [0-t'ah < $t'oh_1$] they believe it. *[IA]*

huba't'en lhai *V* [0-lai] they are busy. *[IA]*

hubadughan *V* [0-ghan < $ghan_1$] we (2) killed them (3+). *[PA]*

hubak'i *N* their mother's sister.

hubanelhu'it *V* [l-'it] I am anxious. *[IA]*

hubanghan *V* [0-ghan < $ghan_1$] he/she killed them (3+). *[PA]*

hubanilh'en *V* [lh-'en < $'en_1$] they are looking at them. *[IA]*

hubat'en lhai *V* [0-lhai] he/she is busy. *[IA]*

hube'indak *N* their flower.

hube'lunatdultsiyan *N* their great-grandfather.

hube'ook'et'as *N* their cook stove.

hube'ook'utudai *N* their table.

hube'ustes N their mattress.

hube'ustlen *N* their subject, servant.

hubeguz *PPC* between them.

hubeghoosusgoo *V* [0-goo < goo_1] I passed them driving. *[PA]*

hubesghan *V* [0-ghan < $ghan_1$] I killed them (3+). *[PA]*

hubhulh 'utet'ah *V* [0-t'ah < $t'oh_2$] they will go by plane. *[FA]*

huboona *N* their older brother.

hubu dani *V* [0-ni] he/she told them. *[PA]*

hoonyan *N* elder.

hubu- *PREFIX* third person duo-plural possessor for class 1 nouns.

hubu'atke *N* their wives.

hububa *N* their father.

hububat *N* their mittens.

hubube'uzdla *N* their cupboard, dresser.

hububez *N* their mother-in-law.

hububezan *N* their paternal aunt.

hububoo *N* their rectums.

hububoos *N* their cat.

hububun *N* their roof.

hububunt'ah *N* their attic.

hububut *N* their stomachs.

hububuttl'ah *N* the bottoms of

their stomachs.

hubuchai *N* their grandchild.

hubuchaicho *N* their great-grandchild.

hubuchaike *N* their grandchildren.

hubuchak *N* their ribs.

hubuchets'un *N* their tailbones.

hubuch'i' *PRO* theirs.

hubuda *N* their lips.

hubudadeltel *N* their porch.

hubudadentanghunez'ai *N* their doorknob.

hubudadentanghunezdla *N* their doorknobs.

hubudadent'az *N* their window.

hubudadent'azts'asdla *N* their curtains.

hubudagha *N* their moustaches.

hubudati *N* their doorway.

hubudedulti *V* [lh-ti < ti_2] we (2) honour them. *[IA]*

hubudelhti *V* [lh-ti < ti_2] he/she honours them. *[IA]*

hubudes *N* their lungs.

hubudesyaz *N* their younger sister.

hubudulhti *V* [lh-ti < ti_2] you (2+) respect them. *[IA]*

hubudune delts'i *N* their living room.

hubuduneti kw'uts'uda *N* their reclining chair.

hubudusneke *N* their ancestors.

hubudzi *N* their hearts.

hubudzidool *N* their arteries.

hubudzik'ut *N* their chest.

hubuchai *N* their grandchild.

hubudzoozt'an *N* their shirt.

hubugan *N* their arms.

hubugha *PPC* because of them.

hubughe *N* their brother-in-law, sister-in-law.

hubughez *N* their testicles.

hubughezzus *N* their scrotums.

hubughoo *N* their teeth.

hubughootsun *N* their gums.

hubughundan *N* their son-in-law.

hubugwaz *N* their nephew/niece.

hubugwut *N* their knees.

hubugwutduts'un *N* their kneecaps.

hubugwutt'uk *N* the backs of their knees.

hubugwutyuschun *N* their shins.

hubuhanghan *V* [0-ghan < $ghan_1$] they killed them (3+). *[PA]*

hubuhunilh'en *V* [lh-'en < $'en_1$] they are taking a deliberate look at each other. *[IA]*

hubujook *N* their toque.

hubukanuta *V* [0-ta < ta_1] he/she is looking for them. *[IA]*

hubukechun *N* their legs.

hubukechunnah *N* their ankles.

hubukechunt'ak *N* the back of their legs.

hubukelagi *N* their toenails.

hubukelamai *N* their toes.

hubukelamaicho *N* their big toes.

hubukelamaiyaz *N* their little toes.

hubukelamaiyazts'unts'e *N* their fourth toes.

hubukelamaiyez *N* their long toes.

hubukelamaiyezts'unts'e *N* their middle toes.

hubukelascho *N* their big toes.

hubukelatsul *N* their heels.

hubukengi *N* their toenails.

hubukent'ak *N* the tops of their feet.

hubukentsul *N* their heels.

hubuketl'ah *N* their soles.

hubuketl'i *N* their toe spaces.

hubuketsul *N* their heels.

hubukoo *N* their house.

hubukuike *N* their husbands.

hubuk'i *N* their pelvic regions.

hubuk'its'un *N* their hipbones.

hubukw'uts'uda *N* their chair.

hubukw'uts'udacho *N* their easy chair.

hubukw'uts'udayez *N* their chesterfield.

hubukw'uz *N* their kidneys.

hubula *PPC* helping them. (1) **Hubula** 'ut'en. 'He helps them.'

hubulachunah *N* their wrists.

hubulagi *N* their fingernails.

hubulak'et *N* their palms.

hubulasge *N* their fingers.

hubulasts'ah *N* their pinkies.

hubulat'ah *N* their wrists.

hubulat'ak *N* the backs of their hands.

hubulatl'i *N* their finger spaces.

hubulik *N* their dog.

hubulikcho *N* their big dog.

hubulili *N* their bed.

hubulilik'usulhchooz *N* their sheet.

hubuloo *N* their mother.

hubuludab *N* their table.

hubuludi *N* their tea.

hubuluz *N* their urine.

hubuluzkah *N* their urinary tract.

hubuluzzus *N* their bladders.

hubulh *PPC* with them. (1) **Hubulh** hunilch'e. 'He is mad at them.'

hubulh 'et'ah *V* [0-t'ah < t'oh$_2$] they are going by plane. *[IA–prog]*

hubulh ne'uzt'a V [0-t'a < t'oh$_2$] they have gone in a loop by plane. *[PA]*

hubulhtus *N* their sister.

hubulhutsin *N* their brother.

hubumoodih *N* their boss.

hubumusdus *N* their cow

hubuna *N* their eyes.

hubunabagha *N* their eyelashes.

hubunalhti *N* their blanket.

hubunats'osdooz *N* their eyebrows.

hubunen *N* their faces.

hubuninchustah *N* their sinus

cavities.

hubuninchuz *N* their thumbs.

hubunints'uzti *N* their elbows.

hubunul'en-i *N* their television.

hubunut'uk *N* their muscles.

hubustocho *N* their big stove, heater.

hubutai *N* their uncle other than mother's brother, step-father.

hubutesk'et *N* their bed, bedroom.

hubutoo *N* their water.

hubutususyin *V* [0-yin < yin_1] I stand amongst them. *[IA]*

hubut'ak *N* their backs.

hubut'akts'un *N* their backbones.

hubut'eke *N* their friend.

hubut'ukts'un *N* their collarbone.

hubutl'ah *N* their bums.

hubutl'asus *N* their pants, dress.

hubutl'egha *N* their pubic hair.

hubutl'et *N* their groins.

hubutl'uz *N* their gall bladders.

hubutsah *PPC* before them, older than them.

hubutsakelhgwus *N* their navels.

hubutsan *N* their feces.

hubutsanbayah *N* their outhouse.

hubutse' *N* their daughter.

hubutsi *N* their heads.

hubutsi'alh *N* their pillow.

hubutsidah *N* the tops of their heads.

hubutsigha *N* their head hair.

hubutsinghai *N* their brains.

hubutsint'ak *N* the backs of their heads.

hubutsints'un *N* their skulls.

hubutsit'ak *N* the backs of their heads.

hubutsitl'ah *N* the backs of their heads.

hubutsiyan *N* their grandfather.

hubutsiyancho *N* their great-grandfather.

hubutsiyanyaz *N* their younger grandfather.

hubutsoo *N* their maternal grandmother.

hubutsoocho *N* their maternal great-grandmother.

hubutsoola *N* their tongues.

hubutsooyaz *N* their younger grandmother.

hubutsukw *N* their penises.

hubutsul *N* their rectums, anuses.

hubuts'ahbul *N* their cowboy hat.

hubuts'anghan *V* [0-ghan < $ghan_1$] we (3+) killed them (3+). *[PA]*

hubuts'ik *N* their intestines.

hubuts'ikcho *N* their large intestines.

hubuts'ilchun *N* their necks.

hubuts'ilchunt'ak *N* the napes of their necks.

hubuts'oo *N* their breasts.

hubuts'oola *N* their nipples.

hubuts'uyi ghutna-a *N* their kitchen.

hubuts'uyibesula *N* their food cupboard.

hubuwuda *N* their calves.

hubuwus *N* their shoulders.

hubuwuz *N* their thighs.

hubuwuzts'un *N* their femurs.

hubuyah *N* their house.

hubuyas'at *N* their daughter-in-law.

hubuyat *N* their older sister.

hubuyats'e *N* their daughter.

hubuye' *N* their son.

hubuyeztli *N* their horse.

hubuyun *N* their floor.

hubuyunlatel *N* their flooring.

hubuyunsubul *N* their floor covering.

hubuyunsulchooz *N* their floor covering.

hubuyunts'un *N* their backbones.

hubuyust'e *N* their bodies.

hubuz'e *N* their mother's brother.

hubuzaz *N* their father-in-law.

hubuzdelhti *V* [lh-ti < ti_2] we (3+) honour them. *[IA]*

hubuzek *N* their mouths.

hubuzesdak *N* their throats.

hubuzesdak'al *N* their uvulas.

hubuzesdalos *N* their uvulas.

hubuzesdamai *N* their uvulas.

hubuzesdazum *N* their tonsils.

hubuzets'un *N* their chins.

hubuzkai *N* their blood.

hubuzkaich'ooz *N* their veins.

hubuzkaich'oozcho *N* their artery.

hubuzkeh *N* their children.

hubuzool *N* their wind pipes.

hubuzoolts'un *N* their larynges.

hubuzumcho *N* their pancreases.

hubuzut *N* their livers.

hubuzuz *N* their hide.

hudelhts'i *V* [l-ts'i < $ts'i_1$] they (3+) are sitting. *[IA]*

hudich'ah *PRO* themselves, by themselves.

hudughai *V* [0-ghai < $ghai_1$] they are hairy. *[IA]*

hudunughai *V* [0-ghai < $ghai_1$] they have facial hair. *[IA]*

huhoont'i *V* [0-t'i] they are happy. *[IA]*

hukhuna *V* [0-na < na_1] they are alive. *[IA]*

hukwa'nuszun *V* [0-zun < zun_1] I want (st). [abs: wh] *[IA]* (1) 'Utesyilh **hukwa'nuszun.** 'I want to eat.' (2) Na'tesdilh **hukwa'nuszun.** 'I want to eat.'

hukwunilhdzut *V* [d-zut] you (1) think about it. *[IA]*

hukwuninta *V* [0-ta < ta_1] you (1) are looking for (st). [abs: wh] *[IA]*

hukwunintsel *V* [0-tsel] he/she chopped a hole in it. [abs: wh] *[PA]*

hukwunusta *V* [0-ta < ta_1] I am looking for (st). [abs: wh] *[IA]*

hukw'elh'az *PPC* after it. [abs: wh]

hukw'unihunilhdzut *V* [lh-zut] they are thinking about it. *[IA]*

hukw'unihuntelhdzut *V* [lh-zut] they will think about it. *[FA]*

hukw'uninalhdzut *V* [lh-zut] you (1) thought about it. *[PA]*

hukw'uninelhdzut *V* [lh-zut] I thought about it. *[PA]*

hukw'uniniduldzut *V* [lh-zut] we (2) are thinking about it. *[IA]*

hukw'uninilhdzut *V* [lh-zut] you (1) are thinking about it. *[IA]*

hukw'unintaduldzut *V* [lh-zut] we (2) will think about it. *[FA]*

hukw'unintalhdzut *V* [lh-zut] you (1) will think about it. *[FA]*

hukw'unintelhdzut *V* [lh-zut] I will think about it. *[FA]*

hukw'uninulhdzut *V* [lh-zut] I am thinking about it. *[IA]*

hukw'units'untelhdzut *V* [lh-zut] we (3+) will think about it. *[FA]*

hukw'uniznalhdzut *V* [lh-zut] we (3+) thought about it. *[PA]*

hukw'uniznilhdzut *V* [lh-zut] we (3+) are thinking about it. *[IA]*

hulan *V* [0-lan] they are numerous. *[IA]*

huldlai *V* [d-lai] they have many. *[IA]*

hulhda *N* yesterday.

hulhda whutsah da *N* day before yesterday.

hulhgha *N* evening.

hunduda *V* [0-da] they are sick. *[IA]*

hunesjan *V* [d-yan < yan_1] they have aged. *[PA]*

hunilch'e *V* [l-ch'e < $ch'eh_1$] he is mad, angry. *[IA]*

hunlhuch'e *V* [l-ch'e < $ch'eh_1$] he/she got mad. *[PA]* (1) Bulh **hunlhuch'e.** 'He got mad at him.'

hunlhuch'e *V* [l-ch'e < $ch'eh_1$] I am angry. *[IA]*

hunuljut *V* [l-jut < $joot_1$] they are afraid. *[IA]*

hunulwus *V* [l-wus] they are hot. *[IA]*

huske *V* [0-ke] he is angry. *[IA]*

hutedulh *V* [0-dulh < dil_1] they (3+) will walk. *[FA]* (1) Fraser Lake ts'i **hutedulh.** 'They will go to Fraser Lake.'

hutegooh V [0-gooh < goo_1] they will drive. *[FA]* (1) Lheidli ts'i **hutegooh.** 'They will drive to Lheidli.'

hutejun *V* [d-yun] they are going to sing. *[FA]*

hutekelh *V* [0-kelh < ke_2] they will go by boat. *[FA]*

hutelgih *V* [l-gih < $gaih_1$] they (2) will run. *[FA]*

hutelwus *V* [l-wus] they (3+) will run. *[FA]*

hutetez *V* [0-tez] they will sleep. *[FA]*

huteyi *V* [0-yi < yi_1] they will eat. *[FA]*

huwal *N* general appeal for funds at a balhats. In general, at a balhats, the funds distributed have been raised previously by the clan or family making the distribution. In some circumstances, however, an appeal is made to those present. This is said to be **huwal.** Compare: 'indamanuk.

huwawhulna *V* [l-na < na2] it is difficult. *[IA]*

huwunli *N* security guard.

huwunline *N* watchmen.

huyegus *V* [0-gus] they are dragging it. *[IA–prog]*

huyi *V* [0-yi < yi_1] they eat. *[IA]*

huylhts'ul *V* [lh-ts'ul] they are

broiling it. *[IA]*

huyoodulhkut *V* [lh-kut] they ask him/her. *[IA]*

huyootalh *V* [0-talh] they are kicking it. *[IA]*

huyulhjus *V* [lh-jus < jas$_2$] they hooked it (fish). *[PA]*

huyulhni *V* [lh-ni] they call it. *[IA]*

huyutsa k'ez hik'ut kwun dilhk'ai *N* Indian slow cooker.

huyuzbuz *V* [0-buz] they stretched it. *[PA]*

idulgi *V* [lh-gi] we (2) are drying (st). *[IA]*

huyuzelhghi *V* [lh-ghi < ghi$_1$] they killed it. *[PA]*

huyuztulh *V* [0-tulh] they kicked him/her. *[PA]*

huzke *V* [0-ke < ke$_1$] they (2) are sitting. *[IA]*

huztez *V* [0-tez] they are sleeping. *[IA]*

idudai *V* [0-dai < yi$_1$] we (2) eat. *[IA]*

idudzoo *V* [0-zoo < zoo$_1$] we (2) are good. *[IA]*

idugooh *V* [0-gooh < goo$_1$] we (2) are driving. *[IA–prog]*

idugus *V* [0-gus] we (2) are dragging (st). *[IA–prog]*

idujez *V* [0-yez] we (2) are tall. *[IA]*

idukelh *V* [0-kelh < ke$_2$] we (2) are going by boat. *[IA–progressive]* (1) Lejak ts'i **idukelh.** 'We are going to Lejac by boat.'

idulgi *V* [lh-gi] we (2) are drying (st). *[IA]*

idulgih *V* [l-gih < gaih$_1$] we (2) are running. *[IA–prog]*

idullhah *V* [lh-tlah] we (2) are smearing. [IA] (1) Yeztli yoo be **idullhah.** 'We are smearing the horse with liniment.'

idults'ul *V* [lh-ts'ul] we (2) are broiling (st). *[IA]*

idutez *V* [0-tez] we (2) slept. *[PA]*

ih *PART* question marker. (1) Lhes inyi **ih?** 'Are you eating bread?'

ilgih *V* [l-gih < gaih$_1$] you (1) are running. *[IA–prog]*

ilh'ulh *V* [lh-'ulh < 'alh$_1$] you (1) are chewing. *[IA]* (1) Dunitsun **ilh'ulh.** 'You are chewing moose meat.'

in'en *V* [0-'en < 'en] he/she sees. *[IA]*

-indak *N* flower.

ingooh *V* [0-gooh < goo$_1$] you (1) are driving. [IA–prog]

inkelh *V* [0-kelh < ke$_2$] you (1) are going by boat. *[IA–progressive]*

inle *V* [0-le] he/she was, former, late (deceased). *[PA]*

inle *V* [0-le] used to, he/she was. *[IA]*

inte *V* [0-te] you (1) slept. *[PA]*

inyez *V* [0-yez] you (1) are tall. *[IA]*

inyi *V* [0-yi < yi1] you (1) eat. *[IA]* (1) Lhes **inyi.** 'You (1) are eating bread.' (2) Lhes **inyi** ih? 'Are you eating bread?'

inzoo *V* [0-zoo < zoo$_1$] you (1) are

good. *[IA]*

it’us *V* [0-’us < ’as1] we (2) are walking. *[IA–prog]* (1) Fraser lake ts’i **it’us** lak? ‘Are we going to Fraser Lake?’

jaboon *N* credit. Etymology: From English “jawbone”, which meant “credit” in the colloquial language of the 19th century, probably via Chinook Jargon.

jan *N* old age.

jenyo *N* bull moose.

Joz *N* man’s given name. Etymology: Loan from French *Georges*.

Joz Dayi *N* Chief George, who became the first dayi in 1862.

jook *N* toque.

jus *N* fish hook.

juschun *N* fishing pole.

ka hoosdutun *V* [d-tun] I cling to (st). *[IA]*

ka’adut’en *V* [d-’en < ’en$_2$] we (2) hunted (st). *[PA]*

ka’ant’en *V* [d-’en < ’en$_2$] you (1) hunted (st). *[PA]*

ka’het’en *V* [d-’en < ’en$_2$] they hunted (st). *[PA]*

ka’huninzun *V* [0-zun < zun1] they want (st). *[IA]* (1) ’Utsungui **ka’huninzun.** ‘They want dried meat.’

ka’hutet’ilh *V* [d-’ilh < ’en$_2$] they will hunt (st). *[FA]*

ka’hut’en *V* [d-’en < ’en$_2$] they are hunting (st). *[IA]*

ka’idut’en *V* [d-’en < ’en$_2$] we (2) are hunting (st). *[IA]*

ka’int’en *V* [d-’en < ’en$_2$] you (1) are hunting (st). *[IA]*

ka’nahzun *V* [0-zun < zun1] you (2+) want (st). *[IA]* (1) Dant’i **ka’nahzun?** ‘What do you want?’

ka’nidudzun *V* [0-zun < zun$_1$] we (2) want (st). *[IA]*

ka’ninzun *V* [0-zun < zun$_1$] he/she wants (st). *[IA]* (1) ’Utsungui **ka’ninzun.** ‘She wants dried meat.’

ka’ninzun *V* [0-zun < zun$_1$] you (1) want (st). *[IA]* (1) Dant’i **ka’ninzun?** ‘What do you want?’ (2) Ts’uyi **ka’ninzun?** ‘Do you want food?’

ka’nuszun *V* [0-zun < zun$_1$] I want (st). *[IA]* (1) ’Utsungui **ka’nuszun.** ‘I want dried meat.’ (2) Too **ka’nuszun.** ‘I want water.’

ka’tadut’ilh *V* [d-’ilh < ’en$_2$] we (2) will hunt (st). *[FA]*

ka’tant’ilh *V* [d-’ilh < ’en$_2$] you (1) will hunt (st). *[FA]*

ka’ts’et’en *V* [d-’en < ’en$_2$] we (3+) hunted (st). *[PA]*

ka’ts’ut’en *V* [d-’en < ’en$_2$] we (3+) are hunting (st). *[IA]*

ka’uneh *V* [0-neh] he/she is going after (hunting, trapping). *[IA]* (1) Dune gah **ka’uneh.** ‘The man is going after rabbits.’

ka’uzninzun *V* [0-zun < zun$_1$] we (3+) want (st). *[IA]* (1) ’Utsungui **ka’uzninzun.** ‘We want dried meat.’

ka’uztet’ilh *V* [d-’ilh < ’en$_2$] we (3+) will hunt (st). *[FA]*

Kaoosdli *N* man’s given name. **Kaoosdli** became keyoh whuduchun in 1834. He was the father of Joz Dayi.

kasdzoon *N* tamarack.

kayanahudlishun *N* face-saving song.

ke *PP* depending on for subsistance.

-ke *N* foot, feet.

ke'et *ADV* barefoot.

-kechun *N* leg.

-kechunnah *N* ankle.

-kechunt'ak *N* back of the legs.

kedeldzolh *V* [l-d-zolh] he/she is dragging his feet. *[IA–prog]*

kegon *N* shoe.

kegudsatalh *N* tramp.

kejudustl'ah *N* bum.

-kelagi *N* toenail.

-kelamai *N* toe.

-kelamaicho *N* big toe.

-kelamaiyaz *N* little toe. This refers to the outermost toe, the leftmost toe on the left foot, the rightmost toe on the right foot, not just any small toe.

-kelamaiyazts'unts'e *N* fourth toe. The toe next to the little toe.

-kelamaiyez *N* long toe. This names the toe next to the big toe, not just any long toe.

-kelamaiyezts'unts'e *N* middle toe.

-kelascho *N* big toe. This means the big toe, not just any large toe, that is, the rightmost toe on the left foot or the leftmost toe on the right foot.

-kelatsul *N* heel.

kenaindutsi *V* [d-tsi] you (1) put your shoes on. *[IA]*

-kengi *N* toenail, claw of rear paw.

-kent'ak *N* top of foot.

-kentsul *N* heel.

kesgwut *N* moccasins, tires.

kesgwutcho N mukluks.

kesgwutch'ul *N* old, worn-out moccasins.

ketul *N* socks.

Ketloh *N* Ketlo, a family name. Etymology: Originally a nickname ke dutloh "feet are wet", for a man who always had wet moccasins.

-ketl'ah *N* sole of foot.

-ketl'i *N* the spaces between the toes.

ketl'oolh *N* shoe laces.

ketsih *N* leggings.

-ketsul *N* heel.

keyah *N* territory, trapline, village.

keyah whuduchun *N* hereditary chief.

Keyah Whujut *N* Old Village. The previous site of the village of Stellakoh.

keyoh whuduchun *NP* village leader, the first among equals of the clan heads. Etymology: Literally, "the village stick", after the talking stick he possesses.

ki ut'i *V* [0-t'i < t'i$_2$] she is married. Literally,"she has a husband". *[IA]*

koh *N* river.

kohtaba *N* shore of river.

koo N house. (1) **Koo** usda. 'She is at home.'

koonangus *N* stove poker.

koonghak *N* corner.

koonk'et *ADV* at home.

koonk'etsih *N* Strawberry Blite.

kooz *ADV* within the house.

kooz *N* home village.

-kui *N* husband.

-kuike *N* husbands.

kuninta *V* [0-ta < ta_1] you (1) are looking for. *[IA]* (1) Ts'ekoo **kuninta?** 'Are you looking for women?'

k'a *N* fat.

k'a *N* rifle shell, bullet.

k'adliyaz *ADV* shortly after.

k'ah *N* tracks, footprints.

k'ak *N* fish split once and dried.

-k'al *N* female genitalia.

k'i *N* Paper Birch.

k'an *ADV* now.

k'an 'awet *ADV* right now.

k'an bundada *N* this morning.

k'an hulhgha *N* this evening.

k'andit *ADV* now.

k'anditdzen *N* today.

k'andzen *N* today.

k'ani *N* wood chuck.

K'atsizchun *N* Old Fort Fraser. The area at the base of Fraser Mountain. The site of the original Northwest Company fort and of the grave of the prophet Boba, now Beaumont Provincial Park. Latitude: 54° 3' 34" N. Longitude: 124° 37' 11" W. Etymology: "the base of Smashed Arrows Mountain".

K'atsizyus *N* Fraser Mountain. Latitude: 54° 1' 73" N. Longitude: 124° 37' 39" W. Etymology: "smashed arrows mountain". Huckleberry picking area. Site of the story of the Nadleh woman who gave birth on top of the mountain. Site of battles with tribes from the south.

k'azba *N* Willow Ptarmigan.

k'azus *N* bullet pouch.

k'edlits'ilhba *N* pussy willow.

k'ehududlishun *N* longing song. A song expressing longing for someone.

k'elh'az *PP* after (in time).

-k'elh'ih *PP* less than. This is the pivot for negative comparison. It attaches to the possessive prefixes.

k'emai *N* Saskatoon berries.

k'en dulk'un *N* Red-Osier Dogwood. Locally known as Red Willow.

k'i *N* Paper Birch.

-k'i *N* pelvic region.

K'ibet *N* A low area in a birch stand near Kenny Nooski's house. Etymology: "inside the birch trees". Rabbit-snaring area.

-k'itsun' *N* pelvic bone.

k'o' *N* hunchback.

k'ohch'az N cracklings, Indian popcorn.

-k'oon *N* fish eggs.

k'ooni *N* female fish.

k'udentuk *V* [0-tuk] it is brittle. E.g. a stick or bone. [abs: d] *[IA]*

k'udentuk *V* [0-tuk < tuk_2] it is broken in two. E.g. a stick. [abs: d] *[PA]*

-k'i *N* mother's sister, mother's female first cousin, father's brother's wife, mother's brother's wife, step-mother.

k'ulan *N* The area along the wall opposite the door in a balhats.

k'uneduyus *V* [0-yus] he/she speaks broken Carrier. *[IA]*

k'unih *N* cambium.

k'untuk *V* [0-tuk < tuk_2] he is broke, without money. *[IA]*

k'unuduwul *V* [0-wul] it is brittle, easily smashed. E.g. glass. *[IA]*

k'us *N* Green Alder.

k'us *CONJ* or, either.

k'us hulan *V* [0-lan] most of them. *[IA]*

k'usdudacho *N* couch.

-k'ut *PP* on.

k'uyalhtuk *V* [lh-tuk] he/she speaks (language). *[IA]* (1) Dakelh **k'uyalhtuk.** 'He speaks Carrier.'

k'uztsan *N* apron.

khah *N* Canada Goose.

khahdai *N* Scouring Rush. Etymology: Literally, "goose food".

khahyaz *N* gosling.

khasdzoon *N* Dwarf Maple.

khast'an *N* Fireweed.

khe *N* grease, oil. Poss: ghe.

khe unli *V* [0-li] it is greasy. *[IA]*

khelh *N* load carried on back, pack. This refers to what one carries, not to the container.

khah *N* Canada Goose.

khelht'az *N* Yellow Water Lily.

khi *N* spruce roots.

khidutna *V* [0-na < na_1] we (2) are alive. *[IA]*

khindli *V* [d-li] you (1) are careful (of yourself). *[IA]*

khinyus *N* raft.

khit *N* winter.

khoh *N* Canada goose.

khuna *V* [0-na < na_1] he/she is alive. *[IA]*

khunai *N* animal.

khunaicho *N* elephant, mastodon, prehistoric monster.

khunaichoya *N* Masked Shrew. Etymology: Literally, "louse from prehistoric monster".

khunaiti *N* animal trail.

khunantesja *V* [0-ja] he/she panicked. *[PA]*

khundughun *N* calamity.

khunek *N* word, message, language.

khunek nu'a *N* messenger.

khunik *N* word, language. Poss: ghunik.

khuntusinilhholhdulh *ADV* tentatively.

khusduke *V* [d-ke] he/she is cranky. *[IA]*

khusih *N* squawfish, Northern Pikeminnow.

khusna *V* [0-na < na$_1$] I am alive. *[IA]*

khuz *N* pus.

khuzbat'an *N* plantain.

kwun *N* fire, light, flame.

kwalhcho *N* woodpecker.

kwatedugoo *V* [0-goo < goo$_1$] we (2) are going after it in a car. *[PA]*

kwatedulgai *V* [l-gai < gaih$_1$] we (2) are running after it. *[PA]*

kwatelhugai *V* [l-gai < gaih$_1$] I am running after it. *[PA]*

kwatesgoo *V* [0-goo < goo$_1$] I am going after it in a car. *[PA]*

kwatesya *V* [0-ya] I am going after it. *[PA]*

kwatet'az *V* [0-'az < 'as$_1$] we (2) are going after it. *[PA]*

kwaztelhughaz *V* [l-ghaz] we (3+) are running after it. *[PA]*

kwaztezdel *V* [0-del < dil$_1$] we (3+) are going after it. *[PA]*

kwaztezgoo *V* [0-goo < goo$_1$] we (3+) are going after it in a car. *[PA]*

kwenduda *N* sore feeling.

kwun *N* fire, light, flame.

kwun dohot'en *V* [d-'en < 'en$_1$] it is orange. [abs: wh] *[IA]* Etymology: "it is fire-coloured".

kwun doonat'en *V* [d-'en < 'en$_1$] it is orange. [abs: n] *[IA]* Etymology: "it is fire-coloured".

kwuncho nainkat *N* meteorite shower, shooting stars.

kwunchoyuk *N* hell.

kwuntoo *N* fire water, white man's drink.

kwuntset *N* ember, coal.

kwunyoo *N* matches.

kwusyoo *N* cough medicine.

kwuz oonde *N* snowbird.

Kw'eoodiltsuk *N* A mountain near Nadleh.

kw'ulh *N* boy's penis. Poss: kw'ulh.

kw'unidutal *V* [0-tal < tulh$_1$] we (2) kicked a hole in it. *[PA]*

kw'unustal *V* [0-tal < tulh$_1$] I kicked a hole in it. *[PA]*

kw'us *N* cloud.

kw'us 'uyinla *V* [0-la] he/she has a cold. *[IA]*

kw'us hooni *V* [0-ni] it is cloudy. *[IA]*

kw'usduda *N* chair.

kw'usduda nubalh *N* rocking chair.

kw'usduda t'az whebulh *N* recliner.

kw'usdudacho *N* couch.

kw'usdudayez *N* couch.

kw'usul *N* beads.

kw'ut nabadunelgih *N* helicopter.

kw'ut ts'uzti *N* mattress.

kw'ut yan wheti *N* bridge.

kw'uts'uda *N* chair.

kw'uts'udacho *N* easy chair.

kw'uts'udayez *N* chesterfield. Etymology: Literally, "long chair".

kw'uyukunenkai *N* turtle. Etymology: Literally, "dish turned upside down".

-kw'uz *N* kidney.

kw'uzbenul'en *N* thermometer. Etymology: "the thing by means of which cold is seen".

kw'uzghoo *N* icicle.

la *PP* in aid of, helping.

la *PART* focus. (1) Nts'e **la** nkooz? 'Where are you from?'

-la *PP* in aid of. Used in certain constructions having to do with helping. With most verbs the help must take the form of participating in a joint activity, rather than merely doing something on behalf of someone else. (1) **Sla** 'inja. 'You have helped me.' (2) **Hubula** 'ut'en. 'He helps them.'

lacholbai *N* yarrow plant, milfoil.

-lachunnah *N* wrist.

-lagi *N* fingernail.

lait'ah *V* [0-t'ah] it is weak. This may mean physically weak or may apply to flavour. *[IA]*

lait'i *N* penis.

lak *PART* yes/no question marker. (1) Nadleh ts'i **lak** teskelh? 'Will I go to Nautley by boat?' (2) Nadleh ts'i **lak** uskelh? 'Am I going to Nautley by boat?' (3) Fraser Lake t'si **lak** tat'us? 'Will we go to Fraser lake?'

-lak'et *N* palm of hand.

Landidzen *N* Monday.

landooz *N* Black Cottonwood.

lanezi *NUM* ten *[generic]*.

lanezi 'on'at 'ilhuk'i *NUM* eleven *[generic]*.

lanezi 'on'at skwunlai *NUM* fifteen *[generic]*.

lanezi lanezi *NUM* one hundred *[generic]*.

lanezidun *NUM* ten *[locative]*.

lanezinun *NUM* ten *[human]*.

lanezit *NUM* ten *[multiplicative]*.

laneziwh *NUM* ten *[abstract]*.

lasal *N* shawl. Etymology: Loan from French *le châle* "shawl".

-lasge *N* finger.

-lasgek *N* index finger.

-lasgeyaz *N* small finger.

-lasts'ah *N* little finger/pinkie.

lasyet *N* plate. Etymology: Loan from French *l'assiete* "plate".

-lat'ah *N* wrist.

-lat'ak *N* back of hand.

-latl'i *N* finger spaces. The spaces between the fingers.

latl'us *N* soap, detergent.

-latsuk *N* fingers.

lelwe *N* king. Etymology: Loan from Canadian French *le roi.*

liba *N* yeast bread. Etymology: Loan from French *le pain.*

ligok *N* chicken. Etymology: A loan from French *le coq* "rooster".

ligok hoolht'as *N* hawk.

ligokbayah *N* chicken coop.

ligokghez *N* chicken egg.

lilet *N* milk. Etymology: Loan from Canadian French *le lait.*

lilet sugi *N* powdered milk.

lili *N* bed. Etymology: Loan from French *le lit.*

lili k'ut subal *N* bed spread.

lilik'usulhchooz *N* sheet. Etymology: Literally, "that which, being two-dimensional and flexible,
lies on a bed".

liyab *N* devil. Etymology: Loan from French *le diable.*

Liyubyints'izti *N* Devil's Hairpin. Etymology: Literally, "devil's elbow road".

lizas *N* angel. Etymology: Loan from French *les anges.*

-loo *N* mother. Duoplural: -loo

lubegin *N* bacon.

lubel *N* spade. Etymology: Loan from French *le pelle.*

lubezo *N* poison. Etymology: Loan from French *le poison.*

lubos *N* pocket. Etymology: A loan from French *la poche* "pocket".

lubot *N* cup. Etymology: Loan from French *le pot.*

lubudak *N* potato. Etymology: A loan from Canadian French *la pataque.* ultimately from *patata*, the word of the same meaning in Taino, the Caribbean native language of Haiti.

lubudakt'ooz *N* potato chips. Etymology: "potato peel".

lubuyos *N* haystack. Etymology: a loan from French *le meulon.* the origin of the final s is obscure.

ludab *N* table. Etymology: Loan from French *la table.*

ludab k'usulhchooz *N* table cloth.

ludi *N* tea. Etymology: Loan from French *le thé.*

ludi 'oosa' *N* tea pot.

ludi musjek *N* Labrador Tea.

ludibot *N* tea pot.

lufooset *N* fork. Etymology: A loan from French *la fourchette* "fork".

luga *N* coffee. Etymology: Loan from French *le café.*

luga 'oosa' *N* coffee pot.

luga be'bizih *N* coffee pot.

luga be'udlez *N* coffee pot.

lugar *N* playing cards. Etymology: Loan from French *les cartes.*

luglik'et *N* keyhole.

lugliz *N* church. Etymology: Loan from French *l'église.*

luglok *N* cape.

luglos *N* bell. Etymology: Loan from

French *la cloche.*

lugloo *N* nail. Etymology: Borrowed from French *le clou.*

luhoos *N* fork. Etymology: Loan from French *la fourche.*

Luksilyoo *N* The name of a clan. Its crest is cariboo.

luminis *N* Protestant.

lusel *N* salt. Etymology: Loan from French *le sel.*

lusook *N* sugar. Etymology: Loan from French *le sucre.*

lusooma *N* towel.

lusooni *N* chimney. Etymology: Loan from French *la cheminee* “fireplace, chimney”.

lusoosam *N* turnips. Etymology: Loan from French *le chou de Siam* “kohlrabi” (literally “Thai cabbage”).

luwagun *N* wagon.

-lh *PP* together with.

lh’ek *N* night.

lhaga’as *N* dried Red Laver seaweed. When fresh this is reddish, purple, or greenish and has the consistency of cellophane. When dried it is black and brittle. It grows on rocks in the lower intertidal zone. It was and to some extent still is traded in from the coast; it does not grow in Carrier territory. Etymology: A loan from a Tsimshianic language, most likely Gitksan. Compare Nisga’a [laq’askw] and Coast Tsimshian [laask].

Lhaghahutullhuk *N* An area traditionally used for broad-jumping contests. On the old Nautley Road near Kenny Nooski’s house. Etymology: “where people compete at jumping”.

lhahuntoolhe’en *V* [l-’en < ’en$_1$] they are looking at each other with admiration. *[IA]*

lhahunul’en *V* [d-lh-’en < ’en$_1$] they are looking at each other. *[IA]*

lhai *V* [0-lhai] they (generic) are many. *[IA]* (1) Mai **lhai.** ‘There are lots of berries around.’ (2) Nja chundoo **lhai.** ‘There are many lodgepole pine here.’ (3) Lhi **lhai.** ‘There are a lot of dogs.’

lhane *Q* many *[human]*

lhanun *Q* many *[human].*

lhat *Q* many *[multiplicative],* many times.

lhat-un *Q* many (places).

lhawh *Q* many *[abstract].*

lhbai *N* porcupine quills used for embroidery.

lhda bundada *N* yesterday morning.

lhda hulhgha *N* yesterday evening.

lhe’ulcho *V* [l-cho < cha$_1$] how big it is! *[IA]*

lhe’ultsolyaz *V* [l-tsol < tsol$_1$] how small it is! *[IA]* (1) Tsalhts’ul bulasge **lhe’ultsolyaz** nulh’en. ‘I am looking at a baby whose fingers are oh so small.’

lheghidli *N* confluence of rivers.

lheidli *N* confluence of rivers.

lhelhdun *ADV* different places.

lhelhdun k’un’a *ADV* in many ways.

lhelhyoone *N* different people.

lhena’duduwhults’it *V* [l-ts’it] he/

she is a braggart. *[IA]*

lhes *N* bread.

lhes 'uzus *N* flour sack.

lhes lhuk'i *N* cookie.

lhes nududzaih *N* flour.

lhes sto bez sut'e *N* oven bannock.

lhes sut'e *N* bannock, fry bread.

lhetl'us *N* mud.

lheyatdendi *V* [0-di] it is cloudy. *[IA]*

lheyih *N* scissors.

lhez *N* ashes, dirt, soil, dust.

Lhez Bunk'ut *N* Klez Lake. Latitude: 54° 1' 15" N. Longitude: 124° 43' 43" W. Etymology: "dust lake". A camping area for trappers and travelers on the Cheslatta Trail, which ran along the west side of the lake.

lhghai *V* [lh-ghai < $ghai_1$] he/she is hairy. *[IA]*

lhghun *ADV* frequently.

lhghunilhdukw *ADV* frequently.

lhi *N* dog. Poss: lik. Duoplural: lhi (1) **Lhi** lhai. 'There are a lot of dogs.'

Lhi Tutl'it *N* A rock cut on the railway on the south shore of Fraser Lake. Etymology: "dogs fart." The name originates from dog teams straining to climb the steep hill on the original trail that ran through this site. A netting area for char and whitefish. Spawning char are speared or gaffed from shore. Nets were also set here in winter. The name includes several campsites along the shore at this location.

lhicho *N* big dog.

lhich'ul *N* worthless dog.

lhijut *N* old dog.

lhike *N* dog's paw.

lhike *N* dogs.

lhiluzchun *N* Canada mayflower, False Solomon's Seal, Sugarberry. This species is also known by the older scientific name Smilacina amplexicaulis. Etymology: Literally, "dog urine plant".

lhitsol *N* little dog.

lhits'e *N* beets.

lhits'e *N* bitch, female dog.

lhiya *N* flea. Etymology: "dog tick".

lhiyaz *N* puppy. Duoplural: lhiyazke

lhkut *V* [lh-kut] it is smooth, slippery. *[IA]*

lhk'einla *V* [0-la] he/she put them on each other. Among other things, this can mean that he crossed his feet while walking or running and tripped. [class: mdo-c] *[PA]*

lhk'eti *V* [0-ti] it (road) branches. *[IA]*

lhk'eti *N* fork in road.

lhk'udidun *N* eight *[locative]*.

lhk'udit *NUM* eight *[multiplicative]*.

lhk'udiwh *N* eight *[abstract]*.

lhk'un nasain'aih *N* red sunset.

lhk'ut *PPC* on top of each other.

lhk'utadun *NUM* six *[locative]*.

lhk'utakiwh *NUM* six *[abstract]*.

lhk'utak'i *NUM* six *[generic]*.

lhk'utanun *NUM* six *[human]*.

lhk'utat *NUM* six *[multiplicative]*.

lhk'utat lanezi *NUM* sixty *[generic].*

lhk'utdine *NUM* eight *[human].*

lhk'utdink'i *NUM* eight *[generic].*

lhk'utdit lanezi *NUM* eighty *[generic].*

lhkw'encho *N* White Sturgeon.

Lhkw'encho'an *N* A deep spot in Fraser Lake in front of I.R. 1. Etymology: "sturgeon hole". A netting site for Sockeye salmon. This is where two men once came upon the ts'uniltsih, which transfixed them with its gaze. One died, the other became very ill.

Lhkw'etsilhchola *N* Mouse Mountain. See also: Neyilasts'ah.

Lhkw'etsilhchola *N* the point of land on Fraser Lake below Mouse Mountain.

lhombilh *N* fish net.

lhooh *N* whitefish.

lhook *N* fish.

lhook sugi *N* dried fish.

lhook tazul *N* fish soup.

lhookat *V* [0-kat < kat_3] he/she claps, applauds. *[IA]*

lhookchetl'a *N* fish tail.

lhookghe *N* cod liver oil.

lhookna *N* fish eyes.

lhooktl'uz *N* milt, fish sperm.

lhooktsi *N* fish head.

lhooktsik *N* fish guts.

lhook'icho *N* whale.

lhoombilh *N* fish net.

lhoombilhchun *N* fish net anchor pole.

lhoombilhtse *N* fish net weights.

lhoonde *N* each other's cousin.

lhootyoo *N* ointment. Etymology: "scab medicine".

lhooz *N* perch.

lhtadelnat *V* [l-nat] he/she split (st) into many pieces lengthwise. *[PA]*

lhtah *PPC* among one another.

lhtak'alt'i *NUM* seven *[generic].*

lhtak'alt'idun *NUM* seven *[locative].*

lhtak'alt'inun *NUM* seven *[human].*

lhtak'alt'it *NUM* seven *[multiplicative].*

lhtak'alt'it lanezi *NUM* seventy *[generic].*

lhtak'alt'iwh *NUM* seven *[abstract].*

lhtanahunesdel *V* [d-del < dil_1] they (3+) came back. *[PA]*

lhtanalkat *N* quilt.

lhtananenja *V* [d-ya] you (1) came back. *[PA]*

lhtananesja *V* [d-ya] he/she came back. *[PA]*

lhtananesja *V* [d-ya] I came back. *[PA]*

lhtananet'az *V* [d-'az < $'as_1$] we (2) came back. *[PA]*

lhtanaznesdel *V* [d-del < dil_1] we (3+) came back. *[PA]*

lhti *N* rifle.

lhtudu'alts'e *N* twins.

lhts'eidalnat *V* [l-nat] he/she split it in two lengthwise. *[PA]*

Lhts'umusyoo *N* The name of a clan. It's crests are: owl, grouse, killer whale, quarter moon.

lhu’ool *V* [l-’ool < ’ool$_1$] it has been braided. *[PA]*

lhuhedughan *V* [d-ghan < ghan$_1$] they (3+) killed each other. *[PA]*

lhooktsik *N* fish guts.

lhuhoolkal *V* [l-kal] they collided with one another. *[PA]*

lhuhoonultsasshun *N* shaming song.

lhuhudughan *V* [d-ghan < ghan$_1$] they (3+) kill each other. *[IA]* (1) Nubah **lhuhudughan.** ‘Soldiers kill each other.’

lhuki *V* [l-ki] it (generic) is sweet. *[IA]*

lhuk’i *V* [0-k’i] it tastes good, it is sweet. *[IA]*

lhulcho *ADV* halfway, half.

lhulcho dezbun *V* [0-bun < bun$_1$] it is half full. *[IA]*

lhulh *PPC* together with each other. (1) **Lhulh** ’uhut’en. ‘They are working together.’

lhulhcho *N* triplets.

lhulhtuske *N* sisters, female first cousins. See also: -ulhtus.

lhum *N* a piece of ice.

lhuna *N* captive.

lhuschuz *N* knot.

lhustsis *N* ash.

lhustsis bekuk *N* ash tray.

lhut *N* smoke.

lhut be’hanajut *N* chimney.

lhuyul *V* [l-yul < yul$_1$] it (generic) is white. *[IA]*

lhuz *N* urine, pee. Poss: -luz.

-lhuzkah *N* urinary tract.

lhuzk’et *N* urinary tract.

-lhuzzus *N* bladder (urinary).

madus *N* mattress.

madli *V* [0-dli] he/she is foolish. *[IA]*

mahudli *V* [0-dli] they are foolish. *[IA]*

mai *N* berries.

mai be’ulha *N* jam.

mai nesdliz *N* jam.

mai nezgi *N* dried berries.

mai nudutleh *N* jam.

mai nuhudleh *N* the distribution of fruit at a balhats.

maitoo *N* fruit juice.

maitoo dot’en *V* [d-’en < ’en$_1$] it is purple. Literally, “it is the colour of berry juice”. *[IA]*

maitlus *N* fruit leather.

maiyun *N* a song sung during the distribution of fruit at a balhats. Never sung at a funeral balhats.

mandah *N* canvas. Etymology: Loan from Spanish manta “coarse cotton cloth” introduced by Mexican pack-train men.

masdli *V* [0-dli] I am foolish. *[IA]*

mba *N* your (1) father.

mbalh'i *V* [lh-'i < 'i$_1$] I am waiting for you (1). *[IA]*

mbat *N* your (1) mitts.

mbe *QWH* who?. (1) **Mbe** nyuts'ulhni? 'What is your name?'

mbez *N* your (1) mother-in-law.

mboo' *N* your (1) rectum.

mboos *N* your (1) cat.

mbun *N* your (1) roof.

mbunt'ah *N* your (1) attic.

mbut *N* your (1) stomach.

mbuttl'ah *N* the bottom of your (1) stomach.

mai *N* berries.

me *N* omen.

meljah *N* white person. Etymology: Said to be a loan from Chinook Jargon, but Chinook Jargon source is unknown.

Meljayah *N* Stellakoh village. An old name for Stellakoh village Latitude: 54° 3' 0'' N. Longitude: 124° 55' 0'' W. [disused] Etymology: Literally, "white people's house", because travellers, including white people, used to make camp in this area.

moodih *N* boss, knowledgable person.

mooki *N* banana.

mulla *N* abalone shell, mother-of-pearl.

mus *N* porridge, oatmeal. Etymology: Loan from English mush.

musdus *N* cow. Etymology: Loan from Cree mustos.

Musdus Lhenul'elh *N* See Noonghak. A ford on the Nechako River up-stream of the Nautley River used before there were bridges or ferries. Etymology: "cows swim across".

musdus ts'eke *N* female cow. Duoplural: musdus ts'ekoo

musdus uljas *N* milk cow.

musdusbayah *N* cow barn, stable for cattle.

musdusdune *N* bull.

musdustsun *N* beef.

musdusyaz *N* calf.

musdzi *N* owl.

musdziyaz *N* moth.

n- *PREFIX* second person singular posessor. Attaches to class 1 nouns.

-na *N* eye, headlight.

na'adudai *V* [d-ai < yi$_1$] we (2) ate (u.o.). *[PA–habitual]*

na'eghilh *V* [0-ghilh < ghe$_1$] he/she is packing back. *[IA–progressive]*

na'esdai *V* [d-ai < yi$_1$] I ate (u.o.). *[PA–habitual]*

na'hududelhts'ah *V* [lh-ts'ah] they make noise so as to announce themselves. This is what people might do if they were approaching someone's campsite at dusk. *[IA]*

na'idudai *V* [d-ai < yi$_1$] we (2) eat (u.o.). *[IA–habitual]*

na'indai *V* [d-ai < yi$_1$] you (1) eat (u.o.). *[IA–habitual]*

na'ondai *V* [d-ai < yi$_1$] you (1) may eat (u.o.). *[OA–habitual]*

na'oosdai *V* [0-dai < yi$_1$] I may eat (u.o.). *[OA]* (1) **Na'oosdai** hukwa'nuszun. 'I want to eat.'

na'tadudilh *V* [d-ilh] we (2) will eat (u.o.). *[FA–habitual]*

na'tesdilh *V* [0-dilh] I will eat (u.o.). *[FA]*

na'ts'edai *V* [d-ai < yi$_1$] we (3+) ate (u.o.). *[PA–habitual]*

na'ts'udai *V* [d-ai < yi$_1$] we (3+) eat (u.o.). *[IA–habitual]*

na'ulhgheh *V* [lh-gheh < ghe$_2$] he/she is melting (u.o.). *[IA]*

na'usdai *V* [d-ai < yi$_1$] I eat (u.o.). *[IA–habitual]*

na'uztedai *V* [d-ai < yi$_1$] we (3+) will eat (u.o.). *[FA]*

-nabagha *N* eyelashes.

-nabalos *N* eyelid.

nachanlhuk'a *V* [l-k'a] he/she has not become obese. *[PN]*

nache'untut'ah *N* tree from which rabbit snare is hung.

nach'ulh *N* Black Birch, Water Birch, Western Birch.

nadalhghez *V* [lh-ghez] you (1) felled. *[PA]* (1) Duchun **nadalhghez?** 'Did you fell the tree?'

nade *N* next year.

nadeh *V* [0-deh] it (snow) is falling from the trees. *[IA]* (1) Yus **nadeh.** 'Snow is falling from the trees.'

nadelduz *V* [l-duz] he/she fell down. *[PA]*

nadelhudzo *N* dew.

nadenjak *V* [0-jak] you (1) are dry. *[IA]*

nadesghuz *V* [0-ghuz < ghis$_1$] I capsized, rolled over (in vehicle). *[PA]*

nadesjak *V* [0-jak] they are dry. *[IA]*

nadewe'ts'elts'ah *V* [l-ts'ah] we (3+) make noise so as to announce ourselves. *[IA]*

nadewelhuts'ah *V* [l-ts'ah] I make noise so as to announce myself. *[IA]*

nadezghuz *V* [0-ghuz < ghis$_1$] it capsized, rolled over. *[PA]*

nadilhghis *V* [lh-ghis < ghis$_1$] you (1) fell. *[IA]* (1) Duchun **nadilhghis.** 'Fell the tree.'

nadilhk'aih *V* [lh-k'aih < k'aih$_1$] you (1) ignite. *[IA]* (1) Kwun **nadilhk'aih.** 'Light a fire.'

Nadot'en *N* Babine person, Takla person.

naduji *V* [d-yi < yi$_1$] we (2) ate. *[PA]*

nadukelh *V* [d-kelh < ke$_2$] he/she is going back by boat. *[IA–progressive]*

nadulyul *V* [l-yul] it is dripping. *[IA]*

nadun *NUM* two *[locative]*. (1) **Nadun** hahonkai. 'He dug two holes.'

-nadun *N* cousin.

nadusdujih *V* [d-jih] we (2) are dry. *[IA]*

Nadleh *N* Nautley. (1) **Nadleh** ts'i chaseskelh. 'I am not going to Nautley.'

Nadlehbun *N* Fraser Lake. Latitude: 54° 4' 56" N. Longitude: 124° 44' 55" W. Etymology: "(salmon) run lake".

Nadlehbunk'ut *N* Fraser Lake. Latitude: 54° 4' 56" N. Longitude: 124° 44' 55" W. Etymology: "(salmon) run lake".

Nadlehbunne *N* Fraser Lake people.

Nadlehkoh *N* Nautley River. Drains Fraser Lake into the Nechako River. Latitude: 54° 5' 9" N. Longitude: 124° 35' 44" W. Etymology: "(salmon) run river". Primary fishing station for sockeye salmon.

Nadlehyah *N* Nautley community hall.

nagoostliyaz *ADV* quietly.

-naghoo *N* eyetooth, wisdom tooth.

nah *INT* here you are. This is what you say to someone as you hand him or her something.

nah- *PREFIX* first person dual possessor. Attaches to class 1 nouns.

nah- *PREFIX* second person duo-plural possessor.

nahba 'alha'hoont'ah *V* [0-t'ah < t'oh$_1$] we (2) believe it. *[IA]*

nahba we'hooja *V* [0-ja] we (2) are bewildered. *[IA]*

nahba't'en lhai *V* [0-lai] we (2) are busy. *[IA]*

nahbai *N* weasel. In Carrier territory the weasel usually encountered is the Short-tailed Weasel (Mustela erminea richardsoni). However this term appears to be applicable as well to the Least Weasel (Mustela rixosa) which is more rarely encountered.

nahbaicho *N* ermine.

nahdani *V* [0-ni] he/she told you (2+). *[PA]*

nahdukelh *V* [d-kelh < ke$_2$] you (2+) are going back by boat. *[IA–progressive]*

nahdustl'us *N* your (2+) paper.

nahedukelh *V* [d-kelh < ke$_2$] they are going back by boat. *[IA–progressive]*

nahgan *N* your (2+) arms.

nahghudaninya *V* [0-ya] he entered your (2+) house. *[PA]*

nahghudazdindil *V* [0-dil] we (3+) entered your (2+) house. *[PA]*

nahghundanindil *V* [0-dil] they (3+) entered your (2+) house. *[PA]*

nahkunuta *V* [0-ta < ta$_1$] he/she is looking for us (2). *[IA]*

nahkunuta *V* [0-ta < ta$_1$] he/she is looking for you (2+). *[IA]*

Nahnelt'o *N* The location of the original fish weir on the Nautley River.

nahtoo *N* your (2+) water.

Nahudujun *N* A mountain near Nadleh. Etymology: "they practice singing".

nahulyis *V* [l-yis] they are resting. *[IA]*

naih ba whenant'uk *N* clothes line.

nahunet'a *V* [0-t'a] they got fooled. *[PA]*

nahutet'ah *V* [0-t'ah < t'oh$_2$] they will fly home. *[FA]*

nahutet'ah *V* [0-t'ah < t'oh$_2$] they will fly in a loop. *[FA]*

nahuwhulyeh *V* [l-yeh] they are dancing. *[IA]*

nahzih *PPC* beside us (2).

naidukelh *V* [d-kelh < ke$_2$] we (2) are going back by boat. *[IA–progressive]*

naih *N* clothes.

naih ba whenant'uk *N* clothes line.

naih benadugih *N* clothes dryer.

naih benaldzooh *N* clothes brush.

naih besula *N* dresser.

naih bewhuna'udeh *N* cleaning cloth.

naihjut *N* old clothes.

nailyis *V* [l-yis] you (1) rest. *[IA]*

nailhni *ADV* right.

naindukelh *V* [d-kelh < ke$_2$] you (1) are going back by boat. *[IA–progressive]*

naints'ut *V* [0-ts'ut] it is foggy. *[IA]*

nainul'en *V* [l-'en < 'en$_1$] he/she checks on it. *[IA]*

nainul'ih *V* [l-'ih < 'en$_1$] he/she checks on it. *[IA–customary]*

nainulhtsul *V* [lh-tsul] he/she dampens it. *[IA]*

naiyulh'ool *V* [lh-'ool < 'ool$_1$] she is braiding it. *[IA]* (1) Dutsigha **nayulh'ool.** 'She is braiding her hair.'

naiyulhgheh *V* [lh-gheh < ghe$_2$] he/she is melting it. *[IA]*

naiyulhkut *V* [lh-kut] he/she patches it. *[IA]*

najas *V* [d-yas] it is snowing. *[IA]*

najih *V* [d-yih] it is healing. *[IA]*

nak'ezdulya *N* eyeglasses.

nak'ubul *V* [0-bul < bul$_1$] he/she is blinking. *[IA]*

nak'uz *N* a single eye.

nakwundilhk'aih *V* [lh-k'aih] you (1) ignite again. *[IA]*

nal'oo *N* anchor for rabbit snare.

nalgheh *V* [l-gheh < ghe$_2$] it is melting. *[IA]*

nalts'ut *V* [l-ts'ut] it fell down. *[PA]*

nalhgheh *V* [lh-gheh < ghe$_2$] I am melting (st). *[IA]*

nalhni *ADV* right side.

nalhti *N* blanket.

nalhticho *N* comforter.

nalhtsul *N* reconstituted dried fish. This refers to dried fish that is soaked in water and then boiled.

nalhughen *V* [l-ghen < ghe$_2$] it has melted. *[PA]*

nalhutsul *V* [l-tsul] they are wet. *[IA]*

nananda *V* [0-da] you (1) made a mistake. *[PA]*

nananghiz *V* [0-ghiz < $ghis_1$] it toppled. *[PA]* (1) Duchun **nananghiz.** 'The tree toppled.'

nanehult'ah *V* [l-t'ah < $t'oh_1$] they are two. *[IA]*

naneltsul *V* [l-tsul] you are wet. *[IA]*

nanestl'oo *N* fence.

nanestl'oo kechun *N* fence post.

nanezmaz *N* button.

nanezmazk'et *N* button hole.

nanguz *N* fox.

Nanguz Bunk'ut *N* Peta lake. Latitude: 54° 13' 10" N. Longitude: 124° 47' 32" W. Etymology: "fox lake". Trout fishing. Maxime George had a cabin on the northwest side.

nanilk'a *V* [l-k'a] he/she has become obese again. *[PA]*

nankah *NUM* two *[generic].*

nanlhu'en *V* [l-'en < $'en_1$] he/she checks on (st). *[IA]*

nanlhu'ih *V* [l-'ih < $'en_1$] he/she checks on (st). *[IA–customary]*

nantantelh *V* [0-telh] you (1) are going to sleep. *[FA]*

nantestelh *V* [0-telh < ti_4] I will go to sleep. *[FA]*

nants'untetus *V* [0-tus] we (3+) will sleep. *[FA]*

nanulhtsul *V* [lh-tsul] I dampen (st). *[IA]*

nanun *NUM* two *[human].*

nanut'a *V* [d-'a] he/she is making a mistake. *[IA]*

nanzin *V* [0-zin] you (1) thought that. *[PA]*

nanyoost'en *V* [d-'en < $'en_1$] I may see you (1) again. *[OA]* (1) **Nanyoost'en** si! 'See you again!'

naolyis *V* [l-yis] you (1) rest. *[OA]*

naondle *V* [d-le] you (1) may be. *[OA]*

naoodnilh *N* ghost. In the early post-contact period was also used to refer to Europeans.

nasain'ai hoolk'un *N* red sunset.

nasdukelh *V* [d-kelh < ke_2] I am going back by boat. *[IA–progressive]*

nasdlez *N* white person. Etymology: Literally, "boiled again", that is, "parboiled", in reference to white people's skin colour.

nasdli *V* [0-dli < dli_1] he/she feels cold. *[IA]*

nasusdutsan *V* [d-tsan < $tsan_1$] I have shat. [PA]

nat *NUM* twice, two *[multiplicative].*

nat lanezi *NUM* twenty *[generic].*

natalyis *V* [l-yis] you (1) will rest. *[FA]*

natandilh *V* [0-dilh] you (1) will eat (st). *[FA–habitual]*

natandudluz *V* [d-luz] you (1) will pee. *[FA]*

natandugoo *V* [0-goo] you (1) are driving away. *[FA]*

natandutsun *V* [d-tsun] you (1) will shit. *[FA]*

Nataoozjaz *N* A lake on the south side of I.R. 1. Reported to be a place to avoid due to quicksand and poisonous water. This lake is a feature in the legend of a creature carried there from another small lake on the north side of Sugarloaf Mountain. Etymology: “water seeps into the ground”.

natedudluz *V* [d-luz] they will pee. *[FA]*

natedugooh *V* [d-gooh] they are driving away. *[FA]*

natehdutsun *V* [d-tsun] you (2+) will shit. *[FA]*

natelghen *V* [l-ghen < ghe_2] it will melt. *[FA]*

natesdudluz *V* [d-luz < luz_1] I am going to pee. *[FA]*

natesdughi *V* [d-ghi < ghe_1] he/she is packing back. *[IA]* (1) Ts’uyi **natesdughi.** ‘He is packing food back (to his cabin).’

natesdutsun *V* [d-tsun] I am going to shit. *[FA]*

natesdluz *V* [d-luz < luz_1] I am going to pee. *[FA]*

natet’ah *V* [0-t’ah < $t’oh_2$] they will fly in a loop. *[IA]*

natukwa’ *V* [0-kwa’] he/she burps. *[IA]*

natuskwa’ *V* [0-kwa’] I burp. *[IA]*

nat’ah *N* Spruce Grouse, Fool’s Hen.

Natl’alikoh *N* Sutherland River. Runs in a northwesterly direction into the southeast tip of Babine Lake. Latitude: 54° 29' 21'' N. Longitude: 125° 10' 50'' W. Etymology: “everything swims in it river”. A site for trapping, fishing, and bear hunting.

Natl’ilikoh *N* Sutherland River.

natsauneltsut *N* smallpox.

nats’edukelh *V* [d-kelh < ke_2] we (3+) are going back by boat. *[IA–progressive]*

-nats’osdooz *N* eyebrows.

nats’ulyis *V* [l-yis] we (3+) are resting. *[IA]*

nawniduk’az *V* [0-k’az] it (weather) turned cold. [abs: wh] *[PA]*

nawnilghaz *V* [l-ghaz < $ghaz_1$] it has gotten hot. *[PA]*

nawntelwus *V* [l-wus] it will start to get hot. *[FA]*

nawtelhch’ul *V* [lh-ch’ul < $ch’ul_1$] he/she will plow. [abs: wh] *[FA]*

nawtezo *V* [0-zo] he/she will plow (st:road), grade (st:road). [abs: wh] *[FA]*

nawus *N* soapberry, Indian ice cream.

nawh *NUM* two *[abstract].*

nawhalhch’ul *V* [lh-ch’ul < $ch’ul_1$] he/she plowed. [abs: wh] *[PA]*

nawhe’indak *N* your (2+) flower.

nawheni *PRO* you (2+). (1) **Nawheni** cha tehjun? ‘Are you guys going to sing too?’

nawhenilghen *V* [l-ghen < ghe_2] it has started to melt. *[PA]*

nawhnul’en *V* [l-’en < $’en_1$] he/she checks on (st). [abs: wh] *[IA]*

nawhnul’ih *V* [l-’ih < $’en_1$] he/she checks on (st). [abs: wh] *[IA–customary]*

nawhudani *V* [0-ni] he/she told us (2). *[PA]*

nawhudugus *V* [d-gus] he/she is washing for himself. [abs: wh] *[IA]* (1) Yun **nawhudugus.** ‘She is washing the floor for herself.’

nawhugus *V* [0-gus] he/she is washing. [abs: wh] *[IA]* (1) Yun **nawhugus.** ‘She is washing the floor.’

nawhujas *V* [d-yas] it is snowing. [abs: wh] *[IA]*

nawhulnuk *N* priest, messenger.

nawhulh ’utet’ah *V* [0-t’ah < t’oh$_2$] we (2) will go by plane. *[FA]*

nawhulh ne’uzt’a *V* [0-t’a < t’oh$_2$] we (2) have gone in a loop by plane. *[PA]*

nawhulhtiyaz *V* [lh-ti] it is drizzling. *[IA]*

nawhulhtus *N* your (2+) sister.

nawhulhtsit *V* [lh-tsit] it is drizzling. *[IA]*

nawhulhutet’ah *V* [0-t’ah < t’oh$_2$] we (2) are going by plane. *[IA–prog]*

nawhuzo *V* [0-zo] he/she is grading (st:road), plowing (st:road). [abs: wh] *[IA]* (1) Ti **nawhuzo.** ‘He is grading the road.’

nazde *ADV* blowing from the south.

naznelhtsul *V* [l-tsul] we (3+) are wet. *[IA]*

naztedilh *V* [0-dilh] we (3+) will eat (st). *[FA]*

naztedudluz *V* [d-luz] we (3+) will pee. *[FA]*

naztedugooh *V* [0-gooh] we (3+) are driving away. *[FA]*

naztedutsun *V* [d-tsun] we (3+) will shit. *[FA]*

ncha *V* [0-cha < cha$_1$] it is large. [abs: 0] *[IA]*

nchai *N* your (1) grandchild.

nchaicho *N* your (1) great-grandchild.

nchaike *N* your (1) grandchildren.

nchak *N* your (1) ribs.

nchets’un *N* your (1) tailbone.

nda *ADV* to the east. (1) **Nda** tesyalh. ‘I am going to go east.’

nda *N* your (1) lips.

ndadeltel *N* your (1) porch.

ndadentanghunez’ai *N* your (1) doorknob.

ndadentanghunezdla *N* your (1) doorknobs.

ndadent’az *N* your (1) window.

ndadent’azts’asdla *N* your (1) curtains.

ndagha *N* your (1) moustache.

ndaneltsuk *V* [l-tsuk] there are as many as. *[IA]*

ndati *N* your (1) doorway.

ndaz *V* [0-daz < daz$_1$] it is heavy. *[IA]*

ndazde *ADV* from the east. (1) **Ndazde** nilhts’i. ‘The wind is blowing from the east.’

ndazde nilhts’i *N* east wind.

ndes *N* your (1) lungs.

ndesyaz *N* your (1) younger sister.

ndezuna *ADV* this way.

ndinda *V* [0-da] you (1) are sick. *[IA]*

Ndoni *N* Holy Spirit. Etymology: “above mind”.

Ndonidzen *N* Pentecost. Etymology: “Holy Spirit Day”.

nduchahoonil V [0-nil] it has not happened. *[PN]*

nduda *V* [0-da < da_2] it is sore, he/she is sick. *[IA]* (1) Stsi **nduda.** 'I have a headache.'

ndudaneghunli *N* nurse.

ndukw *V* [0-dukw < $dukw_1$] he/she is short. *[IA]*

ndulhda *V* [lh-da] I cause (s.o.) pain. *[IA]*

neboos *N* our cat.

ndune delts'i *N* your (1) living room.

nduneti kw'uts'uda *N* your (1) reclining chair.

ndunulmulh *V* [l-mulh] it (liquid) is boiling. *[IA]*

nduwhuneh *V* [0-neh] it is happening. *[IA]*

ndzi *N* your (1) heart.

ndzidool *N* your (1) artery.

ndzik'ut *N* your (1) chest.

ndzoozt'an *N* your (1) shirt.

ne dani *V* [0-ni] he/she told us. *[PA]*

ne- *PREFIX* first person plural posssessor. Attaches to class 1 nouns.

ne'atke *N* our wives.

ne'hulooz *V* [0-looz < $looz_1$] they are going around by sleigh. *[IA]*

ne'indak *N* our flower.

ne'lunatdultsiyan *N* our great-grandfather.

ne'ook'et'as *N* our cook stove.

ne'ook'utudai *N* our table.

ne'ughe *V* [0-ghe < ghe_1] he/she is packing around (u.o.). *[IA]*

ne'usghe *V* [0-ghe < ghe_1] I am packing around (u.o.). *[IA]*

ne'ustes *N* our mattress.

ne'ustlen *N* our subject, servant.

neba *N* our father.

neba *PPC* for us.

neba 'alha'hoont'ah *V* [0-t'ah < $t'oh_1$] we (3+) believe it to be true. *[IA]*

neba whe'ooja *V* [0-ja] we (3+) are bewildered. *[IA]*

neba't'en lhai *V* [0-lai] we are busy. *[IA]*

nebe'uzdla *N* our cupboard, dresser.

nebez *N* our mother-in-law.

neboo' *N* our rectums.

neboos *N* our cat.

nebun *N* our roof.

nebunt'ah *N* our attic.

nebut *N* our stomachs.

nebuttl'ah *N* the bottoms of our stomachs.

nechai *N* our grandchild.

nechaicho *N* our great-grandchild.

nechaike *N* our grandchildren.

nechak *N* our ribs.

nechets'un *N* our tailbones.

nech'i' *PRO* ours.

neda *N* our lips.

nedadeltel *N* our porch.

nedadentanghunez'ai *N* our doorknob.

nedadentanghunezdla *N* our doorknobs.

nedadent'az *N* our window.

nedadent'azts'asdla *N* our curtains.

nedagha *N* our moustaches.

nedati *N* our doorway.

nedes *N* our lungs.

nedesyaz *N* our younger sister.

nedochunalhduz *N* clover.

nedotoo *N* liquor, booze, alcohol.

nedotoo yundulhda *V* [lh-da] he/she has a hangover. Literally, "liquor has made him sick". *[IA]*

nedulhdehun *N* magician.

nedune delts'i *N* our living room.

neduneti kw'uts'uda *N* our reclining chair.

nedusneke *N* our ancestors.

nedustl'us *N* our (3+) paper.

nedzi *N* our hearts.

nedzidool *N* our arteries.

nedzik'ut *N* our chest.

nedzoozt'an *N* our shirt.

negan *N* our arms.

neghe *N* our brother-in-law, sister-in-law.

neghez *N* our testicles.

neghezzus *N* our scrotums.

neghoo *N* our teeth.

neghootsun *N* our gums.

neghundan *N* our son-in-law.

neghunik whuts'inli *V* [C-li] watch what you say. *[IA]*

negwaz *N* our nephew/niece.

negwut *N* our knees.

negwutduts'un *N* our kneecaps.

negwutt'uk *N* the backs of our knees.

negwutyuschun *N* our shins.

nehaneztl'oo *V* [0-tl'oo < tl'oo$_1$] they fenced us in. *[PA]*

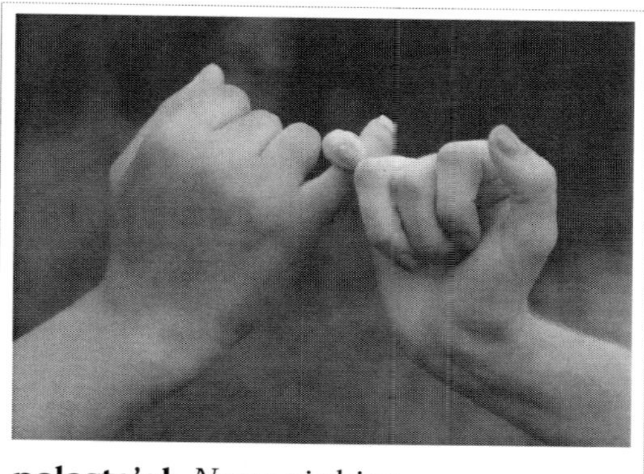

nelasts'ah *N* our pinkies.

nehndunilhtsi *V* [lh-tsi] you (1) put on makeup. *[IA]*

nehndunulhtsi *V* [lh-tsi] they put on makeup. *[IA]*

nehoontun V [0-tun < tun$_1$] he/she holds us. *[IA]* (1) 'Uloo **nehoontun** inle. 'Mother would grasp us.'

nehts'undunultsi *V* [l-tsi] we (3+) put on makeup. *[IA]*

nehu'a *N* lahal.

neilhchuk *N* police officer, RCMP officer.

nejook *N* our toque.

nekahunuta *V* [0-ta < ta$_1$] he/she is

looking for us. *[IA]*

nekechun *N* our legs.

nekechunnah *N* our ankles.

nekechunt'ak *N* the back of our legs.

nekelagi *N* our toenails.

nekelamai *N* our toes.

nekelamaicho *N* our big toes.

nekelamaiyaz *N* our little toes.

nekelamaiyazts'unts'e *N* our fourth toes.

nekelamaiyez *N* our long toes.

nekelamaiyezts'unts'e *N* our middle toes.

nekelascho *N* our big toes.

nekelatsul *N* our heels.

nekengi *N* our toenail.

nekent'ak *N* the tops of our feet.

nekentsul *N* our heels.

neketl'ah *N* our soles.

neketl'i *N* our toe spaces.

neketsul *N* our heels.

nekoo *N* our house.

nekuike *N* our husbands.

nek'i *N* our pelvic regions.

nek'its'un *N* our hipbones.

nekw'uts'uda *N* our chair.

nekw'uts'udacho *N* our easy chair.

nekw'uts'udayez *N* our chesterfield.

nekw'uz *N* our kidneys.

nelachunah *N* our wrists.

nelagi *N* our fingernails.

nelak'et *N* our palms.

nelasge *N* our fingers.

nelasts'ah *N* our pinkies.

nelat'ah *N* our wrists.

nelat'ak *N* the backs of our hands.

nelatl'i *N* our finger spaces.

nelik *N* our dog.

nelikcho *N* our big dog.

nelili *N* our bed.

nelilik'usulhchooz *N* our sheet.

neloo *N* our mother.

neludab *N* our table.

neluz *N* our urine.

neluzkah *N* our urinary tracts.

neluzzus *N* our bladders.

nelh *PPC* with us. (1) **Nelh** hunilch'e. 'He is mad at us.'

nelhtus *N* our sister.

nelhuk'a *V* [l-k'a] he/she is obese, fat. *[IA]*

nelhutsin *N* our brother.

nemoodih *N* our boss.

nemusdus *N* our cow.

-nen *N* face.

nena *N* our eyes.

nenabagha *N* our eyelashes.

nenachailya *V* [l-ya] you (1) have helped us, we (3+) thank you (1). *[PA]*

nenaildzih *N* prehistoric bird. A monster bird that migrates through in Spring and Fall and eats people.

nenalhti *N* our blanket.

nenats'osdooz *N* our eyebrows.

nenen *N* our faces.

nenet’a *V* [d-’a] he/she got fooled. *[PA]*

neni *N* our mind.

neninchustah *N* our sinus cavities.

neninchuz *N* our thumbs.

nenintak *N* our foreheads.

nenints’uzti *N* our elbows.

nenul’en-i *N* our television.

nenut’uk *N* our muscles.

nesih *N* our wall.

nesjan *V* [d-yan < yan_1] he/she has aged, is old. *[PA]*

nesjan *V* [d-yan < yan_1] I am old, have aged. *[PA]*

nestocho *N* our big stove, heater.

netai *N* our uncle other than mother’s brother.

Netats’anghi *N* Sugarloaf Mountain.

netes *N* our knife.

netoo *N* our water.

net’ak *N* our back.

net’akts’un *N* our backbones.

net’eke *N* our friend.

net’es *N* our charcoal.

net’ukts’un *N* our collarbone.

netl’ah *N* our bums.

netl’asus *N* our pants, dress.

netl’egha *N* our pubic hair.

netl’et *N* our groins.

netl’uz *N* our gall bladders.

netsa uneltsut *N* pimple.

netsakelhgwus *N* our navels.

netsan *N* our feces.

netsanbayah *N* our outhouse.

netsawhudelhzulhne *N* seniors.

netse’ *N* our daughter.

netsi *N* our heads.

netsi’alh *N* our pillow.

netsi’ul *N* tamarack. Etymology: Literally “our-head-tree”, so-called because an infusion of the bark is used to treat migraine headache.

netsidah *N* the tops of our heads.

netsigha *N* our head hair.

netsigha bets’ulha *N* hair oil.

netsinghai *N* our brains.

netsint’ak *N* the backs of our heads.

netsints’un *N* our skulls.

netsit’ak *N* the backs of our heads.

netsitl’ah *N* the backs of our heads.

netsiyan *N* our grandfather.

netsiyancho *N* our great-grandfather.

netsiyanyaz *N* our younger grandfather.

netsiyoo *N* headache medicine, aspirin.

netsiyun *N* public song.

netsk’et *N* our bed, bedroom.

netsowhuzulhne *N* elders, seniors.

netsoo *N* our maternal grandmother.

netsoocho *N* our maternal great-grandmother.

netsoola *N* our tongues.

netsooyaz *N* our younger grandmother.

netsudule *N* devil. Etymology: “he/she puts us in the fire”.

netsukw *N* our penises.

netsul *N* our rectums.

nets'ahbul *N* our cowboy hat.

nets'ik *N* our intestines.

nets'ikcho *N* our large intestines.

nets'ilchun *N* our necks.

nets'ilchunt'ak *N* the napes of our necks.

nets'oo *N* our breasts.

nets'oola *N* our nipples.

nets'uyi ghutna-a *N* our kitchen.

nets'uyibesula *N* our food cupboard.

newoontun *V* [0-tun] they hold us. *[IA]*

newuda *N* our calves.

newus *N* our shoulders.

newuz *N* our thighs.

newuzts'un *N* our femurs.

neyas'at *N* our daughter-in-law.

neyat *N* our older sister.

neyats'e *N* our daughter.

neye' *N* our son.

neyeztli *N* our horse.

neyi *N* cannibal. **neyi** are said to be like human beings, but chunky, very strong, and with narrow slitty eyes. They steal people, especially children, and eat them.

Neyilasts'ah *N* Mouse Mountain. See also: Lhkw'etsilhchola. Etymology: Literally, "cannibal's little finger", after the cannibals said once to have lived in a cave on it, and the fact that the bluff looks like a crooked little finger when seen from the West.

neyoona *N* our older brother.

neyulhtus *N* our sister.

neyun *N* our floor.

neyun *N* our song.

neyunlatel *N* our flooring.

neyunsubul *N* our floor covering.

neyunsulchooz *N* our floor covering.

neyunts'un *N* our backbones.

neyust'e *N* our bodies.

nez'e *N* our mother's brother.

Nezanutna *N* given name meaning she moved far away with her chattels.

nezaz *N* our father-in-law.

nezaz *N* our father-in-law.

nezdzak *N* dried berries.

nezek *N* our mouths.

nezek besulhti *N* coffin.

nezesdak *N* our throats.

nezesdak'al *N* our uvulas.

nezesdalos *N* our uvulas.

nezesdamai *N* our uvulas.

nezesdazum *N* our tonsils.

nezets'un *N* our chins.

nezih k'ut *PPC* beside us (3+).

nezkai *N* our blood.

nezkaich'ooz *N* our veins.

nezkaich'oozcho *N* our artery.

nezkeh *N* our children.

nezk'uz *V* [0-k'uz] it is cold. *[IA]*

nezk'uz *V* [0-k'uz < k'uz$_1$] it is cold. *[IA]*

nezool *N* our wind pipes.

nezoolts'un *N* our larynges.

nezumcho *N* our pancreases.

nezut *N* our livers.

nezuz *N* our hide.

ngan *N* your (1) arm.

nghahitelhchulh *V* [lh-chulh < choot$_1$] they will take it from you (1). [FA]

nghate'alh *V* [0-'alh] they are going to feed you. *[FA]*

nghe *N* your (1) brother-in-law, sister-in-law.

nghez *N* your (1) testicles.

nghezzus *N* your (1) scrotum.

nghoo *N* your (1) teeth.

nghootsun *N* your (1) gums.

nghundan *N* your (1) son-in-law.

nghuteschulh *V* [0-chulh < choot$_1$] I am going to take from you (1). *[FA]*

ngwaz *N* my nephew/niece.

ngwut *N* your (1) knee.

ngwutduts'un *N* your (1) kneecap.

ngwutt'uk *N* the back of your (1) knee.

ngwutyuschun *N* your (1) shins.

ni *V* [0-ni] he/she says. *[IA]*

ni *ADV* northward. (1) **Ni** tesyalh. 'I am going to go north.'

-ni *N* mind. Poss: -eni.

ni lhe'nintananesja *V* [d-ya] he/she was turned around. *[PA]*

ni usli *V* [0-li] I am worried. *[IA]*

nidudlat *V* [0-lat] we (2) are floating around. *[IA]*

nidul'en *V* [l-'en] we (2) look at (st). *[IA]*

niduljut *V* [l-jut < joot$_1$] we (2) are afraid. *[IA]*

-nik *N* nostril.

nilghaz *V* [l-ghaz < ghaz$_1$] it is hot. *[IA]*

nilwus *V* [l-wus] you (1) are hot. *[IA]*

nilh'en *V* [lh-'en < 'en$_1$] you (1) look at. *[IA]*

nilhdukw *V* [lh-dukw < dukw$_1$] it is near. *[IA]*

nilhdza *V* [lh-dza < dza$_1$] it is far. *[IA]*

nilhdza beyaztuk *N* telephone.

nilhts'i *V* [lh-ts'i < ts'i$_2$] it (wind) is blowing, it is windy. *[IA]*

nilhts'i *N* wind.

nilhts'i tainya *N* waterspout.

nilhts'idus *N* tornado.

-nimbus *N* cheek.

-nimbusts'un *N* cheek bone.

nin'ai *N* rack for drying food.

ninawhunde *V* [0-de] there are scattered clouds. [IA]

nincha *V* [0-cha < cha$_1$] it is large. [abs: n] *[IA]*

-ninchustah *N* sinus cavity.

-ninchuz *N* thumb.

ningoo *V* [0-goo] you (1) are driving in a loop. *[IA]*

nininle *V* [0-le < le$_1$] you (1) place (st). [class: mdo-c] *[IA]* (1) Soo **nininle.** 'Put them away.'

nink'az *V* [0-k'az] it is cold. *[IA]*

-nint'ak *N* forehead.

nintsi *V* [0-tsi < tsi'$_1$] it is bad. [abs: n] *[IA]* (1) Nukuk **nintsi.** 'The ball is bad.'

-nints'uzti *N* elbow.

ninya *V* [0-ya < ya$_1$] you (1) are walking in a loop. *[IA]*

ninyez *V* [0-yez] it is long. [abs: n] *[IA]*

ninzoo *V* [0-zoo < zoo$_1$] it is good. [abs: n] *[IA]* (1) Nukuk **ninzoo.** 'It is a nice ball. '

nisdutsil *V* [d-tsil] I got a tan. *[PA]*

niti *N* border-crossing, end of road. The point where a trail ends at the frontier between one people's territory and its neighbour's.

nit'as *V* [0-'as < 'as$_1$] we (2) are walking in a loop. *[IA]*

nizde *ADV* from the north. (1) **Nizde** nilhts'i. 'The wind is blowing from the north.'

njahooltel *V* [l-tel < tel$_1$] it is as wide as this. [abs: wh] *[IA]* The standard of comparison is something designated by the speaker using a gesture.

njan *ADV* here.

njaneltsuk *V* [0-tsuk] there are as many as. *[IA]*

njook *N* your (1) toque.

nkechun *N* your (1) leg.

nkechunnah *N* your (1) ankle.

nkechunt'ak *N* the back of your (1) legs.

nkelagi *N* your (1) toenail.

nkelamai *N* your (1) toes.

nkelamaicho *N* your (1) big toe.

nkelamaiyaz *N* your (1) little toe.

nkelamaiyazts'unts'e *N* your (1) fourth toe.

nkelamaiyez *N* your (1) long toe.

nkelamaiyezts'unts'e *N* your (1) middle toe.

nkelascho *N* your (1) big toe.

nkelatsul *N* your (1) heel.

nkengi *N* your (1) toenail.

nkent'ak *N* the top of your (1) foot.

nkentsul *N* your (1) heel.

nketl'ah *N* your (1) sole.

nketl'i *N* your (1) toe spaces.

nketsul *N* your (1) heels.

nkoo *N* your (1) house.

nkooz *N* your (1) home village. (1) Nts'e la **nkooz?** 'Where are you from?' (2) Nts'e **nkooz?** 'Where are you from?'

nkunuta *V* [0-ta < ta$_1$] he/she is looking for you (1). *[IA]*

nk'enentai *V* [0-tai] he/she misses you (1). *[IA]*

nk'esi' *V* [0-tsi' < tsi'$_1$] I love you (1). *[IA]*

nk'i *N* your (1) pelvic region.

nk'its'un *N* your (1) hipbone.

nkwuts'udacho *N* your (1) easy chair.

nkw'uts'uda *N* your (1) chair.

nkw'uts'udayez *N* your (1) chesterfield.

nkw'uz *N* your (1) kidneys.

nlachunah *N* your (1) wrist.

nlagi *N* your (1) fingernail.

nlak'et *N* your (1) palm.

nlasge *N* your (1) finger.

nlasts'ah *N* your (1) pinkie.

nlat'ah *N* your (1) wrist.

nlat'ak *N* the back of your (1) hand.

nlatl'i *N* your (1) finger spaces.

nlik *N* your (1) dog.

nlikcho *N* your (1) big dog.

nlilik'usulhchooz *N* your (1) sheet.

nloo *N* your (1) mother.

nludab *N* your (1) table.

nluz *N* your (1) urine.

nluzkah *N* your (1) urinary tract.

nluzzus *N* your (1) bladder.

nnalhti *N* your (1) blanket.

nninchustah *N* your (1) sinus cavity.

nnintak *N* your (1) forehead.

nnints'uzti *N* your (1) elbow.

noduke *V* [0-ke < ke_2] we (2) may go around in a boat. *[OA]*

noh *N* young woman.

nondeti *N* dinosaur.

nonya *V* [0-ya < ya_1] you (1) may walk in a loop. *[OA]*

noo *N* island.

nooghun *N* embankment along Nechakoh.

noondai *N* arrowhead.

Noonghak *N* A ford on the Nechako River upstream of the Nautley River used before there were bridges or ferries.

Noonghun *N* The embankment along the Nechako River on the east side of the new subdivision by Alana's house.

noosbuz *V* [0-buz < buz_1] I may stretch (st). [abs: n] *[OA]* (1) 'Aw **noosbuz** ghait'ah. 'I can't stretch it.'

noostel *N* wolverine.

Noozkoh *N* Ormond Creek. Latitude: 54° 5' 18" N. Longitude: 124° 44' 52" W. Etymology: "(flowing) from the north creek". There used to be a salmon run, before the Kenney Dam was built.

nsih *N* your (1) wall.

ntai *N* your (1) uncle other than mother's brother.

ntel *V* [0-tel < tel_1] it is wide. [abs: 0] *[IA]*

nt'ak *N* your (1) back.

nt'akts'un *N* your (1) backbone.

nt'eke *N* your (1) friend.

nt'ukts'un *N* your (1) collarbone.

ntl'ah *N* your (1) bum.

ntl'asus *N* your (1) pants, dress.

ntl'egha *N* your (1) pubic hair.

ntl'et *N* your (1) groin.

ntl'uz *N* your (1) gall bladder.

ntsakelhgwus *N* your (1) navel.

ntsan *N* your (1) feces.

ntsanbayah *N* your (1) outhouse.

ntse' *N* your (1) daughter. A man's daughter.

ntsi *N* your (1) head.

ntsi'alh *N* your (1) pillow.

ntsich'a *N* aspirin.

ntsidah *N* the top of your (1) head.

ntsigha *N* your (1) head hair.

ntsinghai *N* your (1) brain.

ntsint'ak *N* the back of your (1) head.

ntsints'un *N* your (1) skull.

ntsit'ak *N* the back of your (1) head.

ntsitl'ah *N* the back of your (1) head.

ntsiyan *N* your (1) grandfather.

ntsiyancho *N* your (1) great-grandfather.

ntsiyanyaz *N* your (1) younger grandfather.

ntsizde *ADV* from the south. (1) **Ntsizde** nilhts'i. 'The wind is blowing from the south.'

ntsol *V* [0-tsol < tsol$_1$] it is small. *[IA]*

ntsolyaz *N* a little bit.

ntsoola *N* your (1) tongue.

ntsooyaz *N* your (1) younger grandmother.

ntsukw *N* your (1) penis.

ntsul *N* your (1) rectum.

nts'ahbul *N* your (1) cowboy hat.

nts'e *QWH* where. (1) **Nts'e** la nkooz? 'Where are you from?'

nts'ezun'a *QWH* how?

nts'ez *QWH* whence?, from where?. (1) **Nts'ez** hainya? 'Where did you come from?'

nts'ik *N* your (1) intestines.

nts'ikcho *N* your (1) large intestine.

nts'ilchun *N* your (1) neck.

nts'ilchunt'ak *N* the nape of your (1) neck.

nts'oh *QWH* how many (years of age). (1) **Nts'oh** nyeyusk'ut? 'How old are you?'

nts'oo *N* your (1) breast.

nts'oola *N* your (1) nipples.

nts'un *ADV* downward. (1) **Nts'un** tesyalh. 'I am going to go south.'

nts'untelhbus *V* [lh-bus < bas$_1$] I will roll (st) to you (1). *[FA]*

nts'untelhchoos *V* [lh-choos < chus$_1$] I will bring you (1) (st). *[FA]*

nts'untelhdoh *V* [lh-doh < do$_1$] I will bring you (1) (st). E.g. loose hay or wool. *[FA]*

nts'untelhdlooh *V* [lh-dlooh] I will lead (s.o.) to you (1). *[FA]*

nts'untes'alh *V* [0-'alh < 'ai$_1$] I will bring you (1) (st). [abs: n] [class: sdo-c] *[FA]*

nts'untesdzih *V* [0-dzih < dzai$_1$] I will bring you (1) (st). E.g. berries, beads, rings. [class: euo-c] *[FA]*

nts'ununtelbus *V* [l-bus] it will roll down. *[FA]*

nts'utelhgooh *V* [lh-gooh < goo$_1$] I will drive (st) to you (1). *[FA]*

nts'utelhtelh *V* [lh-telh < ti$_1$] I will bring you (1) (st). E.g. baby, small animal [class: body-c] *[FA]*

nts'utes'alh *V* [0-'alh < 'ai$_1$] I will bring you (1) (st). [class: sdo-c] *[FA]*

nts'utesgus *V* [0-gus] I will drag to you (1) (st). This includes non-forceful "dragging" such as leading a blind person by the hand. *[FA]*

nts'utesghilh *V* [0-ghilh < ghe$_1$] I will bring you (1) (st) on my back. *[FA]*

nts'uteskalh *V* [0-kalh] I will bring you (1) (st). E.g. a dish of food. [class: coc-c] *[FA]*

nts'utestilh *V* [0-tilh < tan$_1$] I will bring you (1) (st). E.g. a rifle, pole, or stick. [class: lro-c] *[FA]*

nts'utestleh *V* [0-tleh] I will bring you (1) (st). E.g. a handfull of mud. [class: mushy-c] *[FA]*

nts'uyasduk *V* [lh-duk] I am talking to you (1). *[IA]*

nts'uyi ghutna-a *N* your (1) kitchen.

nts'uyibesula *N* your (1) food cupboard.

nubah *N* soldiers.

nubalhtl'ool *N* trawling line.

nuchahateskel *V* [0-kel < ke_2] they are not going to go around by boat. *[FN]*

nuchahidulh *V* [0-dulh < dil_1] they (3+) have not walked in a loop. *[PN]* (1) 'Aw Fraser Lake **nuchahidulh.** 'They have not been to Fraser Lake and back.'

nuchahigooh *V* [0-gooh < goo_1] they have not driven in a loop. *[PN]*

nuchahikel *V* [0-kel < ke_2] they have not gone in a loop by boat. *[PN]*

nuchaigooh *V* [0-gooh < goo_1] he/she has not driven in a loop. *[PN]*

nuchainyal *V* [0-yal < ya_1] you (1) have not walked in a loop. *[PN]*

nuchaiyal *V* [0-yal < ya_1] he/she has not walked in a loop. *[PN]* (1) Dune 'aw 'oh **nuchaiyal.** 'The man has not been walking around.'

nuchasinyah *V* [0-yah < ya_1] you (1) are not walking in a loop. *[IN]*

nuchasit'as *V* [0-'as < $'as_1$] we (2) are not walking in a loop. *[IN]*

nuchasit'az *V* [0-'az < $'as_1$] we (2) have not walked in a loop. *[PN]* (1) 'Aw Fraser Lake ts'i **nuchasit'az.** 'We (2) have not been to Fraser Lake.'

nuchasoduke *V* [0-ke < ke_2] we (2) may not go around in a boat. *[ON]*

nuchasuske *V* [0-ke < ke_2] I am not going in a loop in a boat. *[IN]*

nuchasusyah *V* [0-yah < ya_1] I am not walking in a loop. *[IN]*

nuchasyah *V* [0-yah < ya_1] he/she is not walking in a loop. *[IN]*

nuchasyal *V* [0-yal < ya_1] I have not walked in a loop. *[PN]*

nuchateske *V* [0-ke < ke_2] he/she is not going to go around by boat. *[FN]*

nuchateskel *V* [0-kel < ke_2] he/she is not going to go around by boat. *[FN]*

nuchatesya *V* [0-ya < ya_1] he/she will not walk in a loop. *[FN]*

nuchatuzanya *V* [0-ya < ya_1] you (1) will not walk in a loop. *[FN]*

nuchatuzat'us *V* [0-'us < $'as_1$] we (2) will not walk in a loop. *[FN]*

nuchatuzeske *V* [0-ke < ke_2] I am not going to go around in a boat. *[FN]*

nuchatuzesyal *V* [0-yal < ya_1] I will not walk in a loop. *[FN]*

nuchats'idulh *V* [0-dulh < dil_1] we (3+) have not walked in a loop. *[PN]* (1) 'Aw Fraser Lake ts'i **nuchats'idulh.** 'We have not been to Fraser Lake.'

nuchats'usdil *V* [0-dil < dil_1] we (3+) are not walking in a loop. *[IN]*

nuchaztesdil *V* [0-dil < dil_1] we (3+) will not walk in a loop. *[FN]*

nucha:teske *V* [0-ke < ke$_2$] they are not going to go around by boat. *[FN]*

nucha:teskel *V* [0-kel < ke$_2$] they are not going to go around by boat. *[FN]*[old-fashioned]

nudaih *V* [0-daih < daih$_1$] he/she is dancing. This refers to European styles of dancing, not to traditional dancing. *[IA]*

nudaih ba yah *N* dance hall.

nudubun *N* mercury.

nuduguz *V* [0-guz] it is elastic. *[IA]*

nuduk'ui *N* dragonfly.

nuduldlut *V* [l-dlut] he/she is going in a loop talking loudly. *[IA]* (1) 'Oh **nuduldlut.** 'He is going around talking loudly.'

nuduldlut *V* [l-dlut] he/she is going around talking loudly. *[IA]* (1) 'Oh **nuduldlut.** 'He is going around talking loudly.'

nuduzas *V* [0-zas] he/she is hobo-ing around, messing around. *[IA]* Etymology: Literally, "he/she is abrading".

nugoo *N* sled.

nuhugoo *V* [0-goo] they are driving in a loop. *[IA]*

nuhulhugai *V* [l-gai < gaih$_1$] they (2) ran in a loop. *[PA]*

nuhut'ah *V* [0-t'ah < t'oh$_2$] they are flying in a loop. *[IA]*

nuhuzdil *V* [0-dil < dil$_1$] they (3+) have walked in a loop. *[PA]* (1) Fraser Lake ts'i **nuhuzdil.** 'They have been to Fraser Lake and back.' (2) Fraser Lake ts'i lak **nuhuzdil.** 'Have they been to Fraser Lake and back?'

nuhuzgoo *V* [0-goo < goo$_1$] they have driven in a loop. *[PA]* (1) Lheidli ts'i **nuhuzgoo.** 'They have driven to Lheidli and back.'

nuhuzki *V* [0-ki < ke$_2$] they have gone in a loop by boat. *[PA]*

nuk *ADV* to the west. (1) **Nuk** tesyalh. 'I am going to go west.'

nukats'ulgas *V* [l-gas] we (3+) scurry around. *[IA]*

nukedudeh *V* [0-deh] he/she is migrating. *[IA]*

nukehududeh *V* [0-deh] they are migrating. *[IA]*

nukuk *N* ball.

nukukcho *N* large ball. Roughly, balls the size of a basketball and larger.

nukukyaz *N* small ball. E.g. balls the size of a baseball.

nul'en-i *N* television.

nulai *V* [0-lai < lai$_1$] they are many. [abs: n] *[IA]* (1) Mai **nulai.** 'There are many berries.'

nulat *V* [0-lat] he/she is floating around. *[IA]*

nulch'e' *V* [l-ch'e' < ch'eh$_1$] he/she is pouting. *[IA]*

nuldzus *V* [l-dzus < dzus$_1$] he/she is dancing Indian-style. *[IA]*

nulgaih *V* [l-gaih < gaih$_1$] he/she is running around, he/she runs. *[IA]*

nuljut *V* [l-jut < joot$_1$] he/she is afraid. *[IA]*

nulk'i V [l-k'i] it is sweet, tastes good. [abs: n] *[IA]* (1) Mai **nulk'i.** 'The berries are sweet.'

nulwus *V* [l-wus] it is hot. *[IA]*

nulyul *V* [l-yul < yul$_1$] it is white. [abs: n] *[IA]* (1) Bunen **nulyul.** 'Her face is white.' (2) Snen **nulyul.** 'My face is white.'

nulh'en *V* [lh-'en < 'en$_1$] I am looking at. *[IA]* (1) Tsalhts'ul bulasge lhe'ultsolyaz **nulh'en.** 'I am looking at a baby whose fingers are oh so small.'

nulhghaz *V* [lh-ghaz < ghaz$_1$] he/she heats (st) up. *[IA]*

nulhujut *V* [l-jut < joot$_1$] I am afraid. *[IA]*

nun *N* southward.

nunadidulnih *V* [l-nih < nih$_2$] we (2) bless ourselves. *[IA]*

nunadulnih *V* [l-nih < nih$_2$] he/she blesses himself. *[IA]*

nunadulhunih *V* [l-nih < nih$_2$] I bless myself. *[IA]*

nunahudulnih *V* [l-nih < nih$_2$] they bless themselves. *[IA]*

nunazdulnih *V* [l-nih < nih$_2$] we (3+) bless ourselves. *[IA]*

nundunult'a *V* [l-t'a < 'ai$_1$] he/she is playing baseball. *[IA]*

nunidudzoot *V* [0-zoot < zoot$_1$] we (2) are skating around for pleasure. *[IA]*

nunihuntezut *V* [0-zut < zit$_4$] they will think. *[FA]*

nuninidudzut *V* [0-zut < zit$_4$] we (2) think. *[IA]*

nunininzut *V* [0-zut < zit$_4$] you (1) think. *[IA]*

nunintadudzut *V* [0-zut < zit$_4$] we (2) will think. *[FA]*

nunintanzut *V* [0-zut < zit$_4$] you (1) will think. *[FA]*

nuninteszut *V* [0-zut < zit$_4$] I will think. *[FA]*

nuninuhinzut *V* [0-zut < zit$_4$] they think. *[IA]*

nuninuszut *V* [0-zut < zit$_4$] I think. This describes thinking as a mental activity, not the mere holding of an opinion. *[IA]*

nunulle *V* [l-le] he/she is handling. [abs: n] [class: mdo] *[IA]* (1) Sghez **nunulle.** 'She is handling my testicles.'

nunutsus *V* [0-tsus] he/she whips around. he runs around very fast, frequently changing direction *[IA]*

nus *ADV* forward, ahead.

nus te *ADV* in the future.

nus ts'edulh *V* [0-dulh < dil$_1$] we (3+) are advancing, we (3+) are walking forward. *[IA–prog]*

nus whan'en *V* [0-'en < 'en$_1$] he/she had a premonition. *[FA]*

nus whe'alh V [0-'alh] he/she is procrastinating. *[IA]*

nusah'az *V* [0-'az < 'as$_1$] you (2) have walked in a loop. *[PA]*

nusdaih *V* [0-daih < daih$_1$] I dance. *[IA]*

nusde *ADV* from the west. (1) **Nusde** nilhts'i. 'The wind is blowing from the west.'

nusidugoo *V* [0-goo < goo$_1$] we (2) drove in a loop. *[PA]*

nusiduki *V* [0-ki < ke$_2$] we (2) went in a loop by boat. *[PA]* (1) Lujak ts'i lak **nusiduki?** 'Have we been to Lejac by boat?'

nusidulgai *V* [l-gai < gaih$_1$] we (2) ran in a loop. *[PA]*

nusilgai *V* [l-gai < $gaih_1$] you (1) ran in a loop. *[PA]*

nusingoo *V* [0-gooh < goo_1] you (1) drove in a loop. *[PA]*

nusinya *V* [0-ya < ya_1] you (1) walked in a loop. *[PA]*

nusit'az *V* [0-'az < $'as_1$] we (2) walked in a loop. *[PA]*

nuske *V* [0-ke < ke_2] I am going around by boat. *[IA]*

nusdaih *V* [0-daih < $daih_1$] I dance. *[IA]*

nusuki *V* [0-ki < ke_2] he/she has gone around by boat. *[PA]*

nusulhugai *V* [l-gai < $gaih_1$] I ran in a loop. *[PA]*

nususgoo *V* [0-goo < goo_1] I have driven in a loop. *[PA]* (1) Fraser Lake ts'i **nususgoo.** 'I have been to Fraser Lake by car.' (2) Fraser Lake ts'i lak **nususgoo.** 'Have I driven to Fraser Lake and back?'

nususki *V* [0-ki < ke_2] I have gone by boat in a loop. *[PA]* (1) Nadleh ts'i lak **nususki.** 'Have I been to Nadleh by boat?' (2) Nadleh ts'i **nususki.** 'I have been to Nautley by boat.'

nususya *V* [0-ya < ya_1] I walked in a loop. *[PA]*

nusuya *V* [0-ya < ya_1] he/she walked in a loop. *[PA]*

Nuswhutisdi *N* Peterson's beach and point. Etymology: "a long ways out point".

nusya *V* [0-ya < ya_1] I am walking in a loop. *[IA]*

nutanya *V* [0-ya < ya_1] you (1) will walk in a loop. *[FA]*

nutat'us *V* [0-'us < $'as_1$] we (2) will walk in a loop. *[FA]*

nute *V* [0-te] he/she is dreaming. *[IA]*

nutekelh *V* [0-kelh < ke_2] he/she is going to go around by boat. *[FA]*

nutelhudzulh *V* [l-dzulh] I am going to hunt around (in the bush). *[FA]*

nutesdzulh *V* [l-dzulh] I am going hunting. *[IA]*

nutesya *V* [0-ya < ya_1] I will walk in a loop. *[FA]*

nutet'ah *V* [0-t'ah < $t'oh_2$] he/she will fly in a loop. *[FA]*

nutet'ah *V* [0-t'ah < $t'oh_2$] he/she will fly in a loop. *[FA]*

nuteya *V* [0-ya < ya_1] he/she will walk in a loop. *[FA]*

nut'ah *V* [0-t'ah < $t'oh_2$] he/she is flying in a loop. *[IA]*

nut'ah *V* [0-t'ah < $t'oh_2$] he/she is flying in a loop. *[IA]*

nut'uk *N* muscle.

nuts'udilh *V* [0-dilh < dil_1] we (3+) are walking in a loop. *[IA]*

nuts'ugoo *V* [0-goo] we (3+) are driving in a loop. *[IA]*

nuts'ulhughaz *V* [l-ghaz] we (3+) ran in a loop. *[PA]*

nuts'uzdil *V* [0-dil < dil_1] we (3+)

have walked in a loop. *[PA]* (1) Fraser Lake ts'i **nuts'uzdil.** 'We have been to Fraser Lake and back.' (2) Fraser Lake ts'i lak **nuts'uzdil?** 'Have we been to Fraser Lake?' (3) Nts'e **nuts'uzdil.** 'Where have we been?'

nuts'uzki *V* [0-ki < ke_2] we (3+) have gone in a loop by boat. *[PA]* (1) Stella ts'i **nuts'uzki.** 'We have been to Stella by boat.'

nuwhulyeh *V* [l-yeh] he/she is dancing. This refers to traditional dancing as opposed to European style dancing. *[IA]*

nuya *V* [0-ya] he/she is walking in a loop. *[IA]* (1) Dune 'oh **nuya.** 'The man is walking around.' (2) Soo **nuya.** 'He walks around well.'

nuyeh *V* [0-yeh] he/she (child) is growing. *[IA]*

nuzoot *V* [0-zoot < $zoot_1$] he/she is skating around. *[IA]*

nuztedulh *V* [0-dulh < dil_1] we (3+) will walk in a loop. *[FA]*

nwuda *N* your (1) calf.

nwus *N* your (1) shoulders.

nwuz *N* your (1) thighs.

nwuzts'un *N* your (1) femur.

nyah *N* your (1) house.

Nyan Wheti *N* The trail from Nadleh village to Tsaooche "Sowchea" on Stuart Lake. This trail became the pack trail for the Northwest Company then the Hudson's Bay Company between 1806 and 1906, until it was replaced by the Dog Creek Trail, which skirted Saddle Mountain rather than crossing it. Used for travel, trade, communications, hunting, trap-ping, rabbit snaring, bark and plant gathering. Etymology: "the trail across".

nyas'at *N* your (1) daughter-in-law.

nyat *N* your (1) older sister.

nyats'e *N* your (1) daughter. A woman's daughter.

nye' *N* your (1) son.

nye'elts'ul *V* [l-ts'ul] you (1) are hungry. *[IA]*

nyez *V* [0-yez] he/she is tall, it (generic) is tall, long. *[IA]*

nyeztli *N* your (1) horse.

nyints'iz *N* your (1) elbow.

nyiz *V* [0-yiz] he/she is tall. *[IA]*

nyudani *V* [0-ni] he/she told you (1). *[PA]*

nyun *N* your (1) floor.

nyunlatel *N* your (1) flooring.

nyunsubul *N* your (1) floor covering.

nyunsulchooz *N* your (1) floor covering.

nyunt'sun *N* your (1) backbone.

nyust'e *N* your (1) body.

nzaz *N* your (1) father-in-law.

nzesdak *N* your (1) throat

nzesdak'al *N* your (1) uvula.

nzesdalos *N* your (1) uvula.

nzesdamai *N* your (1) uvula.

nzesdazum *N* your (1) tonsils.

nzets'un *N* your (1) chin.

nzih *PPC* beside you (1).

nzoo *V* [0-zoo < zoo_1] he/she is good, it is pretty. *[IA]* (1) 'Indakya **nzoo.**

'The flowers are pretty.'

nzool *N* your (1) wind pipe.

nzoolts'un *N* your (1) larynx.

nzumcho *N* your (1) pancreas.

nzut *N* your (1) liver.

nzuz *N* your (1) hide.

nyak'i *N* your (1) mother's sister.

nyanus *PPC* greater than you (1). (1) **Nyanus** soo dakelh k'uyalhtuk. 'He speaks Carrier better than you do.' (2) **Nyanus** 'ilyez. 'She is taller than you.'

nye'indak *N* your (1) flower.

nye'lunatdultsiyan *N* your (1) great-grandfather.

nye'ook'et'as *N* your (1) cook stove.

nye'ook'utudai *N* your (1) table.

nye'ustes *N* your (1) mattress.

nye'ustlen *N* your (1) subject, servant.

nyelili *N* your (1) bed.

nyeni *N* your (1) mind.

nyenul'en-i *N* your (1) television.

nyestocho *N* your (1) big stove, heater.

nyeyusk'ut *N* your (1) years, age.

nyoona *N* your (1) older brother.

nyoozi *N* your (1) name.

nyube'uzdla *N* your (1) cupboard, dresser.

nyudusneke *N* your (1) ancestors.

nyulhtus *N* your (1) sister.

nyulhye' *N* your (1) step-son.

nyunilh'en *V* [lh-'en] he/she is looking at you (1). *[IA]*

nyutesk'et *N* your (1) bed, bedroom.

nyutoo *N* your (1) water.

nyuts'ulhni *V* [lh-ni] we (3+) call you (1). *[IA]* (1) Mbe **nyuts'ulhni?** 'What is your name?'

nyuz'e *N* your (1) mother's brother.

nyuzkai *N* your (1) blood.

nyuzkaich'ooz *N* your (1) veins.

nyuzkaich'oozcho *N* your (1) artery.

nyuzkeh *N* your (1) children.

nyuznilh'en *V* [lh-'en] we (3+) are looking at you (1). *[IA]*

odudai *V* [0-dai < yi_1] we (2) eat. *[OA]*

odutez *V* [0-tez] we (2) sleep. *[OA]*

ombuz *V* [0-buz < buz_1] you (1) may stretch. *[OA]* (1) 'Aw **ombuz** ghait'ah. 'You can't stretch it.'

onte *V* [0-te < ti_4] you (1) may sleep. *[OA]*

onyi *V* [0-yi < yi_1] you (1) may eat (st). *[OA]*

otomobil *N* car.

oodainjaz *V* [d-yaz] it is snowed in, blocked by snow. *[PA]* (1) Dati **oodainjaz.** 'The door is blocked by snow.'

oodaningi *V* [0-gi] he/she is hungry, skinny. *[IA]*

oodunaltalh-i *N* firecracker.

oodutalh *V* [0-talh < $tulh_1$] we (2) are kicking (st). *[IA]*

ooka'nidudzun *V* [0-zun] we (2) want it. *[IA]*

ooka'nuszun *V* [0-zun] I want it. *[IA]*

ooka'ts'uninzun *V* [0-zun] we (3+) want it. *[IA]*

ookaih *V* [0-kaih] he/she tries to scoop. [class: coc] *[IA–customary]* (1) Duda be lhukw **ookaih.** 'He (a pelican) scoops up fish in his beak.'

ooket *V* [0-ket < ket_1] he/she is buying (st). *[IA]*

ook'utudai *N* table.

-oona *N* older brother, male first cousin older than ego.

oonats'ilh *V* [0-ts'ilh] he/she closes both eyes and then opens them again. *[IA]*

-oonde *N* male second cousin, any male relative on father's side.

oondunultalh *N* dynamite.

oonidujin *V* [0-yin] we (2) are picking (berries). *[IA]*

oonusyin *V* [0-yin] I am picking (berries). *[IA]*

oonuyin *V* [0-yin] he/she is picking (berries). *[IA]*

oosket *V* [0-ket < ket_1] I am buying (st). *[IA]*

oostalh *V* [0-talh < $tulh_1$] I am kicking (st). *[IA]*

ooteskulh *V* [0-kulh < ket_1] I will buy. *[FA]*

ootsas *V* [0-tsas] he/she has severe pain. *[IA]*

ooyutalhai *N* oranges. Etymology: Literally, "they are juicy inside".

-oozi *N* name.

rice 'uzus *N* rice sack.

s- *PREFIX* first person singular possessor. Attaches to class 1 nouns.

s'at *N* my wife.

sa *ADV* for a long time, slowly.

sa *N* sun.

sa dendi *V* [0-di] it is hot and sunny. *[IA]*

sa dot'en *V* [d-'en < $'en_1$] it is yellow. Literally, "it is the colour of the sun". *[IA]*

oonuyin *V* [0-yin] he/she is picking (berries). *[IA]*

sa ilk'un han'ai *N* red sunrise.

sa k'ah naidutat'en V [d-'en < $'en_1$] the sun is setting. *[IA]*

sa k'ah naidutut'ih *V* [d-'ih < $'en_1$] the sun sets. *[IA–customary]*

sa yenin'ai *N* eclipse. When an eclipse occurs, women must lift their skirts, expose their bare bums, and scratch them. If they do not, they will get the seven-year itch.

Sacho Din'ai *N* December.

Sachonun *N* January.

sadzak'uilke *N* sundogs.

sah *N* gaff hook.

saha'dunet'ah *V* [0-t'ah] he/she

cheats me. *[IA]*

sahke *V* [0-ke < ke$_1$] you (2) are sitting. *[IA]*

sahulhgi *V* [lh-gi] I am amused. *[IA]*

sahunilh'en *V* [lh-'en < 'en$_1$] they are looking at me. *[IA]*

sai *N* sand.

Sainaditi *N* Peterson's Beach Forestry Campsite. A sheltered bay with a sandy beach. A fishing area for netting char and whitefish. Latitude: 54° 6' 1" N. Longitude: 124° 43' 49" W. Etymology: "sand road".

sak'i *N* my mother's sister.

salat t'an *N* dandelion. Etymology: "salad leaf".

sanun *N* month.

sazul *N* heat.

sba *N* my father.

sba *PPC* for me.

sba 'alha'hoont'ah *V* [0-t'ah < t'oh$_1$] I believe it. *[IA]*

sba hooncha *V* [0-cha < cha$_1$] I am taken aback, amazed. *[IA]*

sba huwonenik *V* [0-nik] I am amused. *[IA]*

sba ut'en lhai *V* [0-lai] I am busy. *[IA]*

sba we'hooja *V* [0-ja] I am bewildered. *[IA]*

sbadlo *V* [0-dlo] I am amused. *[IA]*

sbahuwawhulna *V* [l-na < na$_2$] it is difficult for me. *[IA]*

sbalyan *N* Bald Eagle.

sbalh'i *V* [lh-'i < 'i$_1$] he/she is waiting for me. *[IA]*

sbat *N* my mitts.

sbe'uzdla *N* my cupboard, dresser.

sbez *N* my mother-in-law.

sbezan *N* my paternal aunt.

sboo' *N* my rectum.

sboos *N* my cat.

sbun *N* my roof.

sbunt'ah *N* my attic.

sbut *N* my stomach. (1) **Sbut** soo 'uja. 'My stomach is nice and full.'

sbut *PPC* in front of me.

sbuttl'ah *N* the bottom of my stomach.

schahoonzum *V* [0-zum] I am hungry. *[IA]*

schai *N* my grandchild.

schaicho *N* my great-grandchild.

schaike *N* my grandchildren.

schak *N* my ribs.

schas *N* swallow.

schets'un *N* my tailbone.

schul *N* my younger brother.

sch'e' *PRO* mine. (1) **Sch'e'** 'int'ah. 'It is mine.'

sch'i' *PRO* mine.

sda *N* my lips.

Sdabiz *N* a place on the north shore of Fraser Lake.

sdadeltel *N* my porch.

sdadentanghunez'ai *N* my doorknob.

sdadentanghunezdla *N* my doorknobs.

sdadent'az *N* my window.

sdadent'azts'asdla *N* my curtains.

sdagha *N* my moustache.

sdaningi *V* [0-gi] I am hungry/skinny. *[IA]*

sdati *N* my doorway.

sdayun *N* my backbone region.

sdek'a *N* my tobacco.

sdes *N* my lungs.

sdesyaz *N* my younger sister.

sduch'us *V* [d-ch'us] it (rope) has broken. *[PA]*

sdughak'ut *N* peninsula.

sdune delts'i *N* my living room.

sduneti kw'uts'uda *N* my reclining chair.

sdusneke *N* my ancestors.

sdutez *N* hotel.

sdle *N* steelhead salmon.

sdzek hah'elts'ul *V* [l-ts'ul] I have an earache. *[IA]*

sdzi *N* my heart.

sdzidzool *N* my artery.

sdzik'ut *N* my chest.

sdzoozt'an *N* my shirt.

se *N* belt. Poss: ze.

se'ilhdzun *V* [lh-dzun] it is mine. *[IA]*

se'indak *N* my flower.

se'lunatdultsiyan *N* my great-grandfather.

se'ook'et'as *N* my cook stove.

se'ook'utudai *N* my table.

se'ustes *N* my mattress.

se'ustlen *N* my subject, servant.

se'utes *N* my knife.

se'utsun *N* my meat.

sedulghi *V* [lh-ghi < ghi$_1$] we (2) killed (st). *[PA]*

seduljus *V* [lh-jus < jas$_2$] we (2) hooked (fish). *[PA]*

selhghi *V* [lh-ghi < ghi$_1$] I killed (st). *[PA]*

selhkelh *V* [lh-kelh < ke$_2$] he/she is transporting me by boat. *[IA–progressive]*

seni *N* my mind.

sestocho *N* my big stove, heater.

sgan *N* my arm.

Sgookoh *N* The creek that enters Fraser Lake about two miles West of Nadleh along the north shore. Etymology: "worm river'. It used to be a spawning place for gescho.

sgha *PPC* because of me.

sghadutetilh *V* [0-tilh < tan$_1$] he/she will give me (st). [abs: d] [class: lro] *[FA]* (1) Tuz **sghadutetilh.** 'He is going to give me a cane.'

sghainin'ai *V* [0-'ai < 'ai$_1$] he/she gave it to me. [abs: gen] [class: sdo] *[PA]* (1) Lubot **sghainin'ai.** 'He gave me a cup.'

sghainintan *V* [0-tan < tan$_1$] he/she gave it to me. [abs: gen] [class: lro] *[PA]* (1) 'Ulhti **sghainintan.** 'He gave me a rifle.'

sghani'ai *V* [0-'ai < 'ai$_1$] he/she gave me. [class: sdo-gen] *[PA]*

sghaninkai *V* [0-kai < kai$_1$] he/she gave me. [class: coc] *[PA]* (1) Too **sghaninkai.** 'He gave me (a cup of) water.

sghanintan *V* [0-tan < tan$_1$] he/she gave me. [class: lro] *[PA]* (1) 'Uba lhti **sghanintan.** 'Father gave me a rifle.'

sghante'alh *V* [0-'alh < 'ai$_1$] he/she is going to give me. [class: sdo-n] *[FA]* (1) Nukuk **sghante'alh.** 'He is going to give me a ball.'

sghe *N* my brother-in-law, my sister-in-law.

sghe *N* my grease.

sghez *N* my testicles.

sghezzus *N* my scrotum.

sghoo *N* my teeth.

sghootsun *N* my gums.

sghundan *N* my son-in-law.

sgwaz *N* my nephew/niece.

sgwuduts'un *N* my kneecap.

sgwut *N* my knee.

sgwutt'uk *N* the back of my knee.

sgwuttsi *N* my kneecap.

sgwutyuschun *N* my shins.

si *PRO* I, me.

si *PART* apparently. This particle immediately follows the constituent over which it has scope. (1) Za 'et **si** usda. 'Zaa is apparently there.' (2) Za **si** 'et usda. 'Zaa, apparently, is there.'

si *PART* [emphatic particle]. (1) Nanyoost'en **si!** 'See you again!'

sich'ah *ADV* myself, by myself .

siduke *V* [0-ke < ke$_1$] we (2) are sitting. *[IA]*

sidutez *V* [0-tez] we (2) are sleeping. *[IA]*

sih *N* wall.

sinda *V* [0-da < da$_1$] you (1) are sitting. *[IA]*

sinti *V* [0-ti < ti$_4$] you (1) sleep. *[IA]*

sjook *N* my toque.

sjul *N* Golden Eagle.

skanuta *V* [0-ta < ta$_1$] they are looking for me. *[IA]*

skawtezut *V* [0-zut] I fainted. *[PA]*

skechun *N* my leg. (1) **Skechun** deztulh. 'He kicked me in the leg.'

skechunnah *N* my ankle.

skechunt'ak *N* the back of my legs.

skeh *N* children. Singulative: ski.

skehdune *N* boys.

skehts'ekoo *N* girls.

skelagi *N* my toenail.

skelamai *N* my toes.

skelamaicho *N* my big toe.

skelamaiyaz *N* my little toe.

skelamaiyazts'unts'e *N* my fourth toe.

skelamaiyez *N* my long toe.

skelamaiyezts'unts'e *N* my middle toe.

skelascho *N* my big toe.

skelatsul *N* my heel.

skelh'at *N* bed.

skengi *N* my toenail.

Skenicho *N* peak of Greer Mountain. Latitude: 53° 59' 47" N. Longitude: 124° 30' 58" W.

skent'ak *N* the top of my foot.

skentsul *N* my heel.

sketl'ah *N* my sole.

sketl'i *N* my toe spaces.

sketsul *N* my heels.

skitsan *N* mustard. Etymology: Literally, "child shit", presumably after the colour and texture rather than taste.

skoo *N* my house.

ski *N* child. Duoplurals: -skeh, -skehne

ski *N* my husband.

skidune *N* boy. Duoplural: skehdune

skiduneyaz *N* small boy.

skits'eke *N* girl. Duoplural: skehts'ekoo

skits'ekeyaz *N* small girl.

skiyaz *N* small child.

skiza' *N* the child of a noble in the clan system.

skunuta *V* [0-ta < ta_1] he/she is looking for me. *[IA]*

sk'elh'ih *PPC* less than me.

sk'i *N* my pelvic region.

sk'its'un *N* my hipbone.

sk'ut *PPC* on me.

skwunladun *NUM* five *[locative]*.

skwunlai *NUM* five *[generic]*.

skwunlanun *NUM* five *[human]*.

skwunlat *NUM* five *[multiplicative]*.

skwunlat lanezi *NUM* fifty *[generic]*.

skwunlawh *NUM* five *[abstract]*.

skw'uts'uda *N* my chair.

skw'uts'udacho *N* my easy chair.

skw'uts'udayez *N* my chesterfield.

skw'uz *N* my kidneys.

sla *PPC* in aid of me, helping me. (1) **Sla** 'inja. 'You have helped me.'

sla be *PPP* by my hand, manually.

sla inja *VP* you (1) have helped me, thank you. *[PA]*

slachunah *N* my wrist.

slagi *N* my fingernail.

slahja *VP* you (2+) have helped me, thank you. *[PA]*

slaintih *V* [0-tih < tan_1] you (1) give (st) to me. *[IA]*

slak'et *N* my palm.

slasge *N* my finger.

slasgek *N* my index finger.

slasts'ah *N* my pinkie.

slat'ah *N* my wrist.

slat'ak *N* the back of my hand.

slatl'i *N* my finger spaces.

sleghe *N* oolichan grease.

slenyah *N* smoke house.

slik *N* my dog.

slikcho *N* my big dog.

sliktsol *N* my little dog.

slilik'usulhchooz *N* my sheet.

sloo *N* my mother.

sludab *N* my table.

sludi *N* my tea.

sluz *N* my urine.

sluzkah *N* my urinary tract.

sluzzus *N* my bladder.

smoodih *N* my boss.

smusdus *N* my cow.

sna *N* my eye.

snabagha *N* my eyelashes.

snachailya *V* [l-ya] you (1) have helped me, thank you. *[PA]*

slenyah *N* smoke house.

snachalhuya *V* [l-ya] you (2+) have helped me, I thank you (2+). *[PA]*

snadun *N* my cousin.

snalhti *N* my blanket.

snats'osdooz *N* my eyebrows.

snen *N* my face.

sninchustah *N* my sinus cavity.

sninchuz *N* my thumb.

snintak *N* my forehead.

snints'uzti *N* my elbow.

snul'en-i *N* my television.

snut'uk *N* my muscle.

so *N* frost.

soo ADV well, really. (1) Nyanus **soo** dakelh k'uyalhtuk. 'He speaks Carrier better than you do.' (2) **Soo** 'ont'e! 'Be well!' (3) **Soo** nuya. 'He walks around well.'

soo 'a *ADV* quickly.

soo gak *N* nothing.

soo ts'ihun *ADV* definitely.

soocho *ADV* properly. (1) **Soocho** duni'ulh. 'Chew properly.'

soodulhkut *V* [lh-kut] he/she asks me. *[IA]*

soogah *N* sugar. Etymology: Loan from English sugar.

soogah 'uzus *N* sugar sack.

sooh *N* American Robin.

sooltsun *V* [l-tsun] it smellls good. *[IA]*

soona *N* my elder brother.

soonadinja *V* [0-ja] you (1) are dressed up. *[IA]*

soonahduja *V* [0-ja] they are dressed up. *[IA]*

soonazduja *V* [0-ja] we (3+) are dressed up. *[IA]*

sooniya *N* money. Etymology: Loan from Cree soniyaw.

sooniyazuz *N* wallet.

soontun *V* [0-tun] I am being held. *[IA]*

soonuy uti zuz *N* purse.

soonuya *N* money. Etymology: Loan from Cree soniyaw.

soonuyazuz *N* purse.

stai *N* my uncle other than mother's brother, step-father.

Stellakoh *N* Stellakoh River. Latitude: 54° 2' 58" N. Longitude: 124° 53' 33" W.

Stellakoh *N* Stellakoh village. Latitude: 54° 5' 43" N. Longitude: 124° 47' 57" W.

Stellayukh *N* Alexandria. The former Carrier village between Quesnel and Williams Lake.

stesk'et *N* my bed, bedroom.

stisyancho *N* my great-grandfather.

sto *N* stove.

sto lhes sut'e *N* bannock.

stocho *N* big stove, heater. Etymology: sto from English "stove" with suffix cho "big".

stuz *N* my cane. (1) **Stuz** dulyul. 'My cane is white.'

st'ak *N* my back.

st'akts'un *N* my backbone.

st'eke *N* my friend.

stlesicho *N* wasp.

stl'ah *N* my bum. (1) **Stl'ah** yayankat. 'She spanked me.'

stl'asus *N* my pants, dress.

stl'egha *N* my pubic hair.

stl'et *N* my groin.

stl'oolh *N* my rope.

stl'uz *N* my gall bladder.

stsah *PPC* before me.

stsakelhgwus *N* my navel.

stsan *N* my feces.

stsanbayah *N* my outhouse.

stse' *N* my (man's) daughter. A man's daughter.

stsi *N* my head. (1) **Stsi** nduda. 'I have a headache.'

stsi nduda *V* [0-da < da_2] I have a headache. Literally, "my head is sore". *[IA]*

stsi'alh *N* my pillow.

stsidah *N* the top of my head.

stsigha *N* my head hair.

stsinghai *N* my brain.

stsint'ak *N* the back of my head.

stsints'un *N* my skull.

stsit'ak *N* the back of my head.

stsitl'ah *N* the back of my head.

stsiyan *N* my grandfather.

stsiyanyaz *N* my younger grandfather.

stsoo *N* my maternal grandmother.

stsoocho *N* my maternal great-grandmother.

stsoola *N* my tongue.

stsooyaz *N* my younger grandmother.

stsukw *N* my penis.

stsul *N* my rectum.

sts'ahbul *N* my cowboy hat.

sts'ik *N* my intestines.

sts'ikcho *N* my large intestine.

sts'ilchun *N* my neck.

sts'ilchunt'ak *N* the nape of my neck.

sts'oo *N* my breast.

sts'oola *N* my nipples.

sts'udunik *V* [0-nik] I am disliked. Literally, "one dislikes me". *[IA]*

sts'un *PPC* to me. (1) **Sts'un** yailhtuk. 'Call me (on the telephone).'

sts'uyi ghutna-a *N* my kitchen.

sts'uyibesula *N* my food cupboard.

sts'uz *N* house fly.

sts'uzts'uz *N* fly eggs.

subah *PART* I wonder. (1) Nts'e **subah** uztezdil. 'I wonder where we are going.'

sudani *V* [0-ni] he/she told me. *[PA]*

suhoontun *V* [0-tun] they are holding me. *[IA]*

sul *N* heat waves.

sulili *N* my bed.

sulusook *N* my sugar.

sulh *PPC* with me. (1) **Sulh** 'ut'en. 'He is working with me.'

sulh 'et'ah *V* [0-t'ah < t'oh$_2$] I am going by plane. *[IA–prog]* (1) Vancouver ts'i **sulh et'ah.** 'I am flying to Vancouver.'

sulh 'utet'ah *V* [0-t'ah < t'oh$_2$] I will go by plane. *[FA]*

sulh cha'est'ah *V* [0-t'ah < t'oh$_2$] I am not going by plane. *[IN–prog]*

sulh cha'test'ah *V* [0-t'ah < t'oh$_2$] I will not go by plane. *[FN]*

sulh ne'uzt'a *V* [0-t'a < t'oh$_2$] I have gone in a loop by plane. *[PA]*

sulh whechait'ah *V* [0-t'ah < t'oh$_2$] I have not set off by plane. *[PN]*

sulhjus *V* [lh-jus < jas$_2$] I hooked (st:fish). *[PA]*

sulhkulh *V* [lh-kulh] it collided with (st). *[PA]*

sulhtus *N* my sister.

sulhtsulyaz *V* [lh-tsul] it is damp. *[IA]*

sulhts'i *N* pepper.

sulhutsin *N* my brother.

sulhye' *N* my step-son.

sum *N* star.

sum nainkat *N* shooting stars, meteorites.

sum nakat *N* shooting stars, meteorites.

sutilh *N* my basket.

sumcho *N* morning star. Etymology: Literally, "big star".

sumcho nainkat *N* meteorite shower, shooting stars.

Sumdi *N* Saturday.

sundulcho *V* [l-cho < cha$_1$] he/she is as big as me. *[IA]*

suneztulh *V* [0-tulh] he/she kicked me in the face, she is sexually attracted to me. *[PA]*

sunilh'en *V* [lh-'en < 'en$_1$] he/she is looking at me. *[IA]*

sus *N* black bear.

sus *N* bridge.

sus'an *N* black bear den.

susda *V* [0-da < da$_1$] I am sitting. *[IA]* (1) Njan **susda.** 'I am here.'

susdli *V* [0-dli < li$_3$] I have become. *[PA]*

susghe *N* black bear grease.

susih *N* my wall.

susk'a *N* black bear fat.

susmai *N* Bearberry, Black Twinberry.

Susti *N* Gully trail to Nadleh River used by black bears going to feed on spawning salmon. Portions are now under the band office. In historic times used for access to the river for fishing, hauling water, etc. Etymology: "black bear trail".

susti *V* [0-ti < ti$_4$] I am sleeping. *[IA]*

sustsun *N* black bear meat.

susya *N* a thick black caterpillar. If you kill one of these you will kill a bear in the fall. You should make motions of eating it.

susyaz *N* little black bear.

sushas *N* my grizzly bear.

sutilh *N* my basket.

sutoo *N* my water.

suz'e *N* my mother's brother.

suze *N* my belt.

suzekw *N* my saliva.

suzih *PPC* beside me.

suzkai *N* my blood.

suzkaich'ooz *N* my veins.

suzkaich'oozcho *N* my artery.

suzkeh *N* my children.

suzt'ukw V [0-t'ukw < $t'ukw_1$] it has sucked on me. *[PA]* (1) Hoot'ukw **suzt'ukw.** 'A leech has sucked on me.'

swuda *N* my calf.

swus *N* my shoulders.

susyaz *N* little black bear.

swuz *N* my thighs.

swuzts'un *N* my femur.

syah *N* my house.

syas'at *N* my daughter-in-law.

syat *N* my older sister.

syats'e *N* my daughter. A woman's daughter.

sye' *N* my son.

sye'elts'ul *V* [l-ts'ul] I am hungry. *[IA]*

syesdak nez'ai *N* my adenoids.

syeztli *N* my horse.

syiz dilhdukw *V* [lh-dukw < $dukw_1$] I am short of breath. *[IA]*

syiz lhutulh *V* [l-tulh] I am out of breath. *[PA]*

syun *N* my floor.

syun *N* my song.

syunlatel *N* my flooring.

syunsubul *N* my floor covering.

syunsulchooz *N* my floor covering.

syunt'ak *N* the small of my back.

syunts'un *N* my backbone.

syust'e *N* my body.

szaz *N* my father-in-law.

szek *N* my mouth.

szesdak *N* my throat.

szesdak'al *N* my uvula.

szesdalos *N* my uvula.

szesdamai *N* my uvula.

szesdazum *N* my tonsils.

szets'un *N* my chin.

szool *N* my wind pipe.

szoolts'un *N* my larynx.

szumcho *N* my pancreas.

szut *N* my liver.

szuz *N* my hide.

shas *N* grizzly bear. Poss: **shas.**

Shasti *N* A trail (draw) west of Serina Greene's house just north of the north end of the Nautley River bridge. This is where grizzly bears came down to feed on spawning salmon. Etymology: "grizzly bear trail".

shasti *N* The deep gully West of Nadleh.

shen *N* summer.

Shen Whuniz *N* July. Etymology: "mid-summer".

shih yuzelhghi *V* [lh-ghi < ghi_1] he/she is fatigued. *[IA]*

shun *N* song. Poss: yun.

shun hadan'ai *V* [0-'ai] he/she broke into song. *[PA]*

taba *N* shore, beach.

tachek *N* river mouth.

tadez'ai *N* pond, puddle.

tadudai *V* [0-dai < yi_1] we (2) will eat. *[FA]*

tadugooh *V* [0-gooh < goo_1] we (2) will drive. *[FA]*

tadukelh *V* [0-kelh < ke_2] we (2) will go by boat. *[FA]* (1) Lejak ts'i **tadukelh.** 'We will go to Lejac by boat.'

tadulgih *V* [l-gih < $gaih_1$] we (2) will run. *[FA]*

tadun *NUM* three *[locative].*

tadunet'ai *N* cliff that descends into water.

tadutnai *N* bar. Etymology: "(where) one drinks liquor".

tah *N* mountain peaks. *[archaic]*

tah *PP* among.

tahoosde *V* [0-de] they are thirsty. *[IA]*

tahutnai *V* [d-nai < nai_1] they drink liquor. *[IA]*

-tai *N* father's brother, father's sister's husband, mother's sister's husband, father's male first cousin, step-father.

tak'i *NUM* three *[generic].*

takhulh *V* [0-khulh] it (body of water) is deep. *[IA]*

talgih *V* [l-gih] you (1) will run. *[FA]*

talgih *V* [l-gih < $gaih_1$] you (1) will run. *[FA]*

talook *N* salmon.

Talook Ooza *N* August.

talook sugi *N* dried salmon.

talook ts'it *N* snail.

talukwya *N* slug. Etymology: Literally, "salmon louse".

Talhedayui *N* A lake on the south side of I.R. 1. Reported to be a place to avoid due to quicksand and poisonous water.

talhlhoh *V* [lh-tloh] he/she is smearing. *[IA]*

Talht'en N Tahltan person.

Talhtsi Ooza *N* April.

talhutloh *V* [l-tloh] it has been smeared. *[PA]*

talh uki *V* [l-ki] it is fresh from the store. *[IA]* Etymology: Literally, this means "liquid is sweet/tasty". The meaning has presumably been generalized from this.

tangooh *V* [0-gooh < goo_1] you (1) will drive. *[FA]*

tanun *NUM* three *[human].*

tanyalh *V* [0-yalh] you (1) will walk. *[FA]*

tanyi *V* [0-yi < yi] you (1) will eat. *[FA]*

Tanyiz Bun *N* Tatin Lake. Latitude: 54° 8' 25'' N. Longitude: 125° 3' 27''

W. Etymology: “long water lake”. Trout fishing area.

taoodetnuk *V* [d-nuk] it (wh-class) is noisy. *[IA]*

taoonde *V* [0-de] you (1) are thirsty. *[IA]*

taoosde *V* [0-de] I am thirsty. *[IA]*

Tasyen *N* ’uza name. Formerly held by the late Peter George. Etymology: Many people believe that this name means “stands in water”, referring indirectly to a person having short legs, so that he looks like he is standing in water. However, according to elder Bernadette McQuarrie, the sister and daughter of holders of the name, the true meaning is “he/she stands on the high peaks”, that is, “he/she is outstanding”. She says that it is a contraction of tah usyen, where tah is an old word meaning “high peaks”.

tat *NUM* three [multiplicative], thrice.

tat lanezi *NUM* thirty *[generic]*.

tatetnilh *V* [d-nilh < nai_1] they will drink liquor. *[FA]*

tatnilh *V* [d-nilh] you (1) will drink liquor. *[FA]*

Tat’at *N* The eddy at the junction of the Nautley and Nechako rivers. It is not as distinct due to the reduction of the flows since the construction of the Kenney Dam.

tat’us *V* [0-’us < $’as_1$] we (2) will walk. *[FA]* (1) Fraser Lake ts’i **tat’us** lak? ‘Will we go to Fraser Lake?’

tatsi *N* waves on water.

Tats’utnai *N* Morris Creek.

tawdetnuk *V* [d-nuk] it is noisy. *[IA–wh]*

tawh *NUM* three *[abstract]*.

tawhulhti *V* [lh-ti < ti_3] it is raining into water. [abs: wh] *[IA]*

taztenilh *V* [d-nilh] we (3+) will drink liquor. *[FA]*

tazul *N* soup, stew.

tazul bedzuz *N* soup bowl.

tazul benuduka *N* ladle.

te’ninla V [0-la] he/she set net. *[PA]* (1) Tun yah **te’ninla.** ‘He set net under the ice.’

te’ninzun *V* [0-zun < zun_1] he/she is merciful. *[IA]*

te’nusla *V* [0-la] I set net. *[PA]* (1) Lhook ka **te’nusla.** ‘I set net for fish.’

te’teslilh *V* [0-lilh] I will set net. *[FA]* (1) Lhook ka **te’teslilh.** ‘I will set net for fish.’

te’usdle *V* [0-le] I set net. *[IA]* (1) Lhook ka **te’usle.** ‘I set net for fish.’

tegooh *V* [0-gooh < goo_1] he/she will drive. *[FA]*

teh gwuzeh *N* bluejay.

tehchunultsi *N* leech.

tehduchun *N* deadhead.

tehdulh *V* [0-dulh] you (3+) will walk. *[FA]*

Tehdune Yananestez *N* Smooth flat slabs of rock along the shores of I.R. 2 just east of Vinnedge Cove on Fraser Lake. The site of a legend in which a large man sleeping on the rocks was surprised and dove into the water, never to reappear.

The stones are said to resemble men wearing cloaks. Etymology: Literally, "underwater people sleep collectively".

tehgwuzeh *N* Kingfisher.

tehhaoolts'eh *N* A kind of white clay found on the bottom of lakes, used for washing and for chinking cabins. It is said to leave the skin of women who wash with it silky.

tehhoongook *N* Blue Heron.

tehjun *V* [d-yun < yun] you (2+) will sing. *[FA]*

Tehk'eilchuk *N* Vanderhoof. Latitude: 54° 1' 3" N. Longitude: 124° 0' 27" W. Etymology: "bottom of the river takes over".

tehnoo *N* reef.

tehul'en *V* [l-'en] they are wretched, poor. *[IA]*

tehyi *V* [0-yi < yi$_1$] you (2+) will eat. *[FA]* (1) Tehchunultsi 'i cha **tehyilh.** 'You will also eat leeches.'

teidul'en *V* [l-'en] we (2) are wretched, poor. *[IA]*

tejun *V* [d-yun < yun$_1$] he/she will sing. *[FA]*

tekelh *V* [0-kelh < ke$_2$] he/she will go by boat. *[FA]*

tel'en *V* [l-'en] he/she is wretched, poor. *[IA]*

telh'en *V* [lh-'en < 'en$_1$] he/she caught sight of. *[PA]*

telhjoos *N* mink.

telhu'en *V* [l-'en] I am wretched, poor. *[IA]*

telhudzool *V* [l-d-zool] it is bloated. E.g., a carcass. *[PA]*

tejun *V* [d-yun < yun$_1$] he/she will sing. *[FA]*

telhudzulh *V* [l-dzulh] I am going to go hunting. *[FA]*

telhugih *V* [l-gih < gaih$_1$] I will run. *[FA]*

tenadahdli *V* [0-dli] you (2+) pray. *[IA]*

tenadindli *V* [0-dli] you (1) pray. *[IA]*

tenadondlih *V* [0-dlih < dli$_2$] you (1) may pray. *[OA]*

tenadudli *V* [0-dli < dli$_2$] he/she prays. *[IA]*

tenadusdli *V* [0-dli < dli$_2$] I pray. *[IA]*

tenazdoodli *V* [0-dli < dli$_2$] we (3+) may pray. *[OA]*

tenazdudli *V* [0-dli < dli$_2$] we (3+) pray. *[IA]*

tes *N* knife.

tes hoongwun *N* jerry can.

tesgooh *V* [0-gooh < goo$_1$] I will drive. *[FA]* (1) Fraser Lake ts'i **tesgooh.** 'I will drive to Fraser Lake.' (2) Fraser Lake ts'i lak **tesgooh?** 'Will I drive to Fraser Lake?'

tesjas *V* [lh-jas < jas$_1$] I am going to milk. *[FA]* (1) Musdus **telhjas.** 'I am going to milk the cows.'

tesjaz *V* [0-jaz] I am going to ejaculate. *[FA]*

tesjaz *V* [0-jaz] it (fish) is laying eggs. *[PA]*

tesjun *V* [d-yun] I will sing. *[FA]* (1) Nyushun be **tesjun.** 'I am going to sing your song.'

teskelh *V* [0-kelh < ke$_2$] I will go by boat. *[FA]* (1) Nadleh ts'i lak **teskelh?** 'Will I go to Nautley by boat?' (2) Nadleh ts'i **teskelh.** 'I will go to Nautley by boat.'

tesk'et *N* bed, bedroom.

testilh *N* can.

tesyalh *V* [0-yalh < ya$_1$] I will walk. *[FA]*

tesyi *V* [0-yi < yi$_1$] I will eat. *[FA]*

teyalh *V* [0-yalh < ya$_1$] he/she is going to walk. *[FA]*

teyi *V* [0-yi < yi$_1$] he/she will eat (st). *[FA]*

tezts'i *V* [0-ts'i < ts'i$_2$] the wind has begun to blow. *[PA]*

tezya *V* [0-ya] he/she is walking. This means that he has already begun to walk. It contrasts with teyalh, which means that at some point in the future he is going to walk somewhere. *[PA]*

ti *N* handle.

ti *N* road, highway, trail.

tihunindil *V* [0-dil < dil$_1$] they (3+) exited. *[PA]*

tilh *N* basket. Poss: tilh.

tinch'a'unt'en *V* [d-'en < 'en$_2$] he/she feels guilty. *[IA]*

tininya *V* [0-ya < ya$_1$] he/she exited. *[PA]*

tinit'az *V* [0-'az < 'as$_1$] we (2) exited. *[PA]*

tintah *N* bush, forest.

tintulhi *N* coyote.

tinusya *V* [0-ya < ya$_1$] I exited. *[PA]*

tiznindil *V* [0-dil < dil$_1$] we (3+) exited. *[PA]*

too *N* water.

too benuduka *N* dipper.

too busk'ut *N* bank of body of water.

too dulk'un N "mercury". This refers to the "mercury" in a thermometer, that is, to the red-colored alcohol. By extension, it might be used in reference to actual mercury in a mercury thermometer. It does not refer to the metal mercury in general. Etymology: "red liquid".

too hubanghan *V* [0-ghan < ghan$_1$] they (3+) drowned. *[PA]*

too lhuk'i *N* pop. Etymology: "sweet water".

too nezk'uz *N* cold water.

too ulhtus *N* hard liquor. Etymology: "strong water".

too undunulmulh *N* Bromo selzer.

too yuzelhghi *V* [lh-ghi < ghi$_1$] he/she drowned. *[PA]*

too yuzulhghe *V* [l-ghe < ghi$_1$] he/she is drowning. *[IA]*

toobets'ukaih *N* water dipper.

toobets'utnai *N* drinking glass.

toobus *N* river bank.

toobust'ah *N* underpart of overhanging river bank.

toodzoot *N* rain coat. Etymology: "water coat".

took'et *N* water hole.

took'oh *N* river.

toonahdilh *V* [0-dilh] you (3+) take a bath. *[IA]*

toonaingus *V* [0-gus] you (1) wash. *[IA]*

toonidulte *V* [lh-te] we (2) give (s.o.) a bath. *[IA]*

tooninya *V* [0-ya] you (1) take a bath. *[IA]*

toonit'az *V* [0-'az < 'as$_1$] we (2) take a bath. *[IA]*

toonuhudelh *V* [0-delh] they (3+) take a bath. *[IA]*

toonuhuyulhte *V* [l-tre] they give him/her a bath. *[IA]*

toonulhte *V* [lh-te] I give (s.o.) a bath. *[IA]*

toonusya *V* [0-ya] I take a bath. *[IA]*

toonuts'udilh *V* [0-dilh] we (3+) take a bath. *[IA]*

toonuts'ulhte *V* [lh-te] we (3+) give (s.o.) a bath. *[IA]*

toonuya *V* [0-ya] he/she takes a bath. *[IA]*

toonuyulhte *V* [lh-te] he/she gives him/her a bath. *[IA]*

tooznal'en *N* mirror.

tuchus *N* fist.

tughul *N* drum.

tujulh *V* [d-yulh < yul$_3$] it spurts, gushes. *[IA]* (1) 'A buzkai **tujulh.** 'He is quick to anger.'

tukooh *V* [0-kooh] he/she is gagging. *[IA]*

tulalhgus *N* cross.

tulk'us *N* grasshopper.

tullhuk *V* [l-lhuk] it (animal) bucks. *[IA]*

tun *N* ice over a surface. This describes ice on a lake or road as opposed to pieces of ice. (1) **Tun** yah te'ninla. 'He set net under the ice.'

tusih *ADV* probably, maybe.

tut'oh *N* switch (for whipping).

tutl'it *V* [0-tl'it < tl'it$_1$] he/she farts. *[IA]*

tuz *N* cane.

tuzniz *N* midnight.

t'acho *N* mallard duck.

-t'ak *N* back.

-t'ak yusts'i *N* lower back.

-t'akts'un *N* spine, backbone.

t'alint'ah *V* [0-t'ah] it is pale. *[IA]*

-t'an *N* leaf.

t'an bedezdla *N* vase.

t'an behanuyeh *N* flower pot.

t'an dohot'en *V* [d-'en < 'en$_1$] it is green. [abs: wh] *[IA]* Etymology: "it is leaf-coloured".

t'an dot'en *V* [d-'en < 'en$_1$] it is green. *[IA]* Etymology: "it is leaf-coloured".

t'an doonat'en *V* [d-'en < 'en$_1$] it is green. [abs: n] *[IA]* Etymology: "it is leaf-coloured".

t'angoo *N* caterpillar.

t'antet'ot *V* [0-t'ot < t'ot$_1$] he/she will be smoking marijuana. *[FA]*

-t'ayukts'un *N* breast bone.

t'az *N* bat.

t'edinil *V* [0-nil] he/she became quiet. *[PA]*

t'edukoo *N* Irregular plural of t'et, q.v.

t'edusnih *V* [0-nih] I am quiet. *[IA]*

t'eh *N* raw meat.

t'eke *N* friend, relative.

t'es *N* charcoal.

t'et *N* young woman. Duoplurals: t'edukoo, t'eduke

t'etch'ul *N* unfavoured daughter.

t'etyaz *N* favorite daughter.

t'ewdusnih *V* [0-nih] it is quiet. [abs: wh] *[IA]*

t'ughus *N* Trembling Aspen. Locally known as Poplar.

-t'ukts'un *N* collar bone.

tl'abesu'ai *N* shorts.

tl'adak *N* summit.

tl'ah *N* bay.

-tl'ah *N* buttock, bum.

tl'asus *N* pants, dress.

-tl'egha *N* pubic hair.

-tl'et *N* groin.

-tl'it *N* groin, crotch, exhaust pipe.

tl'o *N* grass, hay.

tl'o bedudut'as *N* swather (machine for mowing hay).

tl'o musdus *N* buffalo. Etymology: "grass cow".

tl'obayah *N* barn. Etymology: "house for grass".

tl'ok'et *N* meadow.

tl'ok'ut *N* meadow, pasture.

tl'ok'uzih *N* Common Cattail.

Tl'otelk'et *N* A point on the east end of Fraser Lake on the south side of the Nautley River where it flows out of Fraser Lake. Etymology: "wide sedge grass place".

tl'otelk'ut *N* meadow.

tl'otsun *N* onion.

tl'oolh *N* rope.

tl'ughus *N* snake.

-tl'us *N* fish slime.

-tl'uz *N* gall bladder.

tsalhts'ul *N* baby.

-tl'uz *N* milt.

tl'uzni *N* male fish.

tl'uztse *N* gall stone.

tsa *N* beaver.

tsabai *N* Dolly Varden.

tsabilh *N* beaver snare.

tsacheschun *N* wild carrct.

tsacho *N* big beaver.

tsachun *N* food cache.

tsadzi *N* clock, watch.

tsagwulht'ai *N* butterfly.

tsah *PP* before, preceding.

tsaholhgus *V* [lh-gus] it is dark. [abs: wh] *[IA]*

tsaikaih *N* dried beaver meat.

-tsakelhgwus *N* navel, belly button.

tsaken *N* beaver lodge.

Tsak'et Whucho *N* A large depression on the shortcut road on I.R. 1, where a mass grave was made after a battle between Nadleh people and an invading war party from a tribe to the south. Etymology: "big graveyard".

tsak'etdin'ai *N* prisoner.

tsaluk *N* squirrel.

tsalhtse *N* high-bush cranberries.

tsalhts'ul *N* baby.

tsalhts'ulyaz *N* little baby.

tsan *N* breechcloth.

tsan *N* feces.

tsan *N* shit, feces.

tsan bayah *N* toilet.

tsanbayah *N* bathroom, outhouse.

tsandelh *N* widow, widower.

tsasdlimai *N* Northern Black Currant. Etymology: Literally, "frog berry".

tsati *N* large beaver.

Tsat'en *N* Beaver Indian.

tsatsun *N* beaver meat.

tsatsun sugi *N* dried beaver meat.

Tsayoo *N* The name of a clan. It's crests are: beaver, eagle.

tse *N* stone, rock.

tse dezk'un *N* coal. Etymology: "rock that burns".

tse toos'ai *N* weight at end of fish net.

-tse' *N* man's daughter.

tse'an *N* cave.

tse'ul *N* Arctic sagebrush.

tseba *N* hearth.

tsebalyan *N* bald eagle.

tsehudilhts'ai *V* [lh-ts'ai] they hoop and holler so that the mountains resound. *[IA]*

tsek'az *N* grindstone.

tsek'et *N* muskrat.

tsek'etken *N* muskrat lodge.

tsela Dujoz *N* A large rock on the North bank of the Nautley River below Matthew Ketlo's house. Said to resemble a rock that used to be at Stellako, later removed when the road was built, that Joz Dayi used to stand on to deliver speeches. In the 1940s when Leah Patrick's son Raymond was born, he/she was taken directly after his birth while still wet and rubbed on this rock to give him the strength of the rock. This site is associated with the legend of 'Usdas and the Virgin.

tseldzook *N* comb.

tselgai *N* obsidian.

tselk'eh *N* eel.

Tselhk'az *N* An old village site near the current Endako mine. This is evidently the same as the name of Burns Lake. Probably a reference to the former village of the Burns Lake Band, said by others to have been closer to Burns Lake, near the current site of Babine Forest

Products. Etymology: Literally, "hone".

tselhk'azbunk'ut *N* Burns Lake. Latitude: 54° 12' 35" N. Longitude: 125° 38' 30" W.

tselhk'azkoh *N* the river that flows into Burns Lake.

tselhk'azyus *N* the mountain outside Burns Lake.

tselhts'aichoyus *N* An important moose and deer hunting area. Etymology: "big axe bowl mountain", after a marshy area lying in a bowl-shaped depression with a rock formation resembling a stone axe rising in the middle.

tselhutulh *N* A distinct vertical crack in the lakeshore rocks on the North shore of Fraser Lake near Tehdune Yananestez. Etymology: "cracked rock".

Tsenabulh'a Bunk'ut *N* Leg Lake. A crescent-shaped expansion of Smith Creek on the east side of Lily Lake Road, east of Chowsunkut Lake. Latitude: 53° 59' 33" N. Longitude: 124° 31' 18" W. Etymology: "rock goes around lake". Material harvesting and trapping area. The meadows around this lake were hayed by Maxime George and other Nadleh people prior to their pre-emption by settlers.

tsentoo *N* sweat.

tseskaik'ejil *N* cinnabar.

tsetoos'ai *N* fish net anchor.

tsetsilh *N* axe.

tsets'ai *N* plate.

tsets'ai besula *N* dish cupboard.

tsewhehudilhkw'ah *N* three-way fork in road.

Tseyaz *N* Lejac. Latitude: 54° 3' 49" N. Longitude: 124° 44' 18" W.

tseyaz *N* gravel.

Tsezoolh Dezdla *N* Some wave-formed rocks along the shore in Gala Bay. Etymology: "hollow stones lie there". An area for netting whitefish and char.

tsezul *N* A site located below Basghelh up behind Martin Louie's house. This was the location of men's ceremonial and medicinal sweat lodges. Pits still remain from medicinal bough/steam beds and presumably from firepits. Prior to recent times, the last one to use this site was Seymour Thomas. Morris Louie packed water to a small log sweat lodge for him. Etymology: "sweat lodge".

tsezul ba yah *N* sweat lodge.

-tsi *N* head.

-tsi dunuts'un *V* [0-ts'un] to be stubborn. *[IA]* This literally means "X's head is hard", so it is conjugated for "subject" by possessive prefixation of **tsi** "head". (1)

tsi'alh *N* pillow. Etymology: Literally, "head carrier".

tsi'alhyaz *N* cushion. Etymology: "small pillow".

-tsidah *N* top of the head.

-tsigha *N* hair of head.

-tsigha lhu'ool *N* hair braid.

tsinezdelya *N* scarf.

-tsinghai *N* brain.

tsinkwun *V* [0-kwun] he/she is bald. *[IA]*

tsintel *N* ling cod.

Tsintel Ooza *N* March.

-tsint'ak *N* back of the head.

-tsints'un *N* skull.

tsis *N* Canadian River Otter.

tsitelhya *V* [l-ya] it bolted. E.g. a frightened animal. *[PA]*

-tsit'ak *N* back of the head.

-tsitl'ah *N* back of the head.

tsitsis *N* dandruff.

tsits'un *N* cranium, skull.

-tsiyan *N* grandfather.

-tsiyancho *N* great-grandfather.

-tsiyanyaz *N* younger grandfather.

-tsol *SUFFIX* small.

-tsoo *N* maternal grandmother.

-tsoocho *N* maternal great-grandmother.

-tsoola *N* tongue.

tsootsi *N* spider.

tsootsibilh *N* cobweb, spider web.

-tsooyaz *N* younger grandmother.

-tsukw *N* penis.

-tsul *N* anus, rectum.

tsul'aintal *V* [0-tal < $tulh_1$] you (1) kick (him) up the ass. *[IA]*

tsulhcho *N* huckleberries.

tsun *N* dirt.

-tsun *N* flesh.

tsunah *N* orphan.

tsuntsi *N* Douglas Fir.

tsunts'alh *N* spoon.

tsuyi ba yah *N* food storage building.

tsuz *N* firewood. (1) Tsuz dughut. 'He is sawing wood.'

ts'ah *N* hat, cap.

ts'ahbal *N* sombrero. Etymology: "round hat".

ts'ahbul *N* cowboy hat, wide-brimmed hat.

ts'al *N* Common Red Sphagnum moss. Locally known as Diaper Moss. Used for diapers, sanitary napkins, etc.

ts'anikwa *N* caterpillar. Dialect: Nadleh specific.

ts'ante'ez *ADV* after all.

ts'antez *V* [0-tez] we (3+) slept. *[PA]*

ts'anyi *V* [0-yi < yi_1] we (3+) ate. *[PA]*

ts'edulh *V* [0-dulh < dil_1] we (3+) are walking. *[IA–prog]* (1) Fraser Lake ts'i lak **ts'edulh?** 'Are we going to Fraser Lake?' (2) Nts'e **ts'edulh?** 'Where are we going?'

ts'egus *V* [0-gus] we (3+) are dragging (st). *[IA–prog]*

ts'eke *N* woman. Duoplural: ts'ekoo

ts'eke tsalhts'ul *N* baby girl.

ts'ekecho *N* large woman, respected woman.

ts'ekelh *V* [0-kelh < ke_2] we (3+) are going by boat. *[IA–prog]* (1) Stella ts'i **ts'ekelh.** 'We are going to Stella by boat.' (2) Stella ts'i lak **ts'ekelh?** 'Are we going to Stella by boat?'

ts'eketi *N* very respected female elder. Duoplural: ts'ekooti

ts'i *N* canoe.

ts'eketsi' *V* [0-tsi' < tsi'$_1$] she is ugly. *[IA]*

ts'ekeyaz *N* girl.

ts'ekeyez *N* tall woman. Duoplural: ts'ekooyez

ts'ekeza' *N* female noble in clan system.

ts'ekezoo *V* [0-zoo] she is beautiful. *[IA]* This is applicable only to women.

ts'ekoo *N* women. This is the irregular plural of ts'eke.

ts'ekoo tl'asus *N* panties. Etymology: "women's bum covering".

ts'ekooti *N* Irregular plural of ts'eketi.

ts'ekootoo *N* wine. Etymology: "women's water".

ts'ekooyaz *N* girls.

ts'elwus *V* [l-wus] we (3+) are running. *[IA–prog]*

Ts'esdlol *N* A little mountain on the south side of the Nechako River aross from the mouth of Tats'utnai (Morris Creek). Etymology: "sasquatch", after an event involving a sasquatch that took place here.

ts'eslol *N* sasquatch. Not to be confused with neyi, sasquatch are different from human beings. Their bodies are covered with long hair. They are said to eat plants and fish, not meat, and are not considered dangerous to people.

ts'i *N* canoe.

ts'i *PP* to, from. (1) Nadleh **ts'i** lak teskelh? 'Will I go to Nautley by boat?'

ts'ihna *N* hornet.

ts'ihnus hoolh'i *V* [lh-'i] he/she is honest. *[IA]*

-ts'ik *N* intestines, guts.

-ts'ikcho *N* large intestine, bowel.

-ts'ilchun *N* neck.

-ts'ilchunt'ak *N* nape of the neck.

ts'inyez *V* [0-yez] we (3+) are tall. *[IA]*

ts'inzoo *V* [0-zoo < zoo$_1$] we (3+) are good. *[IA]*

ts'it *INT* don't touch it!

ts'itukdukw *N* skirt.

ts'iyane *N* all *[human]*, everyone.

ts'oh *N* hat, cap.

ts'oh *N* spear.

ts'oo *N* Spruce.

-ts'oo *N* breast.

ts'oobez *N* Black Spruce.

ts'oobuz *V* [0-buz < buz$_1$] we (3+) may stretch. *[OA]*

ts'oodulhkut *V* [lh-kut] we (3+) ask (s.o.). *[IA]*

-ts'oola *N* nipple.

ts'ooladel *N* spruce bark.

Ts'oon'ai Bunk'ut *N* Etcho Lake.

A small lake just east of Ormond Lake. Latitude: 54° 10' 27" N. Longitude: 124° 38' 41" W. Etymology: "spruce drying rack lake". It is suggested, however, that this is a corruption of Wuts'in'ai "always in shadow". Area used for trapping beaver and fishing for trout. The original trail ran along the northwest side.

ts'ootalh *V* [0-talh] we (3+) are kicking (st). *[IA]*

ts'ootsun *N* Subalpine Fir. Locally known as Balsam Fir.

ts'ootsunt'ooz *N* Subalpine Fir bark, Balsam Fir bark.

ts'ooyi *V* [0-yi < yi_1] we (3+) may eat (st). *[OA]*

ts'oozus *N* bra.

ts'udelhts'i *V* [l-ts'i < $ts'i_1$] we (3+) are sitting. *[IA]*

ts'udunughai *V* [0-ghai < $ghai_1$] we (3+) have facial hair. *[IA]*

ts'uhoont'i *V* [0-t'i] we (3+) are happy. *[IA]*

ts'ukhuna *V* [0-na < na_1] we (3+) are alive. *[IA]*

ts'ulki' *V* [l-ki'] it is bitter, bad-tasting. *[IA]*

ts'ulhjus *V* [lh-jus < jas_2] we (3+) hooked (fish). *[PA]*

ts'ulhts'ul *V* [lh-ts'ul] we (3+) are broiling (st). *[IA]*

ts'un *PP* to. (1) **Sts'un** yailhtuk. 'Call me (on the telephone).'

-ts'un *N* paternal grandmother.

ts'uncho *N* swan.

-ts'uncho *N* paternal great-grandmother.

ts'unduda *V* [0-da] we (3+) are sick. *[IA]*

ts'unesjan *V* [d-yan < yan_1] we (3+) have aged. *[PA]*

ts'uniltsih *N* A monstrous animal found in Fraser Lake. Those who see it die.

ts'uninzun *V* [0-zun < zun_1] we (3+) think that. *[IA]*

ts'unk'ut *N* graveyard.

ts'untulwis *N* pill.

ts'unulwus *V* [l-wus] we (3+) are hot. *[IA]*

ts'unulhbuz *N* Northern Flying Squirrel.

ts'unulhduz *N* hummingbird.

ts'usgak *N* chickadee.

ts'utesjun *V* [d-yun] we (3+) will not sing. *[FN]* (1) 'Aw **ts'utesjun.** 'We are not going to sing.'

ts'uteyi *V* [0-yi < yi_1] we (3+) will eat. *[FA]*

ts'utni *V* [d-ni] we (3+) call (st). *[IA]*

ts'utooba *V* [0-ba < ba_2] we (3+) may go to war. *[OA]*

ts'ut'ot *N* tobacco. Etymology: Zero-nominalization of "we smoke".

ts'uwhuljos *N* chipmunk.

ts'uyi *V* [0-yi < yi_1] we (3+) eat. *[IA]*

ts'uyi *N* food.

ts'uyi bayah *N* cafe.

ts'uyi besula *N* food cupboard.

ts'uyi ghaghuna *N* chef.

ts'uyi ghutna-a *N* kitchen.

ts'uztez *V* [0-tez] we (3+) are sleeping. *[IA]*

ts'uztez ba whuz'ai *N* bedroom.

Ts'uzyen Hoh Ts'utsun nun *N* February.

ugooh *V* [0-gooh < goo_1] he/she is driving. *[IA–prog]*

uhkelh *V* [0-kelh < ke_2] you (2+) are going by boat. *[IA–progressive]*

uhyez *V* [0-yez] you (2+) are tall. *[IA]*

uhzoo *V* [0-zoo < zoo_1] you (2+) are good. *[IA]*

ukelh *V* [0-kelh < ke_2] he/she is going by boat. *[IA–progressive]*

ulchan *V* [l-chan < $chan_1$] she is pregnant. *[IA]*

ulgih *V* [l-gih < $gaih_1$] he/she is running. *[IA–progressive]*

ulhgoos *V* [lh-goos < $goos_1$] he/she crunches in his mouth. *[IA–customary]* (1) Dulagi **ulhgoos.** 'He bites his fingernails.' (2) Dughoo **ulhgoos.** 'He grinds his teeth.'

ulhlhah *V* [lh-tlah] he/she is greasing. *[IA]* (1) 'Uzuz **ulhlhah.** 'She is greasing a hide.'

ulhtus *V* [lh-tus < tus_2] it is strong. This can describe either physical strength or the strength of a beverage *[IA]*

-ulhtus *N* sister, female first cousin, co-wife.

-ulhtuske *N* sisters, female first cousins. This is the plural of -ulhtus.

ulht'ukw *V* [lh-t'ukw] he/she is sucking. *[IA]* (1) Duninchuz **ulht'ukw.** 'He is sucking his thumb.'

ulhts'ul *V* [lh-ts'ul] I am broiling (st). *[IA]*

ulhuchan *V* [l-chan < $chan_1$] I am pregnant. *[IA]*

ulhuchan *V* [l-chan < $chan_1$] you (2+) are pregnant. *[IA]*

ulhugih *V* [l-gih < $gaih_1$] I am running. *[IA–prog]*

-ulhutsin *N* brother, male first cousin.

-ulhye' *N* step-son.

undendut'en *V* [d-'en] you (1) are working hard. *[IA]*

unli *V* [0-li < li_3] he/she is. *[IA]* (1) Bughoo dzohtsulyaz tink'us **unli.** 'She has offset teeth.'

unna *N* your (1) eye.

unnabagha *N* your (1) eyelashes.

unnats'osdooz *N* your (1) eyebrows.

unnen *N* your (1) face.

unninchuz *N* your (1) thumb.

unnut'uk *N* your (1) muscle.

unsutesna *V* [0-na] he/she abandoned me. *[PA]*

untesna *V* [0-na] I abandoned (s.o.). *[PA]*

unzek *N* your (1) mouth.

usda *V* [0-da < da_1] he/she is sitting, is located. *[IA]* (1) Koo **usda.** 'She is at home.'

usgooh *V* [0-gooh < goo_1] I am driving. *[IA–prog]* (1) Fraser Lake ts'i **usgooh.** 'I am driving to Fraser Lake.' (2) Fraser Lake ts'i lak **usgooh?** 'Am I driving to Fraser Lake?'

usgus *V* [0-gus] I am dragging (st). *[IA–prog]*

usjiz *V* [d-yiz < yis_1] I am breathing. *[IA]* (1) 'Awhuz **usjiz.** 'I'm still breathing.'

usjun *V* [d-yun < yun$_1$] I sing. *[IA]*

uskelh *V* [0-kelh < ke$_2$] I am going by boat. *[IA–progressive]* (1) Nadleh ts'i lak **uskelh?** 'Am I going to Nautley by boat?' (2) Nadleh ts'i **uskelh.** 'I am going to Nautley by boat.'

uski be'nunulbaz *N* baby buggy.

uskiyaz *N* baby.

uslez *V* [0-lez] I am stewing (st), boiling (st). *[IA]*

usnedut'en *V* [d-'en] we are working hard. *[IA]*

ussun *N* my flesh.

ustun *V* [0-tun < tun$_2$] it froze, was killed by frost. *[PA]*

ust'as *V* [0-t'as] I am mowing. *[IA]* (1) Tl'o **ust'as.** 'I am mowing hay.'

usyalh *V* [0-yalh < ya$_1$] I am walking. *[IA–progressive]*

usyez *V* [0-yez] I am tall. *[IA]*

usyi *V* [0-yi < yi$_1$] I eat (st). *[IA]*

usyiz *V* [0-yiz] I am tall. *[IA]*

usyus *V* [0-yus] I crush (st). *[IA]*

uszoo *V* [0-zoo < zoo$_1$] I am good. *[IA]*

uyalh *V* [0-yalh < ya$_1$] he/she is walking. *[IA–progressive]* (1) 'Azdeh **uyalh.** 'He is coming.'

uyi *V* [0-yi < yi$_1$] he/she is eating. *[IA]* **(1)** Besk'i bet **uyi.** 'A seagull is eating a char.' (2) Ligok **uyi.** 'He is eating a chicken.'

uzdani *ADJ* deceased, the late. Follows the head noun.

uzdelhti *V* [lh-ti < ti$_2$] we (3+) honour (st). *[IA]*

uzdelhts'i *V* [lh-ts'i < ts'i$_1$] we (3+) are sitting . *[IA]*

uznanzin *V* [0-zin] we (3+) thought that. *[PA]*

uznudaih *V* [0-daih < daih$_1$] we (3+) are dancing. *[IA]*

uznuljut *V* [l-jut < joot$_1$] we (3+) are afraid. *[IA]*

uztedulh *V* [0-dulh < dil$_1$] we (3+) will walk. *[FA]* (1) Fraser Lake ts'i **uztedulh.** 'We will go to Fraser Lake.' (2) Fraser Lake ts'i lak **uztedulh?** 'Will we go to Fraser Lake?'

uztekelh *V* [0-kelh < ke$_2$] we (3+) will go by boat. *[FA]* (1) Stella ts'i **uztekelh.** 'We will go to Stella by boat.'

uztelwus *V* [l-wus] we (3+) will run. *[FA]*

uztelwus *V* [l-wus] we (3+) will run. *[FA]*

uztetez *V* [0-tez] we (3+) will sleep. *[FA]*

uztezdil *V* [0-dil < dil$_1$] we (3+) have started on foot. *[PA]* (1) Fraser Lake ts'i **uztezdil.** 'We are going to Fraser Lake.' (2) Nts'e subah **uztezdil.** 'I wonder where we are going.'

uzulh *V* [0-zulh] it (siren, whistle) is blowing. *[IA]*

wa'oodudli *V* [0-li] we (2) are careful with it. [*IA*]

wa'oosdli *V* [0-li] I am careful with it. *[IA]*

wa'ts'ooli *V* [0-li] we (3+) are careful with it. *[IA]*

wachahoolnah *V* [l-nah] it is easy.

[IA]

Wanderdi *N* Friday.

wasi *N* lynx.

wate'nuszun *V* [0-zun < zun_1] I pity him/her. *[IA]*

wedlew *N* sandpiper.

Wilyum sugi *N* Drywilliam Lake. The small lake on the south side of Highway 16 between Fort Fraser and Lejac. Latitude: 54° 3' 25" N. Longitude: 124° 41' 0" W. Etymology: Literally, "dry William", after Roy Nooski's grandfather William, who was thin. (Thin people are often referred to in Carrier as "dry".)

-wuda *N* calf of leg.

-wus *N* shoulder.

wusbuz *V* [0-buz < buz_1] I may stretch (st). [abs: gen] *[OA]* (1) 'Aw **wusbuz** ghait'ah. 'I can't stretch it.'

wusyi *V* [0-yi < yi_1] I may eat (st). *[OA]* **(1)** Dunitsung **wusyi** hukwa'nuszun. 'I want to eat moose meat.' (2) Yoobeduyun dunitsun **wusyi** sulhni. 'The doctor told me to eat moose meat.'

wuyi *V* [0-yi < yi_1] he/she may eat (st). *[OA]*

-wuz *N* thigh, upper leg.

-wuzts'un *N* femur, thigh bone.

whechahigooh *V* [0-gooh < goo_1] they did not set off driving. *[PN]* (1) Burns Lake ts'i **whechahigooh.** 'They did not set off driving to Burns Lake.'

whechaikel *V* [0-kel < ke_2] he/she did not set off by boat. *[PN]*

whechaiyal *V* [0-yal < ya_1] he/she did not set off walking. *[PN]*

whechaskel *V* [0-kel < ke_2] I have not set off by boat. *[PN]* (1) Nadleh ts'i **whechaskel.** 'I have not set off for Nautley by boat.'

wuyi *V* [0-yi < yi_1] he/she may eat (st).

whedunukw *V* [0-nukw] it is flashing. *[IA]*

wheguz *PPC* between them *[areal].*

whehandil *V* [0-dil < dil_1] they (3+) have set off on foot. *[PA]* (1) Fraser Lake ts'i **whehandil.** 'They have left for Fraser Lake.' (2) Fraser Lake ts'i lak **whehandil?** 'Have they left for Fraser Lake?'

whehangoo *V* [0-goo < goo_1] they have set off driving. *[PA]* (1) Lheidli ts'i **whehangoo.** 'They have set off driving to Lheidli.'

whehanki *V* [0-ki < ke_2] they left by boat. *[PA]*

wheidugoo *V* [0-goo < goo_1] we (2) set off driving. *[PA]*

wheiduki *V* [0-ki < ke_2] we (2) set off by boat. *[PA]*

wheinki *V* [0-ki < ke_2] he/she set off by boat. *[PA]*

wheinya *V* [0-ya < ya_1] he/she set off walking. *[PA]*

wheit'az *V* [0-'az < $'as_1$] we (2) set off

walking. *[PA]*

whenduneIduz *V* [l-duz] it bounced away. *[PA]*

wheni *PRO* we, us.

whenich'ah *PRO* ourselves.

whenlhujut *V* [0-jut] I am anxious. *[IA]*

whenlhuk'a *V* [l-k'a] I am gaining weight. *[IA]*

whenun *N* hillside.

whesgoo *V* [0-goo < goo_1] I set off by car. *[PA]* (1) Fraser Lake ts'i **whesgoo.** 'I set off my car to Fraser Lake.'

wheski *V* [0-ki < ke_2] I set off by boat. *[PA]* (1) Nadleh ts'i **wheski.** 'I set off for Nautley by boat.'

whets'andil *V* [0-dil < dil_1] we (3+) set off on foot. *[PA]* (1) Fraser Lake ts'i **whets'andil.** 'We set off for Fraser Lake.'

whets'anki *V* [0-ki < ke_2] we (3+) left by boat. *[PA]* (1) Stella ts'i **whets'anki.** 'We left for Stella by boat.'

whu- *PREFIX* areal object of postposition. Used with postpositions beginning with a consonant.

whuch'ahinelhde *V* [lh-de] they earned it. *[PA]*

whuch'ainelhde *V* [lh-de] he/she earned it. *[PA]*

whuch'anidulde *V* [lh-de] we (2) earned it. *[PA]*

whuch'aznelhde *V* [lh-de] we (3+) earned it. *[PA]*

whudah *N* entrance.

whudanzaz *V* [0-zaz] it is barren (nothing grows there). *[IA]*

whudezulh *V* [0-zulh] (time) is passing. *[IA–prog]* (1) Dzen 'a **whudezulh.** 'The days are passing quickly.'

whudintsi' *V* [0-tsi' < tsi'_1] it (the sky) is overcast and stormy. *[IA]* (1) Yat **whudintsi'.** 'It is overcast and stormy.'

whudintsuk *V* [0-tsuk] it (the sky) becomes overcast and stormy. *[IA–customary]* It becomes overcast and cloudy when there is a body in the water. (1) Yat **whudintsuk.** 'It becomes overcast and stormy.'

whudinzoo *V* [0-zoo < zoo_1] it (weather) is good. [abs: wh] *[IA]*

Whududlun Whucho *N* The steep hill on the old road on the south side of the Nechako River leading to 'Indzik'et. Etymology: "big steep place".

whudughai *V* [0-ghai < $ghai_1$] it is hairy. [abs: wh] *[IA]* (1) But'ak **whudughai.** 'His back is hairy.'

whuduldzan *V* [l-dzan < $dzan_1$] it is blue. [abs: wh] *[IA]*

whudulgi *V* [l-gi < gi_1] it is grey. [abs: wh] *[IA]*

whudulk'un *V* [l-k'un < $k'un_1$] it is red. [abs: wh] *[IA]*

whudulhgus *V* [lh-gus < gus_1] it is black. [abs: wh] *[IA]*

whudustl'us *V* [0-tl'us] I paint (st). [abs: wh] *[IA]*

whudutsun *V* [0-tsun] it is dirty. [abs: wh] *[IA]* (1) Ts'eke buyah **whudutsun.** 'The woman's house is dirty.'

whuduts'ul *V* [0-ts'ul] it is brushy. [abs: wh] *[IA]*

whuduts'un *V* [0-ts'un < ts'un$_1$] it is hard. [abs: wh] *[IA]*

whuduts'un suli *V* [0-li] it hardened, became hard. [abs: wh] *[PA]*

whudzaih *V* [0-dzaih] he/she is grasping, acquisitive. *[IA]*

whudzih *N* cariboo. Etymology: "it scrapes the ground", after the way cariboo scrape lichen off the rocks with their hooves.

whuhubunulhtun *V* [lh-tun < tun$_3$] I show them how, explain to them. *[IA]*

whujut *V* [0-jut] it (wh-class) is old. *[IA]* (1) Keyah **whujut.** 'The old village.'

whulai *V* [0-lai] they are numerous. [abs: wh] *[IA]*

whulecho *N* Red Columbine.

whuledzo *N* beaver's ears.

whulutah *ADV* occasionally.

whulyul *V* [l-yul < yul$_1$] it is white. [abs: wh] *[IA]* (1) Syah **whulyul.** 'My house is white.'

Whulhditdzen *N* Thursday.

whulhdulh *N* Indian Hellebore. Locally known as "ginseng", but not at all the same plant.

Whulhnatdzen *N* Tuesday.

whulhtat-lunatdultsiyan *N* great-great-grandfather.

whulhtatbulunatdultsiyan *N* his/her great-great-grandfather.

Whulhtatdzen *N* Wednesday.

whulhtathubalunatdultsiyan *N* their great-great-grandfather.

whulhtathubelunatdultsoo *N* their great-great-grandmother.

whulhtatnelunatdultsiyan *N* our great-great-grandfather.

whulhtatnelunatdultsoo *N* our great-great-grandmother.

whulhtatyubulunatdultsoo *N* his/her great-great-grandmother.

whulhtatlunatdultsoo *N* great-great-grandmother.

whulhtatselunatdultsiyan *N* my great-great-grandfather.

whulhtatselunatdultsoo *N* my great-great-grandmother.

whulhtsultah *N* swamp.

whunadlal *V* [0-dlal < dlal$_1$] he/she dreamt about something. *[PA]*

whunandunesti *V* [0-ti < ti$_4$] he/she overslept. *[PA]*

whunduda *V* [0-da < da$_2$] it is sore. [abs: wh] *[IA]*

whunezk'uz *V* [0-k'uz < k'uz$_1$] it is cold. [abs: wh] *[IA]*

whunih *V* [0-nih] he/she is awake. *[IA]*

whunik'az *V* [0-k'az] it is chilly. *[IA]*

whunilghaz *V* [l-ghaz < ghaz$_1$] it is hot. *[IA]*

whuniz *N* centre.

whunulh'a *N* flax-like plant. Used for making fishnets.

whunulhghaz *V* [lh-ghaz < ghaz$_1$] he/she heats (st). [abs: wh] *[IA]*

whunulhtun *V* [lh-tun < tun$_3$] I show how, explain. *[IA]*

whunyodulh'eh *V* [lh-'eh < 'eh$_1$]

he/she is teaching you (1). *[IA]*

whusainya *V* [0-ya] he/she has arrived. *[PA]*

whusanasasdughi *V* [d-ghi < ghe_1] I packed (st) back home. *[PA]*

whusanayedughi *V* [d-ghi < ghe_1] he/she packed it back home. *[PA]*

whusodulh'eh *V* [lh-'eh < $'eh_1$] he/she is teaching me. *[IA]*

whusunulhtun *V* [lh-tun < tun_3] he/she shows me how, explains to me. *[IA]*

whutah *PPC* among them (areas).

whutsah *PPC* before it *[areal]*. (1) Dzetniz **whutsah** 'Before noon.'

whutsootsi *N* spider.

whuts'oodutnik *V* [d-nik] it is boring. *[IA]*

whuyats'alhti *V* [lh-ti < ti_1] we (3+) buried. [class: body] *[PA]*

whuyodulh'eh *V* [lh-'eh < $'eh_1$] he/she is teaching him/her. *[IA]*

whuyudilhti *V* [lh-ti] he/she is buried here. *[IA]*

whuz *ADV* over there.

whuzainli *N* seep. A place where water seeps out of the ground.

whuzdli *V* [0-dli < li_3] he/she was born. *[PA]*

whuzih *PPC* beside it *[areal]***.**

whuzk'uz *V* [0-k'uz] it is cold. *[IA]*

whuzun'a *ADV* in that manner, thus.

ya *N* louse.

yaba *N* horizon, far away place, overseas. Etymology: Literally, "edge of the sky".

yabatseh *N* thread.

yabatsoh *N* needle. Etymology: "spear for lice".

yadanwul *V* [0-wul] it is smashed into many pieces. *[PA]*

yadatuk *V* [0-tuk < tuk_2] it has broken into many pieces. *[PA]*

yah *N* house, home.

yah *PP* under. (1) Tun **yah** te'ninla. 'He set net under the ice.'

yahainli *V* [0-li < li_1] it is flowing out of. *[IA]* (1) Dzulh **yahainli.** 'It (water) is flowing out of the mountain.'

yahalhduk *V* [lh-duk < duk_1] they spoke. *[PA]*

yahukih *V* [0-kih < k_2] they land in a boat. *[IA]*

yahulhduk *V* [lh-duk < duk_1] they are speaking. *[IA]*

yahutelhduk *V* [lh-duk < duk_1] they will speak. *[FA]*

yaidantsel *V* [0-tsel] he/she chopped it to pieces. *[PA]*

yaidukih *V* [0-kih < k_2] we (2) land in a boat. *[IA]*

yaiduldukh *V* [lh-duk < duk_1] we (2) speak. *[IA]*

yaidlane *N* the dead.

yailhduk V [lh-duk < duk_1] you (1) speak. **[IA]** (1) Sts'un **yailhtuk.** 'Call me (on the telephone).'

yakih *V* [0-kih < k_2] he/she lands in a boat. *[IA]*

yak'umai *N* low-bush cranberries, bog cranberries.

yak'unulh'a *N* Labrador Tea. Etymology: "it grows in swamps".

Yak'usda *N* God.

Yak'usda bughunek k'une 'ut'en-un *N* Christian person. Etymology: "one who acts in accordance with the word of God".

yak'ut *N* swamp.

yak'uz *N* heaven.

yakwut *N* attic.

yalhduk *V* [lh-duk < duk_1] I speak. *[IA]*

yalhtuk *V* [lh-tuk] he/she speaks. *[IA]*

yalhtsul *N* High-Bush Blueberries.

yanilhghel *N* week.

-yas'at *N* daughter-in-law.

yaskih *V* [0-kih < k_2] I land in a boat. *[IA]*

yat *N* the sky, cloud.

-yat *N* older sister.

yataduldduk *V* [l-duk < duk_1] we (2) will speak. *[FA]*

yatalhduk *V* [lh-duk < duk_1] you (1) will speak. *[FA]*

yatelhduk *V* [lh-duk < duk_1] I will speak. *[FA]*

yatoo *N* ocean.

yat'en *N* angel.

yat'enne *N* angels.

yats'alhduk *V* [lh-duk < duk_1] we (3+) spoke. *[PA]*

-yats'e *N* woman's daughter.

yats'it *N* dry cow moose.

yats'ukeh *V* [0-keh < ke_2] we (3+) land in a boat. *[IA]*

yats'ulhduk *V* [lh-duk < duk_1] we (3+) speak. *[IA]*

Yausilh *N* Crowe Canyon. A canyon on the Nechako River below the Old Crowe homestead at the junction of Dog Creek Road and Sutherland Forest Road. A large eddy here collects floating logs and driftwood.

yawtelhah *V* [0-lhah] it is a frosty clear night. *[IA]*

yawhudoos *N* northern lights, aurora borealis.

yayankat *V* [0-kat < kat_3] he/she slapped, spanked it thoroughly. *[PA]* (1) Stl'ah **yayankat.** 'She spanked me.'

-yaz *SUFFIX* small, little. Attaches to nouns and yields a new noun.

-yaz *N* woman's son.

yazai *N* clear sky.

yaztelhduk *V* [lh-duk < duk_1] we (3+) will speak. *[FA]*

yaztsih *ADV* across.

-ye' *N* man's son.

ye'indak *N* his/her flower.

ye'ubul *N* jump rope.

ye'ulht'ukw *V* [lh-t'ukw < $t'ukw_1$] she is breast-feeding him/her. *[IA]*

yeda' *ADV* last time.

yegus *V* [0-gus] he/she is dragging it. *[IA–prog]*

yenube *N* fin. Etymology: "that by means of which it swims".

yenuljut *V* [l-jut < $joot_1$] he/she is afraid of it. *[IA]*

yesjiz *N* carburetor. Etymology: Literally, "that by means of which it breathes".

yests'e *N* deer.

Yests'e Bunk'ut *N* Stag Lake. Latitude: 54° 1' 48" N. Longitude: 124° 39' 24" W. Etymology: "deer lake". Beaver and muskrat trapping area.

yests'etsun *N* venison, deer meat. Etymology: "deer meat".

yests'ezuz *N* deer skin.

yeyulhdzeh *V* [lh-dzeh < $dzeh_1$] he/she is pouring it into it. *[IA]*

yeztli *N* horse.

yeztlibayah *N* stable for horses. Etymology: "house for horses".

yeztlik'ah *N* horse tracks.

yeztliyaz *N* colt.

-yih k'ut nez'ai *N* Adam's Apple, larynx.

-yints'iz *N* elbow.

Yoh Whuyiz *N* A ceremonial style longhouse once located in the general area now occupied by Allan Nooski's and Alec George's houses. Etymology: "long house".

yoo *N* medicine.

yoo besula *N* medicine cabinet.

yoo'az nilhts'i *N* south wind.

yoobayah *N* health centre.

yoobeduyun *N* doctor.

Yoodughi *N* God.

yoodulhkut *V* [lh-kut] he/she asks him/her. *[IA]*

Yoonoo Bunk'ut *N* Oona Lake. Latitude: 54° 10' 53" N. Etymology: "medicine island lake". Area for fishing for char, trout, and whitefish.

yootalh *V* [0-talh] he/she is kicking it. *[IA]*

yests'e *N* deer.

yootsiz nilhts'i *N* north wind.

yootsiz tezts'i *V* [0-ts'i < $ts'i_2$] a blizzard has begun. *[PA]*

yooya susli *V* [0-li] I am embarassed. *[IA]*

yooyooz *N* whistling. (1) **Yooyooz** be yalhtuk. 'He speaks by whistling.'

yubuz *V* [0-buz < buz_1] he/she is stretching it. *[IA]* (1) Dunizuz **yubuz.** 'He is stretching the moosehide.'

yudani *V* [0-ni] he/she told him/her. *[PA]*

yudats'ein'ai *V* [0-'ai] he/she is making an effort. *[IA]* (1) Soo **yudats'ein'ai.** 'He's really making a lot of effort.'

yugha'ooli *V* [0-li] he/she is careful with it. *[IA]*

yughadzedzi *V* [0-dzi] he/she is caressing her. *[IA]*

yughaghuna *V* [0-na] he/she is bothering him/her. *[IA]*

yughatni *V* [d-ni] he/she is advising him/her. *[IA]*

yughulait'ah *V* [0-t'ah] it is too hard for him/her. *[IA]*

yughunli *V* [0-li] he/she is guarding him/her. *[IA]*

yughusda *V* [0-da < da_1] he/she is married to her, she is married to him/her. *[IA]*

yuk 'ududildzun *V* [l-d-zun] he/she is humble. *[IA]*

yuk 'utez'ai *V* [0-'ai < $'a_3$] the sun is setting. *[IA]*

yukaih *V* [0-kaih < kai_1] he/she is dipping it up. *[IA]*

yukunuta *V* [0-ta < ta_1] he/she is looking for him/her. *[IA]*

yuk'une'ut'en *V* [d-'en < $'en_2$] he/she obeys him/her. *[IA]*

yuk'untal *V* [0-tal < $tulh_1$] he/she kicked a hole in it. *[PA]*

yulh *PPC* with him/her. (1) **Yulh** hunilch'e. 'He is mad at him.'

yulhguk *V* [lh-guk < guk_1] he/she is massaging it, rubbing it. *[IA]*

yulhjas *V* [lh-jas < jas_1] he/she is milking it. *[IA]*

yulhjus *V* [lh-jus < jas_2] he/she hooked it (fish). *[PA]*

yulhkulh *V* [lh-kulh] it collides with it. *[IA]*

yulhlhah *V* [lh-tlah] he/she is smearing it. *[IA]* (1) Duyeztli yoo be **yulhlhah.** 'He is smearing his horse with salve.'

yulhna *V* [lh-na] he/she captures him/her . *[IA]*

yulhtus *N* his/her sister.

yulhts'ul *V* [lh-ts'ul] she is broiling it. *[IA]*

yun *N* floor, ground.

-yun *N* backbone.

yun dot'en *V* [d-'en < $'en_1$] it is brown. *[IA]* Etymology: "it is earth-coloured".

Yun Dulhdum *N* A location on Dog Creek Road just before Tachakoh "Three Mile Creek". Etymology: "the earth resounds".

yun k'ut *PPP* on earth.

yun nughutnah *N* earthquake.

yun nulh'as *N* bug.

yun subal *N* linoleum, carpet. Any floor covering.

yun sulchooz *N* floor covering. Etymology: This literally means "something two-dimensional and flexible that lies upon the floor."

yun sulhchooz *N* carpet.

yunak'ult'ah *V* [l-t'ah] she blinks at her in dislike. *[IA]*

yundunulhmulh *V* [lh-mulh < $mulh_1$] he/she is boiling it (water). *[IA]*

yundzihtel *N* floor boards.

yuneztulh *V* [0-tulh] he/she kicked him in the face, she is sexually attracted to him/her. *[PA]*

yunka dune *N* Aboriginal person.

yunkaot'en *N* Indigenous person.

yunkawhut'en *N* Indigenous person.

yunlatel *N* flooring.

yunlhutel *N* rug.

yunsubul *N* floor covering. Etymology: This literally means "something that has been unrolled onto the floor."

yuntah *N* trapline.

-yunt'ak *N* the small of the back.

-yunts'un *N* spine, backbone.

yunulh'uz *V* [lh-'uz < tl'es$_1$] he/she is hammering it. *[IA]*

yunulhlhus *V* [lh-tlus] he/she is kneading it. *[IA]*

yus *N* snow.

yus *N* wolf.

yus bulh nilhts'i *V* [lh-ts'i < ts'i$_2$] it is blowing snow. *[IA]*

yusi *N* elk. The North American elk was formerly considered to be a subspecies of Cervus elaphus, but as of 2004 biologists consider the North American elk and the European elk to be different species.

Yuske'nighus *N* February.

yusk'ut *N* age, year. (1) Nts'oh **nyeyusk'ut?** 'How old are you?'

yusk'ut *PPP* on the snow.

yusk'ut nugoo *N* skidoo, snow machine.

-yust'e *N* body.

yusts'i *ADV* lower.

yusts'i but'ak *N* his/her lower back.

yusts'i hubuta'k *N* their lower backs.

yusts'i net'ak *N* our lower backs.

yusts'i nt'ak *N* your (1) lower back.

yusts'i st'ak *N* my lower back.

yutoo *N* his/her water.

Yutunuyeh *N* **'uza**, *q.v.,* name. Currently held by George George, Sr. Etymology: Literally, "he/she grew up among the highest mountains".

yutse *V* [0-tse] it is barking. *[IA]*

yuts'uwhulh'ai *V* [lh-'ai] he/she owes him (st). *[IA]*

yuze'unih *V* [0-nih] he/she is choking him/her. *[IA]*

yuze'usnik *V* [0-nik] he/she choked him/her. *[PA]*

yuzelhghi *V* [lh-ghi < ghi$_1$] he/she killed it. *[PA]*

-z'e *N* uncle (mother's brother), mother's male first cousin.

Za *N* man's given name. Etymology: Borrowed from French *Jean.*

Zagali *N* man's given name. Zagali became keyoh whuduchun around 1800. He was the father of Kaoosdli.

-zaz *N* father-in-law.

ze *PART* only. (1) Nukuk sghante'alh. 'He is going to give me a ball.'

-zek *N* mouth.

-zek dusts'ai *N* roof of the mouth.

-zek hoontsi' *V* [0-tsi' < tsi'$_1$] to talk dirty. *[IA]* This is literally "X's mouth is dirty". The "subject" is therefore indicated by possessive prefixation of **-zek** "mouth", e.g. **buzek hoontsi'** "his mouth is dirty", i.e. "he/she talks dirty".

-zek onduk *N* roof of the mouth.

-zek taoodetnuk *V* [d-nuk] to be quarrelsome. *[IA]* This is literally "X's mouth is noisy". The "subject" is therefore indicated by possessive prefixation of **-zek** "mouth", e.g. **buzek taoodetnuk** "his mouth is noisy", i.e. "he/she is quarrelsome".

-zekw *N* saliva.

-zesdak *N* throat (interior).

-zesdak'al *N* uvula.

-zesdalos *N* uvula.

-zesdamai *N* uvula.

-zesdazum *N* tonsils.

-zet *N* female cross cousin.

-zets'un *N* chin.

zih *PP* beside. (1) Ti **zih.** 'Beside the road.'

-zkai *N* blood.

-zkaich'ooz *N* vein.

-zkaich'oozcho *N* artery.

-zool *N* wind pipe, trachea.

Zooldubak *N* Easter. Etymology: Loan from French *jour de paques.*

-zoolts'un *N* larynx, Adam's apple.

-zumcho *N* pancreas.

zus *N* pouch.

-zut *N* liver.

-zuz *N* hide, skin.

yus *N* snow.

ENGLISH TO DAKELH

abalone shell. *N* mulla.

abandoned (s.o.), I. *V* untesna. *[PA]* [0-na]

abandoned me, he/she. *V* unsutesna. *[PA]* [0-na]

Aboriginal person. *N* yunka dune.

above. *ADV* 'onduk.

absent, he/she is. *V* hoolah. *[IA]* [0-lah]

aches, it. *V* bulh'ults'ulh. [abs: 0] *[IA]* [l-ts'ulh]

aches, it. *V* bulh'unults'ulh. [abs: n] *[IA]* [l-ts'ulh]

acorn. *N* 'angwul.

acquisitive, he/she is. *V* whudzaih. *[IA]* [0-dzaih]

across. *ADV* yaztsih.

Adam's Apple. *N* -yih k'ut nez'ai.

Adam's apple. *N* -zoolts'un.

adenoids, my. *N* syesdak nez'ai.

adolescent, female. *N* t'et. Duoplurals: t'edukoo, t'eduke

adolescent, male. *N* chilh. Duoplurals: chilhuka, chilhke

advancing, we (3+) are. *V* nus ts'edulh. *[IA–prog]* [0-dulh < dil$_1$]

advising him/her, he/she is. *V* yughatni. *[IA]* [d-ni]

afraid, he/she is. *V* nuljut. *[IA]* [l-jut < joot$_1$]

afraid, I am. *V* nulhujut. *[IA]* [l-jut < joot$_1$]

afraid, they are. *V* hunuljut. *[IA]* [l-jut < joot$_1$]

afraid, we (2) are. *V* niduljut. *[IA]* [l-jut < joot$_1$]

afraid, we (3+) are. *V* uznuljut. *[IA]* [l-jut < joot$_1$]

afraid of it, he/she is. *V* yenuljut. *[IA]* [l-jut < joot$_1$]

afraid of it, I am. *V* benlhujut. *[IA]* [l-jut < joot$_1$]

afraid of it, they are. *V* hiyenuljut. *[IA]* [l-jut < joot$_1$]

afraid of it, we (2) are. *V* beniduljut. *[IA]* [l-jut < joot$_1$]

afraid of it, we (3+) are. *V* beznuljut. *[IA]* [l-jut < joot$_1$]

after, shortly. *ADV* k'adliyaz.

after (in time). *PP* k'elh'az.

after all. *ADV* ts'ante'ez.

after him/her. *PPC* buk'elh'az.

after it. *PPC* hukw'elh'az. [abs: wh]

afternoon. *N* dzetniz hukw'elh'az.

again. *ADV* doocha.

age. *N* yusk'ut.

age, his. *N* beyusk'ut. Literally, "his winters".

age, your (1). *N* nyeyusk'ut.

aged, he/she has. *V* nesjan. *[PA]* [d-yan < yan$_1$]

aged, I have. *V* nesjan. *[PA]* [d-yan < yan$_1$]

aged, they have. *V* hunesjan. *[PA]* [d-yan < yan$_1$]

aged, we (3+) have. *V* ts'unesjan. *[PA]* [d-yan < yan$_1$]

ahead. *ADV* nus.

aid of, in. *PP* -la. Used in certain constructions having to do with helping. With most verbs the help must take the form of participating in a joint activity, rather than merely doing something on behalf of someone else.

acorn. *N* 'angwul.

ailment. *N* dada.

airplane. *N* benuts'ut'ah.

alcohol. *N* nedotoo.

Alder, Green. [Alnus viridis] *N* k'us.

Alexandria. *N* Stellayukh. The former Carrier village between Quesnel and Williams Lake.

Alf Lake. *N* Chunbaz Bunk'ut.

alive, he/she is. *V* khuna. *[IA]* [0-na < na$_1$]

alive, I am. *V* khusna. *[IA]* [0-na < na$_1$]

alive, they are. *V* hukhuna. *[IA]* [0-na < na$_1$]

alive, we (2) are. *V* khidutna. *[IA]* [0-na < na$_1$]

alive, we (3+) are. *V* ts'ukhuna. *[IA]* [0-na < na$_1$]

all [human]. *N* ts'iyane.

almost. *ADV* 'ankw'us.

alright. *INT* a'ah.

also. *PART* cha.

always. *ADV* 'awhulyiz.

American Crow. [Corvus brachyrhynchos] *N* datsan.

American Robin. [Turdus migratorius] *N* sooh.

am, I. *V* 'ust'ah. *[IA]* [0-t'ah]

amazed, I am. *V* sba hooncha. *[IA]* [0-cha < cha$_1$]

among. *PP* tah.

among one another. *PPC* lhtah.

among them. *PPC* butah.

among them (areas). *PPC* whutah.

amused, I am. *V* dlonus'i. *[IA]* [0-'i]

amused, I am. *V* sahulhgi. *[IA]* [lh-gi]

amused, I am. *V* sba huwonenik. *[IA]* [0-nik]

amused, I am. *V* sbadlo. *[IA]* [0-dlo]

ancestors. *N* dusneke.

ancestors, his/her. *N* budusneke.

ancestors, my. *N* sdusneke.

ancestors, our. *N* nedusneke.

ancestors, their. *N* hubudusneke.

ancestors, your (1). *N* nyudusneke.

anchor, fish net. *N* tsetoos'ai.

anchor for rabbit snare. *N* nal'oo.

and. *CONJ* 'ink'ez.

angel. *N* lizas.

angel. *N* yat'en.

angels. *N* yat'en-ne.

Angly Creek. *N* Bundzikoh. The creek flowing from Angly Lake to Oona Lake.

Angly Lake. *N* Bundzi Bunk'ut.

angry, he/she is. *V* hunilch'e. *[IA]* [l-ch'e < ch'eh$_1$]

angry, he is. *V* huske. *[IA]* [0-ke]

angry, I am. *V* hunlhuch'e. *[IA]* [l-ch'e < ch'eh$_1$]

animal. *N* khunai.

animal trail. *N* khunaiti.

ankle. *N* -kechunnah.

ankle, his/her. *N* bukechunnah.

ankle, my. *N* skechunnah.

ankle, your (1). *N* nkechunnah.

ankles, our. *N* nekechunnah.

ankles, their. *N* hubukechunnah.

ant. *N* 'andih.

ant hill. *N* 'andihken.

antler. *N* -de'.

antlers. *N* 'ude. Indefinitely possessed form.

anus. *N* -boo'. [babytalk]

anus. *N* -tsul.

anuses, their. *N* hubutsul.

anxious, I am. *V* hubanelhu'it. *[IA]* [l-'it]

anxious, I am. *V* whenlhujut. *[IA]* [0-jut]

aorta. *N* 'udzich'ooz.

apparently. *PART* si. This particle immediately follows the constituent over which it has scope.

applauds, he/she. *V* lhookat. *[IA]* [0-kat < kat$_3$]

April. *N* Talhtsi Ooza.

apron. *N* k'uztsan.

Arctic sagebrush. [Artemisia frigida] *N* tse'ul.

are, they. *V* 'uhint'ah. *[IA]* [0-t'ah < t'oh$_1$]

are, you (1). *V* 'int'ah. *[IA]* [0-t'ah]

are, you (1). *V* 'int'ah. *[IA]* [0-t'ah < t'oh$_1$]

areal object of postposition. *PREFIX* whu-. Used with postpositions beginning with a consonant.

arm. *N* -gan.

arm, his/her. *N* bugan.

arm, my. *N* sgan.

arm, your (1). *N* ngan.

arm pit. *N* -chak'ests'ah.

antlers. *N* 'ude.

arms, our. *N* negan.

arms, their. *N* hubugan.

arms, your (2+). *N* nahgan.

around. *ADV* 'oh.

around. *ADV* hoh.

around it. *PPC* bunat.

arrived, he/she has. *V* whusainya. *[PA]* [0-ya]

arrowhead. *N* noondai.

arteries, our. *N* nedzidool.

arteries, their. *N* hubudzidool.

artery. *N* -dzidool.

artery. *N* -zkaich'oozcho.

artery, his/her. *N* budzidool.

artery, his/her. *N* buzkaich'oozcho.

artery, my. *N* sdzidzool.

artery, my. *N* suzkaich'oozcho.

artery, our. *N* nezkaich'oozcho.

artery, their. *N* hubuzkaich'oozcho.

artery, your (1). *N* ndzidool.

artery, your (1). *N* nyuzkaich'oozcho.

ash. *N* lhustsis.

Ash, Mountain. [Sorbus sitchensis] *N* ch'ok.

ash tray. *N* lhustsis bekuk.

ashes. *N* lhez.

ask (s.o.), we (3+). *V* ts'oodulhkut. *[IA]* [lh-kut]

ask him/her, they. *V* huyoodulhkut. *[IA]* [lh-kut]

asks him/her, he/she. *V* yoodulhkut. *[IA]* [lh-kut]

asks me, he/she. *V* soodulhkut. *[IA]* [lh-kut]

aspirin. *N* netsiyoo.

aspirin. *N* ntsich'a.

ate, he/she. *V* anyi. *[PA]* [0-yi < yi_1]

ate, I. *V* esyi. *[PA]* [0-yi < yi_1]

ate, they. *V* hanyi. *[PA]* [0-yi < yi_1]

ate, we (2). *V* naduji. *[PA]* [0-yi < yi_1]

ate, we (3+). *V* ts'anyi. *[PA]* [0-yi < yi_1]

ate, you (1). *V* anyi. *[PA]* [0-yi < yi_1]

ate (u.o.), I. *V* na'esdai. *[PA–habitual]* [d-ai < yi_1]

ate (u.o.), we (2). *V* na'adudai. *[PA–habitual]* [d-ai < yi_1]

ate (u.o.), we (3+). *V* 'uts'anyi. *[PA]* [0-yi < yi_1]

ate (u.o.), we (3+). *V* na ts'edai. *[PA–habitual]* [d-ai < yi_1]

attic. *N* bunt'ah.

attic. *N* yakwut.

attic, his/her. *N* bubunt'ah.

attic, my. *N* sbunt'ah.

attic, our. *N* nebunt'ah.

attic, their. *N* hububunt'ah.

attic, your (1). *N* mbunt'ah.

attracted (sexually) to him/her, she is. *V* yuneztulh. *[PA]* [0-tulh]

attracted (sexually) to me, she is. *V* suneztulh. *[PA]* [0-tulh]

August. *N* Talook Ooza.

aunt, his/her paternal. *N* bubezan.

aunt, my paternal. *N* sbezan.

aunt, paternal. *N* -bez.

aunt, their paternal. *N* hububezan.

aunt (father's sister). *N* 'ubezan.

aunt (father's sister). *N* -bezan.

aunt (mother's sister, father's brother's wife, mother's brother's wife). *N* -k'i.

aunt (mother's sister). *N* -ak'i.

aunt (mother's sister), his/her. *N* bak'i.

aunt (mother's sister), my. *N* sak'i.

aunt (mother's sister), someone's. N 'ak'i.

aunt (mother's sister), their. *N* hubak'i.

aunt (mother's sister), your (1). *N* nyak'i.

aurora borealis. *N* yawhudoos.

autumn. *N* dak'et.

avalanche. *N* 'unt'oh.

awake, he/she is. *V* whunih. *[IA]* [0-nih]

awl. *N* 'utsoh.

axe. *N* tsetsilh.

Babine person. *N* Babine whut'en.

Babine person. *N* Nadot'en.

baby. *N* tsalhts'ul.

baby. *N* uskiyaz.

baby, little. *N* tsalhts'ulyaz.

baby boy. *N* dune tsalhts'ul.

baby buggy. *N* uski be'nunulbaz.

baby girl. *N* ts'eke tsalhts'ul.

baby-sit (s.o.), they. *V* ghuhinli. *[IA]* [0-li]

bachelor, he/she is. *V* bu'at hoolah. *[IA]* [0-lah] Literally, "he/she has no wife".

back. *N* -t'ak.

back, his/her. *N* but'ak.

back, his/her lower. *N* yusts'i but'ak.

back, lower. *N* -t'ak yusts'i.

back, my lower. *N* yusts'i st'ak.

back, our. *N* net'ak.

back, the small of my. *N* syunt'ak.

back, the small of the. *N* -yunt'ak.

back, your (1). *N* nt'ak.

back, your (1) lower. *N* yusts'i nt'ak.

back, my. *N* st'ak.

back of hand. *N* -lat'ak.

back pack. *N* 'unizus.

backbone. *N* -yunts'un.

backbone. *N* -t'akts'un.

backbone. *N* -yun.

backbone, his/her. *N* but'akts'un.

backbone, his/her. *N* buyunts'un.

backbone, my. *N* st'akts'un.

backbone, my. *N* syunts'un.

backbone, the area surrounding. *N* -dayun.

backbone, your (1). *N* nt'akts'un.

backbone, your (1). *N* nyunt'sun.

backbone region, my. *N* sdayun.

backbones, our. *N* net'akts'un.

backbones, their. *N* hubut'akts'un.

backbones, their. *N* hubuyunts'un.

backbones, our. *N* neyunts'un.

backs, our lower. *N* yusts'i net'ak.

backs, their. *N* hubut'ak.

backs, their lower. *N* yusts'i hubuta'k.

bacon. *N* gugoos k'atsun.

bacon. *N* lubegin.

bad, it is. *V* dintsi. *[IA]* [0-tsi < tsi'$_1$]

bad, it is. *V* hoontsi'. [abs: wh] *[IA]* [0-tsi' < tsi'1]

bad, it is. *V* nintsi. [abs: n] *[IA]* [0-tsi < tsi'$_1$]

bad, they are. *V* hintsi. *[IA]* [0-tsi < tsi'$_1$]

bad-tasting, it is. *V* ts'ulki'. *[IA]* [l-ki']

bag. *N* 'uzus.

bag, paper. *N* dustl'uszuz.

bailbones, our. *N* nechets'un.

bailing out (st:boat), I am. *V* betaszih. *[IA]* [0-zih]

bait. *N* 'uni.

bald, he/she is. *V* tsinkwun. *[IA]* [0-kwun]

Bald Eagle. [Haliaeetus leucocephalus] *N* sbalyan.

bald eagle. *N* tsebalyan.

ball. *N* nukuk.

ball, large. *N* nukukcho. Roughly, balls the size of a basketball and larger.

ball, small. *N* nukukyaz. E.g. balls the size of a baseball.

Balsam Fir. [Abies lasiocarpa] *N* ts'ootsun. Locally known as Balsam Fir.

Balsam Fir bark. *N* ts'ootsunt'ooz.

banana. *N* mooki.

bank of body of water. *N* too busk'ut.

Bald Eagle. [Haliaeetus leucocephalus] *N* sbalyan.

bannock. *N* lhes sut'e.

bannock. *N* sto lhes sut'e.

bannock, oven. *N* lhes sto bez sut'e.

bar. *N* tadutnai.

barefoot. *ADV* ke'et.

bark. *N* 'ula.

bark. *N* 'ut'ooz.

bark of Balsam Fir. *N* ts'ootsunt'ooz.

bark of Subalpine Fir. *N* ts'ootsunt'ooz.

barking, it is. *V* yutse. [IA] [0-tse]

Barlow Lake. *N* Duk'ai Hooni Bunk'ut.

barn. *N* tl'obayah.

barn, cow. *N* musdusbayah.

barren, it is (nothing grows there). *V* whudanzaz. *[IA]* [0-zaz]

base of, at the. *PP* cheh.

base of a mountain. *N* dzulhcheh.

baseball, he/she is playing. *V* nundunult'a. *[IA]* [l-t'a < 'ai$_1$]

basket. *N* tilh. Poss: tilh.

basket, large bark. *N* chalhyal. A large bark basket suitable for carrying on back, into which smaller baskets are emptied.

basket, my. *N* sutilh.

bassinet. *N* 'uski ts'ai.

bat. *N* t'az.

bath, he/she gives him/her a bath. *V* toonuyulhte. *[IA]* [lh-te]

bath, he/she takes a. *V* toonuya. *[IA]* [0-ya]

bath, I give (s.o.) a. *V* toonulhte. *[IA]* [lh-te]

bath, I take a. *V* toonusya. *[IA]* [0-ya]

bath, they (3+) take a. *V* toonuhudelh. *[IA]* [0-delh]

bath, they give him/her a. *V* toonuhuyulhte. *[IA]* [l-tre]

bath, we (2) take a. *V* toonit'az. *[IA]* [0-'az < 'as$_1$]

bath, we (2) will give someone a. *V* toonidulte. *[IA]* [lh-te]

bath, we (3+) carry a. *V* toonuts'udilh. *[IA]* [0-dilh]

bath, we (3+) give (s.o.) a. *V* toonuts'ulhte. *[IA]* [lh-te]

bath, you (1) take a. *V* tooninya. *[IA]* [0-ya]

bath, you (3+) take a. V toonahdilh. *[IA]* [0-dilh]

bath tub. *N* betoonuts'uya.

bathroom. *N* tsanbayah.

bay. *N* tl'ah.

be, you (1) may. *V* 'ont'e. *[OA]* [0-t'e < t'oh$_1$]

beach. *N* taba.

beads. *N* kw'usul.

beak. *N* -da.

Bearberry. [Lonicera involucrata] *N* susmai.

bear, black. [Ursus americanus] *N* sus.

bear, grizzly. [Ursus arctos] *N* shas. Poss: shas.

bear, my grizzly. *N* sushas.

bear den, black. *N* sus'an.

bear grease. *N* susghe.

beautiful, he/she is. *V* ts'ekezoo. *[IA]* [0-zoo] This is applicable only to women.

beautiful, it is. *V* dinzoo. [abs: d] *[IA]* [0-zoo < zoo$_1$]

beaver. [Castor canadensis] *N* tsa.

beaver, big. *N* tsacho.

Beaver Indian. *N* Tsat'en.

beaver, large. *N* tsati.

beaver lodge. *N* tsaken.

beaver meat. *N* tsatsun.

beaver snare. *N* tsabilh.

beaver's ears. *N* whuledzo.

becalmed, it is. *V* dezghel. *[PA]* [0-ghel < ghel$_2$]

because of. *PP* gha.

because of me. *PPC* sgha.

because of them. *PPC* hubugha.

become, I have. *V* susdli. *[PA]* [0-dli < li$_3$]

bed. *N* lili.

bed. *N* skelh'at.

bed. *N* tesk'et.

bed, his/her. *N* bulili.

bed, his/her. *N* butesk'et.

bed, my. *N* stesk'et.

bed, my. *N* sulili.

bed, our. *N* nelili.

bed, our. *N* netsk'et.

bed, their. *N* hubulili.

bed, their. *N* hubutesk'et.

bed, your (1). *N* nyelili.

bed, your (1). *N* nyutesk'et.

bed spread. *N* lili k'ut subal.

bedroom. *N* tesk'et.

bedroom. *N* ts'uztez ba whuz'ai.

bedroom, his/her. *N* butesk'et.

bedroom, my. *N* stesk'et.

bedroom, our. *N* netsk'et.

bedroom, their. *N* hubutesk'et.

bedroom, your (1). *N* nyutesk'et.

bedwildered, I am. *V* sba we'hooja. *[IA]* [0-ja]

bee. *N* hoolht'o.

beef. *N* musdustsun.

beehive. *N* hoolht'ot'o.

beetle. *N* hoolhkw'ul.

beets. *N* 'buzkai nawhudleh.

beets. *N* lhits'e.

before. *PP* tsah.

before him/her. *PPC* butsah.

before it [areal]. *PPC* whutsah.

before me. *PPC* stsah.

before them. *PPC* hubutsah.

begging, he/she is. *V* datsa. *[IA]* [0-tsa]

Belgatse. *N* Belhk'achek. A former village site on Cheslatta Lake.

believe it, we (2). *V* nahba 'alha'hoont'ah. *[IA]* [0-t'ah < t'oh$_1$]

believe it, I. *V* sba 'alha'hoont'ah. *[IA]* [0-t'ah < t'oh$_1$]

believe it, they. *V* huba 'alha'hoont'ah. *[IA]* [0-t'ah < t'oh$_1$]

believe it to be true, we (3+). *V* neba 'alha'hoont'ah. *[IA]* [0-t'ah < t'oh$_1$]

believes it to be true, he/she. *V* ba 'alha'hoont'ah. *[IA]* [0-t'ah < t'oh$_1$]

bell. *N* luglos.

belly. *N* -but.

belly. *N* -chan.

belly button. *N* -tsakelhgwus.

belt. *N* se. Poss: ze.

belt, my. *N* suze.

be, you (1) may. *V* naondle. *[OA]* [d-le]

berries. *N* mai.

berries, dried. *N* mai nezgi.

berries, dried. *N* nezdzak.

beside. *PP* zih.

beside it [areal]. *PPC* whuzih.

beside me. *PPC* suzih.

beside us (2). *PPC* nahzih.

beside us (3+). *PPC* nezih k'ut.

beside you (1). *PPC* nzih.

between mountains. *ADV* dzulhyeguz.

between them. *PPC* hubeguz.

between them [areal]. *PPC* wheguz.

bewildered, he/she is. *V* ba we'hooja. *[IA]* [0-ja]

bewildered, we (2) are. *V* nahba we'hooja. *[IA]* [0-ja]

bewildered, we (3+) are. *V* neba whe'ooja. *[IA]* [0-ja]

bicycle. *N* benuts'ulgaih.

big. *SUFFIX* -cho.

big, it is. *V* dincha. [abs: d] *[IA]* [0-cha < cha$_1$]

big, they are. *V* hincha. *[IA]* [0-cha < cha$_1$]

big as me, he/she is as. *V* sundulcho. *[IA]* [l-cho < cha$_1$]

big it is!, how. *V* lhe'ulcho. *[IA]* [l-cho < cha$_1$]

bile. *N* dultso.

birch conk. *N* diyooh.

Birch, Black. [Betula occidentalis] *N* nach'ulh.

Birch, Water. [Betula occidentalis] *N* nach'ulh.

Birch, Western. [Betula occidentalis] *N* nach'ulh.

bird. *N* dut'ai.

bitch. *N* lhits'e.

bitter, it is. *V* ts'ulki'. *[IA]* [l-ki']

black, it is. *V* dulhgus. *[IA]* [lh-gus < gus$_1$]

black, it is. *V* dunulhgus. [abs: n] *[IA]* [lh-gus < gus$_1$]

black, it is. *V* whudulhgus. [abs: wh] *[IA]* [lh-gus < gus$_1$]

black bear, little. *N* susyaz.

blackbird, red-winged. [Agelaius phoeniceus] *N* ch'uk.

Black Cottonwood. [Populus balsamifera ssp. trichocarpa] *N* landooz.

Black Gooseberry. [Ribes oxycanthoides] *N* 'indawuz.

Black Spruce. [Picea mariana] *N* ts'oobez.

Black Twinberry. [Lonicera involucrata] *N* susmai.

bladder, gall. *N* -tl'uz.

bladder, his/her. *N* buluzzus.

bladder, my. *N* sluzzus.

bladder, your (1). *N* nluzzus.

bladder (urinary). *N* -lhuzzus.

bladders, our. *N* neluzzus.

bladders, their. *N* hubuluzzus.

blanket. *N* nalhti.

blanket, his/her. *N* bunalhti.

blanket, my. *N* snalhti.

blanket, our. *N* nenalhti.

blanket, their. *N* hubunalhti.

blanket, your (1). *N* nnalhti.

bled to death, he/she. *V* 'uskai yuzelhghi. *[PA]* [lh-ghi < ghi$_1$] Literally, "blood killed him/her ".

bleeding, he/she is. *V* 'uskai suli. *[IA]* [0-li < li$_1$]

bless myself, I. *V* nunadulhunih. *[IA]* [l-nih < nih$_2$]

bless ourselves, we (2). *V* nunadidulnih. *[IA]* [l-nih < nih$_2$]

bless ourselves, we (3+). *V* nunazdulnih. *[IA]* [l-nih < nih$_2$]

bless themselves, they. *V* nunahudulnih. *[IA]* [l-nih < nih$_2$]

blesses himself, he/she. *V* nunadulnih. *[IA]* [l-nih < nih$_2$]

blew in, a draft. *V* danilhk’az. *[PA]* [lh-k’az]

blind, he/she is. *V* chawes’en. *[IN]* [0-’en < ’en$_1$]

blind, I am. *V* chawzes’en. *[IN]* [0-’en < ’en$_1$]

blinking, he/she is. *V* nak’ubul. *[IA]* [0-bul < bul$_1$]

blinks at her in dislike, she. *V* yunak’ult’ah. *[IA]* [l-t’ah]

blizzard has begun, a. *V* yootsiz tezts’i. *[PA]* [0-ts’i < ts’i$_2$]

bloated, it is. *V* telhudzool. *[PA]* [l-d-zool] E.g., a carcass.

blood. *N* ’uzkai. Indefinitely possessed form.

blood. *N* -zkai.

blood, his/her. *N* buzkai.

blood, my. *N* suzkai.

blood, our. *N* nezkai.

blood, their. *N* hubuzkai.

blood, your (1). *N* nyuzkai.

blood sucker. *N* hoot’ukw. Also known as “blood sucker”.

blow, the wind has begun to. *V* tezts’i. *[PA]* [0-ts’i < ts’i$_2$]

blowing, it (siren, whistle) is. *V* uzulh. *[IA]* [0-zulh]

blowing, it (wind) is. *V* nilhts’i. *[IA]* [lh-ts’i < ts’i$_2$]

blowing from the south. *ADV* nazde.

blowing into, it (wind) is. *V* danilhts’i. *[IA]* [lh-ts’i < ts’i$_2$] This refers to a substantial wind blowing in, as through an open window.

blowing into, it (wind) is. *V* dats’i. *[IA]* [lh-ts’i < ts’i$_2$] This refers to a draft, where a little bit of wind blows in, as through a crack.

Blue Grouse. [Falcipennis canadensis] *N* dihcho.

Blue Heron. *N* tehhoongook.

Blueberries, High-Bush. [Vaccinium ovalifolium] *N* yalhtsul.

blue, it is. *V* duldzan. *[IA]* [l-dzan < dzan$_1$]

blue, it is. *V* dunuldzan. [abs: n] *[IA]* [l-dzan < dzan$_1$]

blue, it is. *V* whuduldzan. [abs: wh] *[IA]* [l-dzan < dzan$_1$]

blue, it is not. *V* chanildzan. [abs: n] *[IN]* [l-dzan < dzan$_1$]

blue jeans. *N* doso tl’asus.

bluejay. *N* teh gwuzeh.

board. *N* dzihtel.

boards, floor. *N* yundzihtel.

boat, he/she did not set off by. *V* whechaikel. *[PN]* [0-kel < ke$_2$]

boat, he/she has gone around by. *V* nusuki. *[PA]* [0-ki < ke$_2$]

boat, he/she is not going by. *V* chaskelh. *[IN–prog]* [0-kelh < ke$_2$]

boat, he/she set off by. *V* wheinki. *[PA]* [0-ki < ke$_2$]

boat, he/she will go by. *V* tekelh. *[FA]* [0-kelh < ke$_2$]

boat, he/she will not go by. *V* chateskel. *[FN]* [0-kel < ke$_2$]

boat, I am not going in a loop in. *V* nuchasuske. *[IN]* [0-ke < ke$_2$]

braided hair. *N* -tsigha lhu'ool.

boat, we (2) set off by. *V* wheiduki. *[PA]* [0-ki < ke_2]

bodies, our. *N* neyust'e.

bodies, their. *N* hubuyust'e.

body. *N* 'uyust'e. Indefinitely possessed.

body. *N* -yust'e.

body, his/her. *N* buyust'e.

body, my. *N* syust'e.

body, your (1). *N* nyust'e.

body odour, he/she has. *V* bujootsun. *[IA]* [0-tsun]

boil, it came to a. *V* handunelmulh. *[PA]* [l-mulh]

boiling, it (liquid) is. *V* ndunulmulh. *[IA]* [l-mulh]

boiling (st), I am. *V* uslez. *[IA]* [0-lez]

boiling (u.o.), he/she is. *V* 'undunulhmulh. *[IA]* [lh-mulh < $mulh_1$]

boiling (u.o.), I am. *V* 'undunulhmulh. *[IA]* [lh-mulh]

boiling it (water), he/she is. *V* yundunulhmulh. *[IA]* [lh-mulh < $mulh_1$]

boiling it (water), they are. *V* hindunulhmulh. *[IA]* [lh-mulh < $mulh_1$]

bolted, it. *V* tsitelhya. *[PA]* [l-ya] E.g. a frightened animal.

bones of fish. *N* 'ughak. Indefinitely possessed form.

booze. *N* nedotoo.

border-crossing. *N* niti. The point where a trail ends at the frontier between one people's territory and its neighbour's.

boring, it is. *V* whuts'oodutnik. *[IA]* [d-nik]

born, he/she was. *V* whuzdli. *[PA]* [0-dli < li_3]

boss. *N* moodih.

boss, my. *N* smoodih.

boss, our. *N* nemoodih.

boss, their. *N* hubumoodih.

bossy, he/she is. *VP* dune dunuyuz. *[IA]*

both of them. *V* nanehult'ah. *[IA]* [l-t'ah < $t'oh_1$]

bothering him/her, he/she is. *V* yughaghuna. *[IA]* [0-na]

bottom, a. *N* 'utl'ah. Indefinitely possessed.

bottom, its. *N* buttl'ah.

bounced away, it. *V* whendunelduz. *[PA]* [l-duz]

bouncing up and down, it is. *V* duk duntuldus. *[IA]* [l-dus]

bowel. *N* -ts'ikcho.

bowl, soup. *N* tazul bedzuz.

box. *N* chunkhelh.

box, cardboard. *N* dustl'us chunkelh.

boy. *N* 'uskidune.

boy. *N* duneyaz.

boy. *N* skuidune. Duoplural: skehdune

boy, baby. *N* dune tsalhts'ul.

boy, small. *N* skuiduneyaz.

boys. *N* skehdune.

bra. *N* ts'oozus.

braggart, he/she is a. *V* lhena'duduwhults'it. *[IA]* [l-ts'it]

bragging about himself, he/she is. *V* dena'whudults'it. *[IA]* [l-ts'it]

braided, it has been. *V* lhu'ool. *[PA]* [l-'ool < $'ool_1$]

braided hair. *N* -tsigha lhu'ool.

braiding it, she is. *V* naiyulh'ool. *[IA]* [lh-'ool < $'ool_1$]

brain. *N* -tsinghai.

brain, his/her. *N* butsinghai.

brain, my. *N* stsinghai.

brain, your (1). *N* ntsinghai.

brains, our. *N* netsinghai.

brains, their. *N* hubutsinghai.

brainy, he/she is. *V* beni hooni. *[IA]* [0-ni] Literally, "his mind exists".

branch of conifer. *N* 'ul.

branch of deciduous tree. *N* 'uyooschum.

branches, it (road). *V* lhk'eti. *[IA]* [0-ti]

branches from it, it. *V* buts'ahainalhyi. *[IA]* [l-yi] E.g. a limb from the trunk of a tree.

brassiere. *N* ts'oozus.

bread. *N* lhes.

bread, fry. *N* lhes sut'e.

bread, yeast. *N* liba.

breakable, it is. *V* du'ek. *[IA]* [0-'ek < $'ek_1$]

branches from it, it. *V* buts'ahainalhyi.

breast. *N* 'utsoo. Indefinitely possessed form.

breast. *N* -ts'oo.

breast, his/her. *N* buts'oo.

breast, my. *N* sts'oo.

breast, your (1). *N* nts'oo.

breast bone. *N* -t'ayukts'un.

breast feed him/her, I. *V* be'ulht'uwk. *[IA]* [lh-t'ukw < $t'ukw_1$]

breasts, our. *N* nets'oo.

breasts, their. *N* hubuts'oo.

breast-feeding him/her, she is. *V* ye'ulht'ukw. *[IA]* [lh-t'ukw < $t'ukw_1$]

breath, I am out of. *V* syiz lhutulh. *[PA]* [l-tulh]

breath, I am short of. *V* syiz dilhdukw. *[IA]* [lh-dukw < $dukw_1$]

breathing, I am. *V* usjiz. *[IA]* [d-yiz < yis_1]

breechcloth. *N* tsan.

bridge. *N* kw'ut yan wheti.

bridge. *N* sus.

bring you (1) (st), I will. *V* nts'untelhchoos. *[FA]* [lh-choos < $chus_1$]

bring you (1) (st), I will. *V* nts'untelhdoh. *[FA]* [lh-doh < do$_1$] E.g. loose hay or wool.

bring you (1) (st), I will. *V* nts'untes'alh. [class: sdo-c] [abs: n] *[FA]* [0-'alh < 'ai$_1$]

bring you (1) (st), I will. *V* nts'untesdzih. [class: euo-c] *[FA]* [0-dzih < dzai$_1$] E.g. berries, beads, rings.

bring you (1) (st), I will. *V* nts'utelhtelh. [class: body-c] *[FA]* [lh-telh < ti$_1$] E.g. baby, small animal

bring you (1) (st), I will. *V* nts'utes'alh. [class: sdo-c] *[FA]* [0-'alh < 'ai$_1$]

bring you (1) (st), I will. *V* nts'uteskalh. [class: coc-c] *[FA]* [0-kalh] E.g. a dish of food.

bring you (1) (st), I will. *V* nts'utestilh. [class: lro-c] *[FA]* [0-tilh < tan$_1$] E.g. a rifle, pole, or stick.

bring you (1) (st), I will. *V* nts'utestleh. [class: mushy-c] *[FA]* [0-tleh] E.g. a handful of mud.

bring you (1) (st) on my back, I will. *V* nts'utesghilh. *[FA]* [0-ghilh < ghe$_1$]

bringing (st) to a boil, he/she is. *V* handunelhmulh. *[IA]* [lh-mulh]

brisket. *N* 'ujoohtsun.

brittle, it is. *V* k'udentuk. [abs: d] *[IA]* [0-tuk] E.g. a stick or bone.

brittle, it is. *V* k'unuduwul. *[IA]* [0-wul] E.g. glass.

Brittle Horsehair Lichen. [Bryoria lanestris] *N* dahgha.

Bromo selzer. *N* too undunulmulh.

broad, it is. *V* hoontel. [abs: wh] *[IA]* [0-tel < tel$_1$]

broad, it is. *V* ntel. [abs: 0] *[IA]* [0-tel < tel$_1$]

broad as this, it is. *V* njahooltel. [abs: wh] *[IA]* [l-tel < tel$_1$] The standard of comparison is something designated by the speaker using a gesture.

broiling (st), I am. *V* ulhts'ul. *[IA]* [lh-ts'ul]

broiling (st), we (2) are. *V* idults'ul. *[IA]* [lh-ts'ul]

broiling (st), we (2) are. *V* ts'ulhts'ul. *[IA]* [lh-ts'ul]

broiling it, he/she is. *V* yulhts'ul. *[IA]* [lh-ts'ul]

broiling it, they are. *V* huylhts'ul. *[IA]* [lh-ts'ul]

broke, he/she is. *V* k'untuk. *[IA]* [0-tuk < tuk$_2$]

broken, it (rope) has. *V* sduch'us. *[PA]* [d-ch'us]

broken in two, it is. *V* k'udentuk. [abs: d] *[PA]* [0-tuk < tuk$_2$] E.g. a stick.

broken into many pieces, it has. *V* yadatuk. *[PA]* [0-tuk < tuk$_2$]

broom. *N* betinawhultsas.

brothel. *N* 'ul'en bayah.

brother. *N* -ulhutsin.

brother, his/her. *N* bulhutsin.

brother, his/her older. *N* boona.

brother, my. *N* sulhutsin.

brother, my elder. *N* soona.

brother, my younger. *N* schul.

brother, older. *N* -oona.

brother, our. *N* nelhutsin.

brother, our older. *N* neyoona.

brother, their. *N* hubulhutsin.

brother, their older. *N* huboona.

brother, younger. *N* -chist'loh.

brother, younger. *N* -chul.

brother, your (1) older. *N* nyoona.

brother-in-law, his/her. *N* bughe.

brother-in-law, my. *N* sghe.

brother-in-law, my. *N* sghe.

brother-in-law, our. *N* neghe.

brother-in-law, their. *N* hubughe.

brother-in-law, your (1). *N* nghe.

brown, it is. *V* yun dot'en. [IA] [d-'en < $'en_1$]

brush, clothes. *N* naih benaldzooh.

brushy, it is. *V* whuduts'ul. [abs: wh] *[IA]* [0-ts'ul]

bucks, it (animal). *V* tullhuk. *[IA]* [l-lhuk]

buffalo. [Bison bison] *N* tl'o musdus.

bug. *N* yun nulh'as.

bull. *N* musdusdune.

bullet. *N* k'a.

bullet pouch. *N* k'azus.

bum. *N* -tl'ah.

bum. *N* kejudustl'ah.

bum, his/her. *N* butl'ah.

bum, my. *N* stl'ah.

bum, your (1). *N* ntl'ah.

bumblebee. *N* hoolht'ocho.

bums, our. *N* netl'ah.

bums, their. *N* hubutl'ah.

burbot. [Lota lota] *N* tsintel.

buried, he/she is. *V* 'adilhti. *[PA]* [lh-ti < ti_1]

buried, we (3+). *V* 'ats'alhti. [class: body] *[PA]* [lh-ti < ti_1]

buried, we (3+). *V* whuyats'alhti. [class: body] *[PA]* [lh-ti < ti_1]

buried here, he is. *V* whuyudilhti. *[IA]* [lh-ti]

buried him/her, he/she. *V* 'ayalhti. *[PA]* [lh-ti < ti_1]

buried him/her, they. *V* hidilhti. *[PA]* [lh-ti]

buried in an avalanche or landslide, he/she is. *V* buk'enahoolhah. *[IA]* [0-lhah]

Burns Lake. *N* tselhk'azbunk'ut.

burlap. *N* doso.

burp, I. *V* natuskwa'. *[IA]* [0-kwa']

burp, we (3+). *V* 'uztukw'ak. *[IA]* [0-kw'ak]

burp, you (1). *V* 'utinkw'ak. *[IA]* [0-kw'ak]

burp, you (2+). *V* 'utahkw'ak. *[IA]* [0-kw'ak]

burps, he/she. *V* natukwa'. *[IA]* [0-kwa']

bush. *N* tintah.

bush. *N* duchuntah.

Bushy-tailed Wood Rat. [Neotoma cinerea] *N* dlimcho.

Bushy-tailed Wood Rat. [Neotoma cinerea] *N* dlooncho.

busy, he/she is. *V* hubat'en lhai. *[IA]* [0-lhai]

busy, he/she is. *V* ba't'en lhai. *[IA]* [0-lai]

busy, I am. *V* sba ut'en lhai. *[IA]* [0-lai]

busy, they are. *V* huba't'en lhai. *[IA]* [0-lai]

busy, we (2) are. *V* nahba't'en lhai. *[IA]* [0-lai]

busy, we are. *V* neba't'en lhai. *[IA]* [0-lai]

butter. *N* dooda.

buttery. *N* tsagwulht'ai.

buttock. *N* -tl'ah.

button. *N* nanezmaz.

button hole. *N* nanezmazk'et.

buy, I will. *V* ooteskulh. *[FA]* [0-kulh < ket$_1$]

buying (st), he/she is. *V* ooket. *[IA]* [0-ket < ket$_1$]

buying (st), I am. *V* oosket. *[IA]* [0-ket < ket$_1$]

buying (u.o.), they are. *V* 'uhooket. *[IA]* [0-ket < ket$_1$]

buying (u.o.), we (2) are. *V* 'ooduket. *[IA]* [0-ket < ket$_1$]

buying (u.o.), we (3+) are. *V* 'uts'ooket. *[IA]* [0-ket < ket$_1$]

by himself. *PRO* dich'ah.

by means of. *PP* be.

by my hand. *PPP* sla be.

by myself. *ADV* sich'ah.

by reason of that. *PPC* 'ighun.

cabbage. *N* 'utancho.

cache. *N* hawukaih.

cache, food. *N* tsachun.

cafe. *N* ts'uyi bayah.

calamity. *N* khundughun.

calf. *N* musdusyaz.

calf, my. *N* swuda.

calf, his/her. *N* buwuda.

calf, your (1). *N* nwuda.

calf of leg. *N* -wuda.

call (st), we (3+). *V* ts'utni. *[IA]* [d-ni]

call it, they. *V* huyulhni. *[IA]* [lh-ni]

call you (1), we (3+). *V* nyuts'ulhni. *[IA]* [lh-ni]

calm, it is. *V* dezghel. *[PA]* [0-ghel < ghel$_2$]

calves, our. *N* newuda.

calves, their. *N* hubuwuda.

calves, their. *N* hubuwuda.

cambium. *N* k'unih.

came back, he/she. *V* lhtananesja. *[PA]* [d-ya]

came back, I. *V* lhtananesja. *[PA]* [d-ya]

came back, they (3+). *V* lhtanahunesdel. *[PA]* [d-del < dil$_1$]

came back, we (2). *V* lhtananet'az. *[PA]* [d-'az < 'as$_1$]

came back, we (3+). *V* lhtanaznesdel. *[PA]* [d-del < dil$_1$]

came back, you (1). *V* lhtananenja. *[PA]* [d-ya]

came from, you (1). *V* hainya. *[PA]* [0-ya]

came from, I. *V* hasya. *[PA]* [0-ya]

camera. *N* benek'edugus.

camera. *N* benek'ehugis.

can. *N* testilh.

Canada goose. [Branta canadensis] *N* khoh.

Canada mayflower. [Maianthemum racemosum] *N* lhiluzchun. This species is also known by the older scientific name Smilacina amplexicaulis.

Canadian River Otter. [Lontra canadensis] *N* tsis.

can, jerry. *N* tes hoongwun.

can opener. *N* bedahudut'as.

cancer. *N* dadacho.

cane. *N* tuz.

cane, his/her. *N* butuz.

cane, my. *N* stuz.

canister set. *N* be'usdudzak.

cannibal. *N* neyi. neyi are said to be like human beings, but chunky, very strong, and with narrow slitty eyes. They steal people, especially children, and eat them.

cannibal. *N* dune uyi.

cannot. *PART* ghait'ah. This particle is used to express inability. It follows the optative affirmative form of the verb, which is in turn preceded by the negative particle 'aw.

canoe. *N* ts'i.

canoe, spruce bark. *N* 'ulats'i.

canvas. *N* doso.

canvas. *N* mandah.

canyon. *N* duyuk.

Canyon Creek. *N* Noozkoh.

cap. *N* ts'ah.

cap. *N* ts'oh.

cape. *N* luglok.

capsized, I. *V* nadesghuz. *[PA]* [0-ghuz < $ghis_1$]

capsized, it. *V* nadezghuz. *[PA]* [0-ghuz < $ghis_1$]

captive. *N* lhuna.

captures him/her, he/she. *V* yulhna. *[IA]* [lh-na]

calf. *N* musdusyaz.

car. *N* benuts'ugoo.

car. *N* otomobil.

carbon dioxide. *N* benuts'ugoo tl'it.

carburetor. *N* yesjiz.

card board. *N* dustl'us dutsun.

cards, playing. *N* lugar.

care of (st), they take. *V* ghuhinli. *[IA]* [0-li]

careful (of yourself), you (1) are. *V* khindli. *[IA]* [d-li]

careful with it, he/she is. *V* yugha'ooli. *[IA]* [0-li]

careful with it, I am. *V* wa'oosdli. *[IA]* [0-li]

careful with it, they are. *V* higha'ooli. *[IA]* [0-li]

careful with it, we (2) are. *V* wa'oodudli. *[IA]* [0-li]

careful with it, we (3+) are. *V* wa'ts'ooli. *[IA]* [0-li]

caressing her, he/she is. *V* yughadzedzi. *[IA]* [0-dzi]

cariboo. [Rangifer tarandus] *N* whudzih.

carpet. *N* yun subal. Any floor covering.

carpet. *N* yun sulhchooz.

Carrier person. *N* dakelh

carrot, wild. *N* tsacheschun.

cartilage. *N* 'ughuts.

cartilage. *N* -ghuts.

cased, I. *V* ghanususguz. *[PA–distributive]* [0-guz] This refers to skinning by pulling the carcass out of the hide rather than cutting the hide away with a blade.

cat. *N* boos.

cat, his/her. *N* buboos.

cat, little. *N* boosyaz.

cat, my. *N* sboos.

cat, our. *N* neboos.

cat, their. *N* hububoos.

cat, your (1). *N* mboos.

caterpillar. *N* t'angoo.

caterpillar. *N* ts'anikwa. Dialect: Nadleh specific.

caterpillar, thick black. *N* susya.

cattail plant. [Typha latifolia] *N* tl'ok'uzih.

caught sight of, he/she. *V* telh'en. *[PA]* [lh-'en < 'en$_1$]

cave. *N* 'an.

cave. *N* 'u'an.

cave. *N* tse'an.

cedar. *N* chunzool.

cedar bark. *N* hat'al.

cedar chest. *N* hoongwun.

ceiling. *N* bunt'ah.

centre. *N* whuniz.

centre of lake. *N* bunniz.

chair. *N* kw'usduda.

chair. *N* kw'uts'uda.

chair, easy. *N* kw'uts'udacho.

chair, his/her. *N* bukw'uts'uda.

chair, his/her easy. *N* bukw'uts'udacho.

chair, his/her reclining. *N* buduneti kw'uts'uda.

chair, my. *N* skw'uts'uda.

chair, my easy. *N* skw'uts'udacho.

chair, my reclining. *N* sduneti kw'uts'uda.

chair, our. *N* nekw'uts'uda.

chair, our easy. *N* nekw'uts'udacho.

chair, our reclining. *N* neduneti kw'uts'uda.

chair, reclining. *N* duneti kw'uts'uda.

chair, rocking. *N* kw'usduda nubalh.

chair, their. *N* hubukw'uts'uda.

chair, their easy. *N* hubukw'uts'udacho.

chair, their reclining. *N* hubuduneti kw'uts'uda.

chair, your (1). *N* nkw'uts'uda.

chair, your (1) easy. *N* nkwuts'udacho.

chair, your (1) reclining. *N* nduneti kw'uts'uda.

char. [Salvelinus namaycush] *N* bet.

charcoal. *N* t'es.

charcoal, our. *N* net'es.

cheap, it is. *V* chaditi. *[IN]* [0-ti < ti$_2$]

cheats, he/she. *V* ha'dunet'ah. *[IA]* [0-t'ah]

cheats me, he/she. *V* saha'dunet'ah. *[IA]* [0-t'ah]

checks on (st), he/she. *V* nanlhu'en. *[IA]* [l-'en < 'en$_1$]

checks on (st), he/she. *V* nanlhu'ih. *[IA–customary]* [l-'ih < 'en$_1$]

checks on (st), he/she. *V* nawhnul'en. [abs: wh] *[IA]* [l-'en < 'en$_1$]

checks on (st), he/she. *V* nawhnul'ih. [abs: wh] *[IA–customary]* [l-'ih < 'en$_1$]

checks on it, he/she. *V* nainul'en. *[IA]* [l-'en < 'en$_1$]

checks on it, he/she. *V* nainul'ih. *[IA–customary]* [l-'ih < 'en$_1$]

cheek. *N* -nimbus.

cheek bone. *N* -nimbusts'un.

chef. *N* ts'uyi ghaghuna.

Cherry, Bitter. [Prunus emarginata] *N* dulgoosmai.

Cheslatta village. *N* Chestl'ada.

chickadee. *N* ts'usgak.

chest. *N* -dzik'ut.

chest, cedar. *N* hoongwun.

chest, his/her. *N* budzik'ut.

chest, my. *N* sdzik'ut.

chest, our. *N* nedzik'ut.

chest, their. *N* hubudzik'ut.

chest, your (1). *N* ndzik'ut.

chest of drawers. *N* bena'dukuk.

chesterfield. *N* kw'uts'udayez.

chesterfield, his/her. *N* bukw'uts'udayez.

chesterfield, my. *N* skw'uts'udayez.

chesterfield, our. *N* nekw'uts'udayez.

chesterfield, their. *N* hubukw'uts'udayez.

chesterfield, your (1). *N* nkw'uts'udayez.

chewing, he/she is. *V* duna'ulh. *[IA]* [0-'ulh < 'alh$_1$]

chewing, you (1) are. *V* duni'ulh. *[IA]* [0-'ulh < 'alh$_1$]

chewing, you (1) are. *V* ilh'ulh. *[IA]* [lh-'ulh < 'alh$_1$]

chews (u.o.), he/she. *V* 'a'alh. *[IA–customary]* [0-'alh < 'alh$_1$]

chickadee. *N* ts'usgak.

chicken. *N* ligok.

chicken coop. *N* ligokbayah.

chicken egg. *N* ligokghez.

chief, generally with reference to band chief. *N* dayi.

child. *N* ski.

child. *N* skui. Duoplurals: -skeh, -skehne

child, small. *N* 'uskiyaz.

child, small. *N* skuiyaz.

children. *N* 'uskehne.

children. *N* skeh. Singulative: ski.

children, his/her. *N* buzkeh.

children, my. *N* suzkeh.

children, our. *N* nezkeh.

children, their. *N* hubuzkeh.

children, your (1). *N* nyuzkeh.

chills, he/she has. *V* buts'utelhk'az. *[IA]* [l-k'az]

chilly, it is. *V* whunik'az. *[IA]* [0-k'az]

chimney. *N* lusooni.

chimney. *N* lhut be'hanajut.

chin. *N* -zets'un.

chin, his/her. *N* buzets'un.

chin, my. *N* szets'un.

chin, your (1). *N* nzets'un.

chins, our. *N* nezets'un.

chins, their. *N* hubuzets'un.

chipmunk. *N* ts'uwhuljos.

chipped, it got. *V* hadanwul. *[PA]* [0-wul]

Chokecherry. [Prunus emarginata] *N* dulgoosmai.

choked him/her, he/she. *V* yuze'usnik. *[PA]* [0-nik]

choked on something solid, he/she. *V* 'untelhah. *[PA]* [0-lhah]

choked on somthing liquid, he/she. *V* 'untelhuyiz. *[PA]* [l-yiz]

choking him/her, he/she is. *V* yuze'unih. *[IA]* [0-nih]

chopped a hole in it, he/she. *V* hukwunintsel. [abs: wh] *[PA]* [0-tsel]

chopped it to pieces, he/she. *V* yaidantsel. *[PA]* [0-tsel]

chopping (u.o.), he/she is. *V* 'utselh. *[IA]* [0-tselh]

chopsticks. *N* duchun be'ts'uyi.

Chowsunkut Lake. *N* Chawts'unk'ut.

Christian person. *N* Yak'usda bughunek k'une 'ut'en-un.

church. *N* lugliz.

cinnabar. *N* tseskaik'ejil.

clam. *N* dulkw'ah besk'um.

clan, one's father's. *N* buts'aha'elts'utne.

clan song. *N* dedohshun.

claps, he. *V* lhookat. *[IA]* [0-kat < kat_3]

claw of rear paw. *N* -kengi.

clay, kind of. *N* tehhaoolts'eh. A kind of white clay found on the bottom of lakes, used for washing and for chinking cabins. It is said to leave the skin of women who wash with it silky.

cleaning cloth. *N* naih bewhuna'udeh.

cliff that descends into water. *N* tadunet'ai.

cling to, I. *V* 'awhenadustl'as. *[IA]* [0-tl'as]

cling to (st), I. *V* ka hoosdutun. *[IA]* [d-tun]

clings to, he/she. *V* 'awhenadintl'as. *[IA]* [0-tl'as]

clippers, hair. *N* behananudughas.

clock. *N* tsadzi.

closes both eyes and then opens them again, he/she. *V* oonats'ilh. *[IA]* [0-ts'ilh]

cloth. *N* dech'ulh.

cloth, cleaning. *N* naih bewhuna'udeh.

clothes. *N* naih.

chopping (u.o.), he/she is. *V* 'utselh. *[IA]* [0-tselh]

clothes, old. *N* naihjut.

clothes brush. *N* naih benaldzooh.

clothes chest. *N* bena'dukuk.

clothes dryer. *N* naih benadugih.

clothes line. *N* naih ba whenant'uk.

clothes washer. *N* be'tunadugus.

clotted, it (blood) has. *V* dul suli. *[PA]* [0-li]

cloud. *N* kw'us.

cloud. *N* yat.

cloudy, it is. *V* kw'us hooni. *[IA]* [0-ni]

cloudy, it is. *V* lheyatdendi. *[IA]* [0-di]

clover. [Trifolium species] *N* nedochunalhduz.

club, war. *N* belhyal.

coal. *N* kwuntset.

coal. *N* tse dezk'un.

coat. *N* dzoot.

cobweb. *N* tsootsibilh.

coccyx. *N* -chets'un.

cod liver oil. *N* lhookghe.

coffee. *N* lugafi.

coffee pot. *N* lugafi 'oosa'.

coffee pot. *N* lugafi be'bizih.

coffee pot. *N* lugafi be'udlez.

coffin. *N* dune besulhti.

coffin. *N* nezek besulhti.

cold, he/she feels. *V* nasdli. *[IA]* [0-dli < dli_1]

cold, he/she has a. *V* kw'us 'uyinla. *[IA]* [0-la]

cold, it (weather) turned. *V* nawniduk'az. [abs: wh] *[PA]* [0-k'az]

cold, it got. *V* hoonk'az. [abs: wh] *[PA]* [0-k'az]

cold, it is. *V* nezk'uz. *[IA]* [0-k'uz < $k'uz_1$]

cold, it is. *V* nezk'uz. *[IA]* [0-k'uz]

cold, it is. *V* nink'az. *[IA]* [0-k'az]

cold, it is. *V* whunezk'uz. [abs: wh] *[IA]* [0-k'uz < $k'uz_1$]

cold, it is. *V* whuzk'uz. *[IA]* [0-k'uz]

cold cream. *N* beznulha.

cold water. *N* too nezk uz.

collar bone. *N* -gant'uk.

collar bone. *N* -t'ukts'un.

collarbone, his/her. *N* but'ukts'un.

collarbone, our. *N* net'ukts'un.

collarbone, their. *N* hubut'ukts'un.

collarbone, your (1). *N* nt'ukts'un.

collided with (st), it. *V* sulhkulh. *[PA]* [lh-kulh]

collided with one another, he/she. *V* lhuhoolkal. *[PA]* [l-kal]

collides with it, it. *V* yulhkulh. *[IA]* [lh-kulh]

colon. *N* 'utsutle.

colt. *N* yeztliyaz.

Columbine, Red. [Aquilegia

formosa] *N* whulecho.

comb. *N* tseldzook.

come here!. *INT* 'ane.

come here!. *INT* 'anih.

comforter. *N* nalhticho.

Common Cattail. [Typha latifolia] *N* tl'ok'uzih.

Common Loon. [Gavia immer] *N* dadzi.

Common Raven. [Corvus corax] *N* datsancho.

Common Red Sphagnum moss. [Sphagnum capillaceum] *N* ts'al. Locally known as Diaper Moss.

community hall, Nautley. *N* Nadlehyah.

confluence of rivers. *N* lheghidli.

confluence of rivers. *N* lheidli.

conk, birch. *N* diyooh.

cook stove. *N* 'ook'et'as.

cooker, Indian slow. *N* huyutsa k'ez hik'ut kwun dilhk'ai.

cookie. *N* lhes lhuk'i.

corner. *N* koonghak.

costly, it is. *V* dezti. [IA] [0-ti < ti_2]

cottonwood. [Populus balsamifera ssp. trichocarpa] *N* landooz.

couch. *N* k'usdudacho.

couch. *N* kw'usdudacho.

couch. *N* kw'usdudayez.

cougar. *N* booscho.

cough medicine. *N* kwusyoo.

coughing, he/she is. *V* dulkwus. *[IA]* [l-kwus]

coughing, I am. *V* dulhukwus. *[IA]* [l-kwus]

cousin. *N* -nadun.

cousin, each other's. *N* lhoonde.

cousin, father's female first. *N* -bezan.

cousin, father's male first. *N* -tai.

cousin, female cross. *N* -zet.

cousin, female first. *N* -ulhtus.

cousin, female second older than ego. *N* -dezcho.

cousin, female second younger than ego. *N* -dezyaz.

cousin, male first. *N* -ulhutsin.

cousin, male first older than ego. *N* -oona.

cousin, male second. *N* -oonde.

cousin, mother's female first. *N* -k'i.

cousin, mother's male first. *N* -z'e.

cousin, my. *N* snadun.

cousin, younger male. *N* -chul.

cousins, female first. *N* lhulhtuske.

cousins, female first. *N* -ulhtuske. This is the plural of -ulhtus.

cow. *N* musdus.

cow, female. *N* musdus ts'eke. Duoplural: musdus ts'ekoo.

cow, his/her. *N* bumusdus.

cow, milk. *N* musdus uljas.

cow, my. *N* smusdus.

cow, our. *N* nemusdus.

Cow Parsnip. [Heracleum lanatum] *N* goos.

cow, their. *N* hubumusdus.

cowboy hat. *N* ts'ahbul.

coyote. [Canis latrans] *N* chuntulhi.

coyote. *N* chuntunuye.

coyote. [Canis latrans] *N* tintulhi.

co-wife. *N* -ulhtus.

crackling, it is. *V* dulk'us. *[IA]* [l-k'us] Describes the sound made by a fire or by cracking a joint.

cracklings. *N* k'ohch'az.

cranberries, bog. [Vaccinium oxycoccos] *N* yak'umai.

cranberries, high-bush. [Viburnum edule] *N* tsalhtse.

cranberries, low-bush. [Vaccinium oxycoccos] *N* yak'umai.

crane, sandhill. [Grus canadensis/ Antigone canadensis] *N* dilh.

Crane, Whooping. [Grus americana] *N* dilhtsul.

cranium. *N* tsits'un.

cranky, he/she is. *V* khusduke. [IA] [d-ke]

cream. *N* dalghen.

Creamy Peavine. [Lathyrus ochroleucus] *N* chunalhduz.

credit. *N* jaboon.

Cree person. *N* Dushin.

cross. *N* tulalhgus.

cross-eyed, he/she is. *V* buna duguz. *[IA]* [0-guz]

crotch. *N* -tl'it.

crotch, his/her. *N* butl'it.

crow. [Corvus brachyrhynchos] *N* datsan.

Crowe Canyon. *N* Yausilh. A canyon on the Nechako River below the Old Crowe homestead at the junction of Dog Creek Road and Sutherland Forest Road. A large eddy here collects floating logs and driftwood.

crown of head. *N* 'utsitah. Indefinitely possessed form.

crumple (st), I. *V* dunusyus. [*IA]* [0-yus]

crunches in his mouth, he/she. *V* ulhgoos. *[IA–customary]* [lh-goos < $goos_1$]

crush (st), I. *V* usyus. *[IA]* [0-yus]

crying, they are. *V* ghunadutsa. *[IA]* [0-tsa]

crying, we are. *V* ghunasdutsa. *[IA]* [0-tsa]

crying, you are. *V* ghuninedutsa. *[IA]* [d-tsa]

cup. *N* lubot.

cupboard. *N* be'uzdla.

cupboard, dish. *N* tsets'ai besula.

cupboard, food. *N* ts'uyi besula.

cupboard, his/her. *N* bube'uzdla.

cupboard, my. *N* sbe'uzdla.

cupboard, our. *N* nebe'uzdla.

cupboard, their. *N* hubube'uzdla.

cupboard, your (1). *N* nyube'uzdla.

cured hide. *N* 'uzuz.

curlers, hair. *N* benaznuldooz.

currant, northern black. [Ribes hudsonianum] *N* tsasdlimai.

curtain. *N* dadent'az ts'oh sula.

curtain, window. *N* dadent'azts'asdla.

curtains. *N* dadent'az ts'osdla.

curtains, his/her. *N* budadent'azts'asdla.

curtains, my. *N* sdadent'azts'asdla.

curtains, our. *N* nedadent'azts'asdla.

curtains, their. *N* hubudadent'azts'asdla.

curtains, your (1). *N* ndadent'azts'asdla.

cushion. *N* tsi'alhyaz.

dam. *N* 'ulh.

damp, it is. *V* sulhtsulyaz. *[IA]* [lh-tsul]

dampen (st), I. *V* nanulhtsul. *[IA]* [lh-tsul]

dampens it, he/she. *V* nainulhtsul. *[IA]* [lh-tsul]

dance, I. *V* nusdaih. *[IA]* [0-daih < $daih_1$]

dance hall. *N* nudaih ba yah.

dance in the Indian way, I. *V* 'ulhudzus. *[IA]* [l-dzus < $dzus_1$]

dance in the Indian way, they. *V* 'uhuldzus. *[IA]* [l-dzus < $dzus_1$]

dance with one's partner in balhats, a. *N* 'indumanuk.

dances in the Indian way, he/she. *V* 'uldzus. *[IA]* [l-dzus < $dzus_1$]

dancing, he/she is. *V* nudaih. *[IA]* [0-daih < $daih_1$] This refers to European styles of dancing, not to traditional dancing.

dancing, he/she is. *V* nuwhulyeh. *[IA]* [l-yeh] This refers to traditional dancing as opposed to European style dancing.

dancing, they are. *V* nahuwhulyeh. *[IA]* [l-yeh]

dancing, we (3+) are. *V* uznudaih. *[IA]* [0-daih < $daih_1$]

dancing Indian-style, he/she is. *V* nuldzus. *[IA]* [l-dzus < $dzus_1$]

dandelion. [Taraxacum species] *N* ditnikwun.

dandelion. *N* salat t'an.

dandruff. *N* tsitsis.

dangerous, it is. *V* beoonujut. *[IA]* [0-jut < $joot_1$]

dark, it is. *V* tsaholhgus. [abs: wh] *[IA]* [lh-gus]

daughter, favorite. *N* t'etyaz.

daughter, his/her. *N* butse'.

daughter, her. *N* buts'e.

daughter, man's. *N* -tse'.

daughter, my. *N* syats'e. A woman's daughter.

daughter, my (man's). *N* stse'. A man's daughter.

daughter, our. *N* netse'.

daughter, our. *N* neyats'e.

daughter, their. *N* hubutse'.

daughter, their. *N* hubuyats'e.

daughter, unfavoured. *N* t'etch'ul.

daughter, woman's. *N* -yats'e.

daughter, your (1). *N* ntse'. A man's daughter.

daughter, your (1). *N* nyats'e. A woman's daughter.

daughter-in-law. *N* -yas'at.

daughter-in-law, his/her. *N* buyas'at.

daughter-in-law, my. *N* syas'at.

daughter-in-law, our. *N* neyas'at.

daughter-in-law, their. *N* hubuyas'at.

daughter-in-law, your (1). *N* nyas'at.

day. *N* dzen.

day after tomorrow. *N* bunde ombun.

day before yesterday. *N* hulhda whutsah da.

daytime. *N* dzenes.

dead, the. *N* yaidlane.

dancing, he/she is. *V* nuwhulyeh.

deadfall trap. *N* gooh.

deadhead. *N* tehduchun.

dear. *INT* goolh. Endearment addressed children and younger brothers and sisters.

deceased. *ADJ* uzdani. Follows the head noun.

December. *N* Dzen Dilhdukw.

December. *N* Sacho Din'ai.

deep, it (body of water) is. *V* takhulh. *[IA]* [0-khulh]

deep, it (snow) is. *V* dincha. [abs: d] *[IA]* [0-cha < cha_1]

deer. *N* yests'e.

deer skin. *N* yests'ezuz.

definitely. *ADV* soo ts'ihun.

delusions of grandeur, he/she has. *VP* duba nadesuti. *[IA]*

den. *N* 'an.

den, black bear. *N* sus'an.

denim. *N* doso.

dentalium shell. [Dentalium preciosum] *N* dulhbai.

depending on for subsistance. *PP* ke.

detergent. *N* latl'us.

devil. *N* liyab.

devil. *N* netsudule.

Devil's Club. [Oplopanax horridus] *N* hoolhghulh.

Devil's Hairpin. *N* Liyubyints'izti.

dew. *N* nadelhudzo.

diaper moss. [Sphagnum capillaceum] *N* ts'al. Locally known as Diaper Moss.

did something, he/she. *V* 'uja. *[PA]* [0-ja]

did something, you (1). *V* 'inja. *[PA]* [0-ja]

died, he/she. *V* dazsai. *[PA]* [0-tsai]

died of exposure, he/she. *V* dli yuzelhghi. *[PA]* [lh-ghi < ghi_1] Literally, "feeling cold killed him/her".

different people. *N* lhelhyoone.

different places. *ADV* lhelhdun.

difficult, it is. *V* huwawhulna. [IA] [l-na < na_2]

difficult for me, it is. *V* sbahuwawhulna. *[IA]* [l-na < na_2]

dining room. *N* 'udai-a.

dining room. *N* 'uts'uyi ba whuz'ai.

dinosaur. *N* nondeti.

dinosaur. *N* dondeti.

dipper. *N* betadzih.

dipper. *N* too benuduka.

dipper (for water). *N* toobets'ukaih.

dipping it up, he/she is. *V* yukaih. *[IA]* [0-kaih < kai_1]

dirt. *N* lhez.

dirt. *N* tsun.

dirty, it (generic) is. *V* dutsun. *[IA]* [0-tsun]

dirty, it is. *V* dunutsun. [abs: n] *[IA]* [0-tsun]

dirty, it is. *V* whudutsun. [abs: wh] *[IA]* [0-tsun]

disappeared, he/she. *V* hoolil. *[PA]* [0-lil]

dishonest, he/she is. *V* duzah. *[IA]* [0-zah]

dislike him/her, I. *V* buts'udusnik. *[IA]* [0-nik]

disliked, I am. *V* sts'udunik. *[IA]* [0-nik] Literally, "one dislikes me".

doctor. *N* yoobeduyun.

dog. *N* lhi. Poss: lik. Duoplural: lhi.

dog, big. *N* lhicho.

dog, female. *N* lhits'e.

dog, his/her. *N* bulik.

dog, his/her big. *N* bulikcho.

dog, little. *N* lhitsol.

dog, my. *N* slik.

dog, my big. *N* slikcho.

dog, my little. *N* sliktsol.

dog, old. *N* lhijut.

dog, our. *N* nelik.

dog, our big. *N* nelikcho.

dog, their. *N* hubulik.

dog, their big. *N* hubulikcho.

dog, worthless. *N* lhich'ul.

dog, your (1). *N* nlik.

dog, your (1) big. *N* nlikcho.

dog's paw. *N* lhike.

dogs. *N* lhike.

doing, they are. *V* 'uhut'en. *[IA]* [d-'en < $'en_2$]

Dolly Varden. [Salvelinus malma] *N* tsabai.

don't touch it!. *INT* ts'it.

door. *N* dadentan.

doorknob. *N* dadentanghunez'ai. Duoplural: dadentanghunezdla

doorknob, his/her. *N* budadentanghunez'ai.

doorknob, my. *N* sdadentanghunez'ai.

doorknob, our. *N* nedadentanghunez'ai.

doorknob, their. *N* hubudadentanghunez'ai.

doorknob, your (1). *N* ndadentanghunez'ai.

doorknobs. *N* dadentanghunezdla.

doorknobs, his/her. *N* budadentanghunezdla.

dogs. *N* lhike.

doorknobs, my. *N* sdadentanghunezdla.

doorknobs, our. *N* nedadentanghunezdla.

doorknobs, their. *N* hubudadentanghunezdla.

doorknobs, your (1). *N* ndadentanghunezdla.

doorway. *N* dadentank'et.

doorway. *N* dati.

doorway. *N* datuk.

doorway, his/her. *N* budati.

doorway, my. *N* sdati.

doorway, our. *N* nedati.

doorway, their. *N* hubudati.

doorway, your (1). *N* ndati.

down (of birds). *N* chus.

downpour, there is. *V* chan nainli. *[IA]* [0-li < li$_1$]

downward. *ADV* nts'un.

draft, they caused a. *V* dahalhts'i. *[PA]* [lh-ts'i < ts'i$_2$] This is said of people who allow the wind to enter by leaving a door or window open.

drag to you (1) (st), I will. *V* nts'utesgus. *[FA]* [0-gus] This includes non-forceful "dragging" such as leading a blind person by the hand.

dragging (st), I am. *V* usgus. *[IA–prog]* [0-gus]

dragging (st), we (2) are. *V* idugus. *[IA–prog]* [0-gus]

dragging (st), we (3+) are. *V* ts'egus. *[IA–prog]* [0-gus]

dragging his feet, he/she is. *V* kedeldzolh. *[IA–prog]* [l-d-zolh]

dragging it, he/she is. *V* yegus. *[IA–prog]* [0-gus]

dragging it, they are. *V* huyegus. *[IA–prog]* [0-gus]

dragonfly. *N* nuduk'ui.

draw, I. *V* 'uk'une'ulhtsi. *[IA]* [lh-tsi]

dreaming, he/she is. *V* nute. *[IA]* [0-te]

dreamt about him/her, I. *V* bunasdlal. *[PA]* [0-dlal < dlal$_1$]

dreamt about something, he/she. *V* whunadlal. *[PA]* [0-dlal < dlal$_1$]

dress. *N* tl'asus.

dress, his/her. *N* butl'asus.

dress, my. *N* stl'asus.

dress, pants. *N* netl'asus.

dress, their. *N* hubutl'asus.

dress, your (1). *N* ntl'asus.

dressed up, they are. *V* soonahduja. *[IA]* [0-ja]

dressed up, we (3+) are. *V* soonazduja. *[IA]* [0-ja]

dressed up, you (1). *V* soonadinja. *[IA]* [0-ja]

dresser. *N* be'uzdla.

dresser. *N* naih besula

dried beaver meat. *N* tsaikaih.

dried beaver meat. *N* tsatsun sugi.

dried berries. *N* mai nezgi.

dried berries. *N* nezdzak.

dried meat. *N* 'utsun sugi.

dried salmon. *N* talook sugi.

drink, white man's. *N* kwuntoo.

drink liquor, they. *V* tahutnai. *[IA]* [d-nai < nai$_1$]

drink liquor, they will. *V* tatetnilh. *[FA]* [d-nilh < nai$_1$]

drink liquor, we (3+) will. *V*

taztenilh. *[FA]* [d-nilh]

drink liquor, you (1) will. *V* tatnilh. *[FA]* [d-nilh]

dripping, it is. *V* nadulyul. *[IA]* [l-yul]

drive, he/she will. *V* tegooh. *[FA]* [0-gooh < goo$_1$]

drive, I will. *V* tesgooh. *[FA]* [0-gooh < goo1]

drive, I will not. *V* chatuzesgooh. *[FN]* [0-gooh < goo$_1$]

drive, they will. *V* hutegooh. *[FA]* [0-gooh < goo$_1$]

drive, they will not. *V* chahutesgooh. *[FN]* [0-gooh < goo$_1$]

drive, we (2) will. *V* tadugooh. *[FA]* [0-gooh < goo$_1$]

drive, we (2) will not. *V* chatuzadugoo. *[FN]* [0-goo < goo$_1$]

drive, you (1) will. *V* tangooh. *[FA]* [0-gooh < goo$_1$]

drive, you (1) will not. *V* chatuszangooh. *[FN]* [0-gooh < goo$_1$]

drive (st) to you (1), I will. *V* nts'utelhgooh. *[FA]* [lh-gooh < goo$_1$]

driven in a loop, he/she has not. *V* nuchaigooh. *[PN]* [0-gooh < goo$_1$]

driven in a loop, I have. *V* nususgoo. *[PA]* [0-goo < goo$_1$]

driven in a loop, they have. *V* nuhuzgoo. *[PA]* [0-goo < goo$_1$]

driven in a loop, they have not. *V* nuchahigooh. *[PN]* [0-gooh < goo$_1$]

driving, he/she is. *V* ugooh. *[IA–prog]* [0-gooh < goo$_1$]

driving, he/she is not. *V* chasgooh. *[IN–prog]* [0-gooh < goo$_1$]

driving, I am. *V* usgooh. *[IA–prog]* [0-gooh < goo$_1$]

driving, they are. *V* hegooh. *[IA–prog]* [0-gooh < goo$_1$]

driving, they are not. *V* chahesgooh. *[IN–prog]* [0-gooh < goo$_1$]

drum. *N* tughul.

driving, they have set off. *V* whehangoo. *[PA]* [0-goo < goo$_1$]

driving, we (2) are. *V* idugooh. *[IA–prog]* [0-gooh < goo$_1$]

driving, we (2) are not. *V* chasidugooh. *[IN–prog]* [0-gooh < goo$_1$]

driving, we (2) set off. *V* wheidugoo. *[PA]* [0-goo < goo$_1$]

driving, you (1) are. *V* ingooh. *[IA–prog]* [0-gooh < goo$_1$]

driving, you (1) are not. *V* chasangooh. *[IN–prog]* [0-gooh < goo$_1$]

driving away, they are. *V* natedugooh. *[FA]* [d-gooh]

driving away, we (3+) are. *V* naztedugooh. *[FA]* [0-gooh]

driving away, you (1) are. *V* natandugoo. *[FA]* [0-goo]

driving in a loop, they are. *V* nuhugoo. *[IA]* [0-goo]

driving in a loop, we (3+) are. *V* nuts'ugoo. *[IA]* [0-goo]

driving in a loop, you (1) are. *V* ningoo. *[IA]* [0-goo]

drizzling, it is. *V* nawhulhtiyaz. *[IA]* [lh-ti]

drizzling, it is. *V* nawhulhtsit. *[IA]* [lh-tsit]

drove in a loop, we (2). *V* nusidugoo. *[PA]* [0-goo < goo_1]

drove in a loop, you (1). *V* nusingoo. *[PA]* [0-gooh < goo_1]

drowned, he/she. *V* too yuzelhghi. *[PA]* [lh-ghi < ghi_1]

drowned, they (3+) drowned. *V* too hubanghan. *[PA]* [0-ghan < $ghan_1$]

drowning, he/she is. *V* too yuzulhghe. *[IA]* [l-ghe < ghi_1]

drum. *N* tughul.

drumming, it (Ruffed Grouse) is. *V* 'udutsut. *[IA]* [d-tsut]

drumming, it is. *V* 'utsut. *[IA]* [0-tsut < $tsut_1$] This describes the way a Ruffed Grouse drums with its wings.

dry, they are. *V* nadesjak. *[IA]* [0-jak]

dry, we (2) are. *V* nadusdujih. *[IA]* [d-jih]

dry, you (1) are. *V* nadenjak. *[IA]* [0-jak]

dry cow moose. *N* yats'it.

dryer, clothes. *N* naih benadugih.

drying (st), we (2) are. *V* idulgi. *[IA]* [lh-gi]

drying poles. *N* 'ujooh. The poles from which fish or meat is hung to dry.

Dry William lake. *N* Wilyum sugi. The small lake on the south side of Highway 16 betwen Fort Fraser and Lejac.

Drywilliam Lake. *N* Wilyum sugi. The small lake on the south side of Highway 16 betwen Fort Fraser and Lejac.

duck. *N* dut'ai.

duff. *N* chehhaindzool.

dug (with a shovel), he/she. *V* hahonkai. *[PA]* [0-kai]

Dwarf Maple. [Acer glabrum] *N* khasdzoon.

dynamite. *N* oondunultalh.

Eagle, Bald. [Haliaeetus leucocephalus] *N* sbalyan.

Easter. *N* Zooldubak.

Endakoh River. *N* Endakoh.

Etcho Lake. *N* Ts'oon'ai Bunk'ut. A small lake just east of Ormond Lake.

eagle, bald. *N* tsebalyan.

eagle, golden. [Aquila chrysaetos] *N* sjul.

ear. *N* -dza.

ear canal. *N* -dzek.

earache, he/she has an. *V* budzek hah'elts'ul. *[IA]* [l-ts'ul]

earache, I have an. *V* sdzek hah'elts'ul. *[IA]* [l-ts'ul]

eardrum. *N* -dzek whudadilhchooz.

earlobe. *N* -dzabal.

earned it, he/she. *V* whuch'ainelhde. *[PA]* [lh-de]

earned it, they. *V* whuch'ahinelhde. *[PA]* [lh-de]

earned it, we (2). *V* whuch'anidulde. *[PA]* [lh-de]

earned it, we (3+). *V* whuch'aznelhde. *[PA]* [lh-de]

earning income, he/she is. *V* 'udutl'us. *[IA]* [0-tl'us]

earrings. *N* dzekw'ul.

ears, rabbit. *N* gahdzo.

earth, on. *PPP* yun k'ut.

earth (soil), dust. *N* lhez.

earthquake. *N* yun nughutnah.

earthworm. *N* 'usgoo.

east, from the. *ADV* ndazde.

east, to the. *ADV* nda.

east wind. *N* ndazde nilhts'i.

easy, it is. *V* wachahoolnah. *[IA]* [l-nah]

easy chair. *N* kw'uts'udacho.

eat, he/she does not. *V* chasyi. *[IN]* [0-yi < yi_1]

eat, he/she will not. *V* chatesyi. *[FN]* [0-yi < yi_1]

eat, I do not. *V* chasusyi. *[IN]* [0-yi < yi_1]

eat, I may not. *V* chaoosyi. *[ON]* [0-yi < yi_1]

eat, I will. *V* tesyi. *[FA]* [0-yi < yi_1]

eat, I will not. *V* chatuzesyi. *[FN]* [0-yi < yi_1]

eat, they. *V* huyi. *[IA]* [0-yi < yi_1]

eat, they do not. *V* chahusyi. *[IN]* [0-yi < yi_1]

eat, they will. *V* huteyi. *[FA]* [0-yi < yi_1]

eat, they will not. *V* chahutesyi. *[FN]* [0-yi < yi_1]

eat, we (2). *V* idudai. *[IA]* [0-dai < yi_1]

eat, we (2). *V* odudai. *[OA]* [0-dai < yi_1]

eat, we (2) will. *V* tadudai. *[FA]* [0-dai < yi_1]

eat, we (3+). *V* ts'uyi. *[IA]* [0-yi < yi_1]

eat, we (3+) do not. *V* chats'usyi. *[IN]* [0-yi < yi_1]

eat, we (3+) will. *V* ts'uteyi. *[FA]* [0-yi < yi_1]

eat, we (3+) will not. *V* chaztesyi. *[FN]* [0-yi < yi_1]

eat, you (1). *V* inyi. *[IA]* [0-yi < yi_1]

eat, you (1) will. *V* tanyi. *[FA]* [0-yi < yi_1]

eat, you (1) will not. *V* chatuzanyi. *[FN]* [0-yi < yi_1]

eat, you (2) do not. *V* chasinyi. *[IN]* [0-yi < yi_1]

eat, you (2+). *V* ahyi. *[IA]* [0-yi < yi_1]

eat, you (2+) do not. *V* chazahyi. *[IN]* [0-yi < yi_1]

eat, you (2+) will. *V* tehyi. *[FA]* [0-yi < yi_1]

eat, you (2+) will not. *V* chatuzehyi. *[FN]* [0-yi < yi_1]

eat (st), he/she may. *V* wuyi. *[OA]* [0-yi < yi_1]

eat (st), he/she will. *V* teyi. *[FA]* [0-yi < yi_1]

eat (st), I. *V* usyi. *[IA]* [0-yi < yi_1]

eat (st), I may. *V* wusyi. *[OA]* [0-yi < yi_1]

eat (st), they may. *V* hooyi. *[OA]* [0-yi < yi_1]

eat (st), we (3+) may. *V* ts'ooyi. *[OA]* [0-yi < yi_1]

eat (st), we (3+) will. *V* naztedilh. *[FA]* [0-dilh]

eat (st), you (1) may. *V* onyi. *[OA]* [0-yi < yi$_1$]

eat (st), you (1) will. *V* natandilh. *[FA–habitual]* [0-dilh]

eat (u.o.), he will. *V* ’uteyilh. *[FA]* [0-yilh]

eat (u.o.), I. *V* ’usyi. *[IA]* [0-yi < yi$_1$]

eat (u.o.), I. *V* na’usdai. *[IA–habitual]* [d-ai < yi$_1$]

eat (u.o.), I may. *V* ’oosyi. *[OA]* [0-yi < yi$_1$]

eat (u.o.), I may. *V* na’oosdai. *[OA]* [0-dai < yi$_1$]

eat (u.o.), I will. *V* ’utesyilh. *[FA]* [0-yilh < yi$_1$]

eat (u.o.), I will. *V* na’tesdilh. *[FA]* [0-dilh]

eat (u.o.), we (2). *V* na’idudai. *[IA–habitual]* [d-ai < yi$_1$]

eat (u.o.), we (2) will. *V* na’tadudilh. *[FA–habitual]* [d-ilh]

eat (u.o.), we (3+). *V* ’uts’uyi. *[IA]* [0-yi < yi$_1$]

eat (u.o.), we (3+). *V* na’ts’udai. *[IA–habitual]* [d-ai < yi$_1$]

eat (u.o.), we (3+) will. *V* na’uztedai. *[FA]* [d-ai < yi$_1$]

eat (u.o.), we (3+) will. *V* ’uzteyilh. *[FA]* [0-yilh < yi$_1$]

eat (u.o.), you (1). *V* ’inyi. *[IA]* [0-yi < yi$_1$]

eat (u.o.), you (1). *V* na’indai. *[IA–habitual]* [d-ai < yi$_1$]

eat (u.o.), you (1) may. *V* na’ondai. *[OA–habitual]* [d-ai < yi$_1$]

eat (u.o.), you (1) will. *V* ’utanyilh. *[FA]* [0-yilh < yi$_1$]

eating, he/she is. *V* uyi. *[IA]* [0-yi < yi$_1$]

eating (u.o.), he/she is. *V* ’uyi. *[IA]* [0-yi < yi$_1$]

eating (u.o.), they are. *V* ’uhuyi. *[IA]* [0-yi < yi$_1$]

eating (u.o.), we (2) are. *V* ’iduji. *[IA]* [0-yi < yi$_1$]

eaves. *N* bunt’ah.

echo. *N* dazul.

eclipse. *N* sa yenin’ai.

eddy. *N* ’ok’et.

edge. *N* ba.

eel. *N* tselk’eh.

effort, he/she is making an. *V* yudats’ein’ai. *[IA]* [0-’ai]

egg, chicken. *N* ligokghez.

egg, fish. *N* ’uk’oon.

eggs. *N* ’ughez.

eggs. *N* -ghez.

eggs, ant. *N* ’andihghez.

eggs, fish. *N* -k’oon.

eggs, fly. *N* sts’uzts’uz.

eight [abstract]. *N* lhk’udiwh.

eight [locative]. *N* lhk’udidun.

eight [generic]. *NUM* lhk’utdink’i.

eight [human]. *NUM* lhk’utdine.

eight [multiplicative]. *NUM* lhk’udit.

eighty [generic]. *NUM* lhk’utdit lanezi.

either. *CONJ* k’us.

ejaculate, I am going to. *V* tesjaz. *[FA]* [0-jaz]

elastic, it is. *V* nuduguz. *[IA]* [0-guz]

elbow, my. *N* snints'uzti.

elbow. *N* -nints'uzti.

elbow. *N* -yints'iz.

elbow, his/her. *N* bunints'uzti.

elbow, your (1). *N* nnints'uzti.

elbow, your (1). *N* nyints'iz.

elbows, our. *N* nenints'uzti.

elbows, their. *N* hubunints'uzti.

elder. *N* hoonyan.

elders. *N* netsowhuzulhne.

electrical outlet or plug. *N* 'ana'dutsih.

elephant. *N* khunaicho.

eleven [generic]. *NUM* lanezi 'on'at 'ilhuk'i.

elk. [Cervus canadensis] *N* yusi. The North American elk was formerly considered to be a subspecies of Cervus elaphus, but as of 2004 biologists consider the North American elk and the European elk to be different species.

embankment along Nechakoh. *N* nooghun.

embarassed, I am. *V* yooya susli. *[IA]* [0-li]

ember. *N* kwuntset.

emphatic particle. *PART* si.

end of road. *N* niti. The point where a trail ends at the frontier between one people's territory and its neighbour's.

entered, he/she. *V* daninya. *[PA]* [0-ya < ya$_1$]

entered, I. *V* danusya. *[PA]* [0-ya < ya$_1$]

entered, they (3+). *V* dahunindil. *[PA]* [0-dil < dil$_1$]

entered, we (2). *V* danidut'az. *[PA]* [0-'az < 'as$_1$]

entered, we (3+). *V* daznindil. *[PA]* [0-dil < dil$_1$]

entered your (2+) house, he. *V* nahghudaninya. *[PA]* [0-ya]

entered your (2+) house, they (3+). *V* nahghundanindil. *[PA]* [0-dil]

entered your (2+) house, we (3+). *V* nahghudazdindil. *[PA]* [0-dil]

entrance. *N* dati.

entrance. *N* whudah.

ermine. *N* nahbaicho.

evening. *N* hulhgha.

evening, this. *N* k'an hulhgha.

evening, tomorrow. *N* bunde lhgha.

evening, yesterday. *N* lhda hulhgha.

every morning. *N* bundada totsuk.

everyone. *N* ts'iyane.

excrement. *N* tsan.

exhaust pipe. *N* -tl'it.

exhaust pipe, its. *N* butl'it.

exited, he/she. *V* tininya. *[PA]* [0-ya < ya$_1$]

exited, I. *V* tinusya. *[PA]* [0-ya < ya$_1$]

exited, they (3+). *V* tihunindil. *[PA]* [0-dil < dil$_1$]

exited, we (2). *V* tinit'az. *[PA]* [0-'az < 'as$_1$]

exited, we (3+). *V* tiznindil. *[PA]* [0-dil < dil$_1$]

expensive, it is. *V* dezti. *[IA]* [0-ti < ti$_2$]

expensive, it is not. *V* chaditi. *[IN]* [0-ti < ti_2]

explain, I. *V* whunulhtun. *[IA]* [lh-tun < tun_3]

explain to them, I. *V* whuhubunulhtun. *[IA]* [lh-tun < tun_3]

explains to me, he/she. *V* whusunulhtun. *[IA]* [lh-tun < tun_3]

exposure, he/she died of. *V* dli yuzelhghi. *[PA]* [lh-ghi < ghi_1] Literally, "feeling cold killed him/her ".

exult, they. *V* 'uhoolhkes. *[IA]* [lh-kes]

exults, he/she. *V* 'oolhkes. *[IA]* [lh-kes]

eye. *N* -na.

eye, a single. *N* nak'uz.

eye, his/her. *N* buna.

eye, my. *N* sna.

eye, your (1). *N* unna.

eyebrows. *N* -nats'osdooz.

eyebrows, his/her. *N* bunats'osdooz.

eyebrows, my. *N* snats'osdooz.

eyebrows, our. *N* nenats'osdooz.

eyebrows, their. *N* hubunats'osdooz.

eyebrows, your (1). *N* unnats'osdooz.

eyeglasses. *N* nak'ezdulya.

eyelashes. *N* -nabagha.

eyelashes, his/her. *N* bunabagha.

eyelashes, my. *N* snabagha.

eyelashes, our. *N* nenabagha.

eyelashes, their. *N* hubunabagha.

eyelashes, your (1). *N* unnabagha.

eyelid. *N* -nabalos.

eyes, fish. *N* lhookna.

eyes, our. *N* nena.

eyes, their. *N* hubuna.

eyetooth. *N* 'unaghoo.

eyetooth. *N* -naghoo.

face. *N* -nen.

face, his/her. *N* bunen.

face, my. *N* snen.

face, your (1). *N* unnen.

faces, our. *N* nenen.

faces, their. *N* hubunen.

facial hair, he/she has. *V* dunughai. [abs: n] *[IA]* [0-ghai < $ghai_1$]

facial hair, I have. *V* dunusghai. *[IA]* [0-ghai < $ghai_1$]

facial hair, they have. *V* hudunughai. *[IA]* [0-ghai < $ghai_1$]

facial hair, we (3+) have. *V* ts'udunughai. *[IA]* [0-ghai < $ghai_1$]

fade, it will. *V* be'dutedulh. *[FA]* [0-dulh]

faded, it. *V* be'didil. *[PA]* [C-dil]

fainted, he/she. *V* bukawetezut. *[PA]* [0-zut]

fainted, I. *V* skawtezut. *[PA]* [0-zut]

falcon, peregrine. *N* hoolht'as.

fall. *N* dak'et.

falling from the trees, it (snow) is. *V* nadeh. *[IA]* [0-deh]

False Solomon's Seal. [Maianthemum racemosum] *N* lhiluzchun. This species is also known by the older scientific name Smilacina amplexicaulis.

far, it is. *V* nilhdza. *[IA]* [lh-dza < dza_1]

far, it is not. *V* chanilhdza. *[IN]* [lh-dza < dza_1]

far away place. *N* yaba.

farts, he/she. *V* tutl'it. *[IA]* [0-tl'it < $tl'it_1$]

fast. *ADV* 'a.

fat. *N* k'a.

fat. *N* 'uk'a.

fat, black bear. *N* susk'a.

fat, he/she has become again. *V* nanilk'a. *[PA]* [l-k'a]

fat, he/she is. *V* nelhuk'a. *[IA]* [l-k'a]

fat, he/she is not. *V* chanilk'ah. *[IN]* [l-k'ah]

father. *N* 'uba.

father. *N* -ba.

falcon, peregrine. *N* hoolht'as.

father, his/her. *N* buba.

father, my. *N* sba.

father, our. *N* neba.

father, their. *N* hububa.

father, your (1). *N* mba.

father-in-law. *N* -zaz.

father-in-law, his/her. *N* buzaz.

father-in-law, my. *N* szaz.

father-in-law, our. *N* nezaz.

father-in-law, their. *N* hubuzaz.

father-in-law, your (1). *N* nzaz.

fatigued, he/she is. *V* shih yuzelhghi. *[IA]* [lh-ghi < ghi_1]

February. *N* Ts'uzyen Hoh Ts'utsun nun.

February. *N* Yuske'nighus.

feces. *N* tsan.

feces, his/her. *N* butsan.

feces, my. *N* stsan.

feces, our. *N* netsan.

feces, their. *N* hubutsan.

feces, your (1). *N* ntsan.

feed you, they are going to. *V* nghate'alh. *[FA]* [0-'alh]

feet. *N* -ke.

feet, his/her own. *N* duke.

feet, the tops of our. *N* nekent'ak.

feet, the tops of their. *N* hubukent'ak.

fell, you (1). *V* nadilhghis. *[IA]* [lh-ghis < $ghis_1$]

fell down, he/she. *V* nadelduz. *[PA]* [l-duz]

fell down, it. *V* nalts'ut. *[PA]* [l-ts'ut]

felled, you (1). *V* nadalhghez. *[PA]* [lh-ghez]

female elder, very respected. *N* ts'eketi. Duoplural: ts'ekooti

female fish. *N* k'ooni.

femur. *N* -wuzts'un.

femur, his/her. *N* buwuzts'un.

femur, my. *N* swuzts'un.

femur, your (1). *N* nwuzts'un.

femurs, our. *N* newuzts'un.

femurs, their. *N* hubuwuzts'un.

fence. *N* nanestl'oo.

fence post. *N* nanestl'oo kechun.

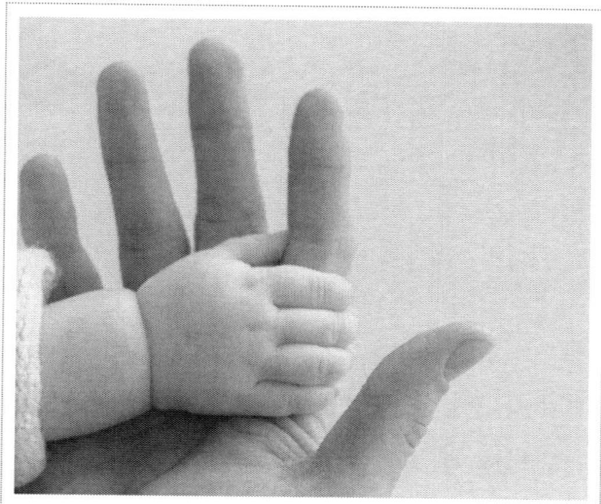

fingers, our. *N* nelasge.

fenced us in, they. *V* nehaneztl'oo. *[PA]* [0-tl'oo < tl'oo$_1$]

ferrets things out, he/she. *V* ha'ulih. *[IA]* [0-lih]

ferrets things out, he/she. *V* ha'ulih. [class: mdo-c-gen] *[IA–customary]* [0-lih]

fertilizer. *N* hanuyeh buts'uyi.

fifteen [generic]. *NUM* lanezi 'on'at skwunlai.

fifty [generic]. *NUM* skwunlat lanezi.

fight, we are going to. *V* beztedulh. *[FA]* [0-dulh]

fight, you (1) are going to. *V* betandulh. *[FA]* [0-dulh]

fight, you (2+) are going to. *V* betehdulh. *[FA]* [0-dulh]

fin. *N* yenube.

finger. *N* -lasge.

finger, his/her. *N* bulasge.

finger, index. *N* -lasgek.

finger, my. *N* slasge.

finger, my index. *N* slasgek.

finger, small. *N* -lasgeyaz.

finger spaces. *N* -latl'i. The spaces between the fingers.

finger spaces, his/her. *N* bulatl'i.

finger spaces, my. *N* slatl'i.

finger spaces, our. *N* nelatl'i.

finger spaces, their. *N* hubulatl'i.

finger spaces, your (1). *N* nlatl'i.

fingernail. *N* 'ulagi.

fingernail. *N* -lagi.

fingernail, his/her own. *N* dulagi.

fingernail, his. *N* bulagi.

fingernail, my. *N* slagi.

fingernail, your (1). *N* nlagi.

fingernails, our. *N* nelagi.

fingernails, their. *N* hubulagi.

fingers. *N* -latsuk.

fingers, our. *N* nelasge.

fingers, their. *N* hubulasge.

finger, your (1). *N* nlasge.

Fir, Douglas. [Pseudotsuga menziesii] *N* tsuntsi.

fire. *N* kwun.

fire water. *N* kwuntoo.

firecracker. *N* oodunaltalh-i.

firewood. *N* tsuz.

Fireweed. [Epilobium angustifolium] *N* khast'an.

first person dual possessor. *PREFIX* nah-. Attaches to class 1 nouns.

first person plural posssessor. *PREFIX* ne-. Attaches to class 1 nouns.

first person singular possessor. *PREFIX* s-. Attaches to class 1 nouns.

fish. *N* lhook.

fish, dried. *N* lhook sugi.

fish split once and dried. *N* k'ak.

fish, female. *N* k'ooni.

fish, male. *N* tl'uzni.

fish, reconstituted dried. *N* nalhtsul. This refers to dried fish that is soaked in water and then boiled.

fish egg. *N* 'uk'oon.

fish eggs. *N* -k'oon.

fish eyes. *N* lhookna.

fish guts. *N* lhooktsik.

fish head. *N* lhooktsi.

fish hook. *N* jus.

fish net. *N* lhombilh.

fish net. *N* lhoombilh.

fish net anchor. *N* tsetoos'ai.

fish net anchor pole. *N* lhoombilhchun.

fish net weights. *N* lhoombilhtse.

fish soup. *N* lhook tazul.

fish spear. *N* dagwut.

fish split once and dried. *N* k'ak.

fish tail. *N* lhookchetl'a.

fish trap. *N* 'uk'oondzai.

fisher. [Martes pennanti] *N* chunihcho.

fishing pole. *N* juschun.

fishing with a hook, he/she is. *V* 'oolhjas. *[IA]* [lh-jas < jas$_2$]

fist. *N* tuchus.

five [abstract]. *NUM* skwunlawh.

five [generic]. *NUM* skwunlai.

five [human]. *NUM* skwunlanun.

five [locative]. *NUM* skwunladun.

five [multiplicative]. *NUM* skwunlat.

flame. *N* kwun.

flammable, it is. *V* dillhuk. *[IA]* [l-lhuk]

flashing, it is. *V* whedunukw. *[IA]* [0-nukw]

flat, it (tire) is. *V* buniltal. *[PA]* [l-tal]

flea. *N* lhiya.

flesh. *N* -tsun.

flesh, my. *N* ussun.

floating around, he/she is. *V* nulat. *[IA]* [0-lat]

floating around, we (2) are. *V* nidudlat. *[IA]* [0-lat]

floats for fish net. *N* dolulh.

floor. *N* yun.

floor, his/her. *N* buyun.

floor, my. *N* syun.

floor, our. *N* neyun.

floor, their. *N* hubuyun.

floor, your (1). *N* nyun.

floor boards. *N* yundzihtel.

floor covering. *N* yun sulchooz.

floor covering. *N* yunsubul.

floor covering, his/her. *N* buyunsubul.

floor covering, his/her. *N* buyunsulchooz.

floor covering, my. *N* syunsubul.

floor covering, my. *N* syunsulchooz.

floor covering, our. *N* neyunsubul.

floor covering, our. *N* neyunsulchooz.

floor covering, their. *N* hubuyunsubul.

floor covering, their. *N* hubuyunsulchooz.

floor covering, your (1). *N* nyunsubul.

floor covering, your (1). *N* nyunsulchooz.

flooring. *N* yunlatel.

flooring, his/her. *N* buyunlatel.

flooring, my. *N* syunlatel.

flooring, our. *N* neyunlatel.

flooring, their. *N* hubuyunlatel.

flooring, your (1). *N* nyunlatel.

flour. *N* lhes nududzaih.

flour sack. *N* lhes 'uzus.

flower. *N* 'indak.

flower. *N* -indak.

flower, his/her. *N* be'indak.

flower, his/her. *N* ye'indak.

flower, his/her own. *N* de'indak.

flower, my. *N* se'indak.

flower, our. *N* ne'indak.

flower, their. *N* hube'indak.

flower, your (1). *N* nye'indak.

flower, your (2+). *N* nawhe'indak.

flower pot. *N* t'an behanuyeh.

flowers. *N* 'indakya.

flowers, its. *N* bindak.

flowing out of, it is. *V* yahainli. *[IA]* [0-li < li_1]

fly, black. *N* dahjolh.

fly, house. *N* sts'uz.

fly home, they will. *V* nahutet'ah. *[FA]* [0-t'ah < $t'oh_2$]

fly in a loop, he/she will. *V* nutet'ah. *[FA]* [0-t'ah < $t'oh_2$]

fly in a loop, he/she will. *V* nutet'ah. *[FA]* [0-t'ah < $t'oh_2$]

fly in a loop, they will. *V* nahutet'ah. *[FA]* [0-t'ah < $t'oh_2$]

fly in a loop, they will. *V* natet'ah. *[IA]* [0-t'ah < $t'oh_2$]

flying in a loop, he/she is. *V* nut'ah. *[IA]* [0-t'ah < $t'oh_2$]

flying in a loop, he/she is. *V* nut'ah. *[IA]* [0-t'ah < $t'oh_2$]

flying in a loop, they are. *V* nuhut'ah. *[IA]* [0-t'ah < $t'oh_2$]

Flying Squirrel. [Glaucomys

sabrinus alp.] *N* ts'unulhbuz.

foam. *N* hawus.

focus. *PART* la.

fog. *N* 'a.

foggy, it is. *V* 'a naints'ut. *[IA]* [0-ts'ut]

foggy, it is. *V* naints'ut. *[IA]* [0-ts'ut]

food. *N* ts'uyi.

food cache. *N* tsachun.

food cupboard, his/her. *N* buts'uyibesula.

food cupboard, my. *N* sts'uyibesula.

food cupboard, our. *N* nets'uyibesula.

food cupboard, their. *N* hubuts'uyibesula.

food cupboard, your (1). *N* nts'uyibesula.

food storage building. *N* tsuyi ba yah.

fooled, he/she got. *V* nenet'a. *[PA]* [d-'a]

fooled, they got. *V* nahunet'a. *[PA]* [0-t'a]

foolish, he/she is. *V* madli. *[IA]* [0-dli]

foolish, I am. *V* masdli. *[IA]* [0-dli]

foolish, they are. *V* mahudli. *[IA]* [0-dli]

Fool's Hen. [Falcipennis canadensis] *N* nat'ah.

foot. *N* -ke.

foot, the top of his/her. *N* bukent'ak.

foot, the top of my. *N* skent'ak.

foot, the top of your (1). *N* nkent'ak.

foot, top of. *N* -kent'ak.

foot path. *N* duneti.

footprints. *N* 'uk'ah.

footprints. *N* k'ah.

for. *PP* ba.

for me. *PPC* sba.

for us. *PPC* neba.

forehead. *N* -nint'ak.

forehead, my. *N* snintak.

forehead, your (1). *N* nnintak.

foreheads, our. *N* nenintak.

forest. *N* tintah.

forest. *N* duchuntah.

fork. *N* be'ooget.

fork. *N* beooduget.

fork. *N* lufooset.

fork. *N* luhoos.

fork in road. *N* lhk'eti.

fork in road, three-way. *N* tsewhehudilhkw'ah.

former. *V* inle. *[PA]* [0-le]

forty [generic]. *NUM* dit lanezi.

forward. *ADV* nus.

four [abstract]. *NUM* diwh.

four [generic]. *NUM* dink'i.

four [human]. *NUM* dinun.

four [locative]. *NUM* didun.

four [multiplicative]. *NUM* dit.

fourth. NUM dit.

fox. *N* nanguz.

frame, stretching. *N* busdi.

Fraser Lake. *N* Nadlehbun.

Fraser Lake. *N* Nadlehbunk’ut.

Fraser Lake people. *N* Nadlehbunne.

Fraser Mountain. *N* K’atsizyus.

freezer. *N* be’utun.

frequently. *ADV* lhghun.

frequently. *ADV* lhghunilhdukw.

fresh from the store, it is. *V* talh. uki. *[IA]* [l-ki]

Friday. *N* ’Utsun Ts’usyih.

Friday. *N* Wanderdi.

fried meat. *N* ’utsun sut’e.

friend. *N* t’eke.

friend, his/her. *N* but’eke.

friend, my. *N* st’eke.

friend, our. *N* net’eke.

friend, their. *N* hubut’eke.

friend, your (1). *N* nt’eke.

frog. *N* dulkw’ah.

from. *PP* ts’i.

from away. *ADV* ’azdeh.

from outside. *ADV* ’azdeh.

from where?. *QWH* nts’ez.

front of me, in. *PPC* sbut.

frost. *N* so.

frost up high. *N* dahzo.

frosty clear night, it is a. *V* yawtelhah. [IA] [0-lhah]

froze, it. *V* ustun. *[PA]* [0-tun < tun$_2$]

fruit juice. *N* maitoo.

fruit leather. *N* maitlus.

fruitful, it is. *V* belhk’a. *[PA]* [l-k’a] Means that an area not only produces a lot but that whatever comes from it is fat and of good quality.

frying pan. *N* be’ut’es.

full of it, I think he is. *V* bech’anuszun. *[IA]* [0-zun < zun$_1$]

fur. *N* ’ugha.

fur. *N* -gha.

fur, his/her own. *N* dugha.

furry, it is. *V* dughai. [abs: 0] *[IA]* [0-ghai < ghai$_1$]

future, in the. *ADV* nus te.

gaff hook. *N* sah.

gagging, he/she is. *V* tukooh. *[IA]* [0-kooh]

gaining weight, I am. *V* whenlhuk’a. *[IA]* [l-k’a]

gall bladder. *N* -tl’uz.

gall bladder, his/her. *N* butl’uz.

gall bladder, my. *N* stl’uz.

gall bladder, your (1). *N* ntl’uz.

gall bladders, our. *N* netl’uz.

gall bladders, their. *N* hubutl’uz.

gall stone. *N* tl’uztse.

gamble, we (3+) will. *V* ’aztelih. *[FA]* [0-lih < lih$_1$]

gamble, you (1) will. *V* ’atanlih. *[FA]* [0-lih < lih$_1$]

gamble, you (2+) will. *V* ’atehlih. *[FA]* [0-lih < lih$_1$]

gambling, he/she is. *V* ’alih. *[IA]* [0-lih < lih$_1$]

gambling, I am. *V* ’aslih. *[IA]* [0-lih]

gambling, they are. *V* ’ahulih. *[IA]* [0-lih]

gambling, we (2) are. *V* ’aidudlih. *[IA]* [0-lih]

gambling, we (3+) are. *V* ’ats’ulih.

[IA] [0-lih]

gave it to me, he/she. *V* sghainin'ai. [class: sdo] [abs: gen] *[PA]* [0-'ai < 'ai$_1$]

gave it to me, he/she. *V* sghainintan. [class: lro] [abs: gen] *[PA]* [0-tan < tan$_1$]

gave me, he/she. *V* sghani'ai. [class: sdo-gen] *[PA]* [0-'ai < 'ai$_1$]

gave me, he/she. *V* sghaninkai. [class: coc] *[PA]* [0-kai < kai$_1$]

gave me, he/she. *V* sghanintan. [class: lro] *[PA]* [0-tan < tan$_1$]

generous, he/she is. *V* bula ncha. *[IA]* [0-cha < cha$_1$] Literally, "his hand is large".

generous, he/she is. *V* danzoo. *[IA]* [0-zoo < zoo$_1$]

genitalia, female. *N* -k'al.

ghost. *N* dunezoolh.

ghost. *N* naoodnilh. In the early post-contact period was also used to refer to Europeans.

giant bird, extinct. *N* dune oolhdzai.

ginseng. [Veratrum viride] *N* whulhdulh. Locally known as "ginseng", but not at all the same plant.

girl. *N* skuits'eke. Duoplural: skehts'ekoo

girl. *N* ts'ekeyaz.

girl, baby. *N* ts'eke tsalhts'ul.

girl, small. *N* skuits'ekeyaz.

girls. *N* skehts'ekoo.

girls. *N* ts'ekooyaz.

give (st) to me, you (1). *V* slaintih. *[IA]* [0-tih < tan$_1$]

give me. *INT* de'. In spite of the curtness suggested by the English translations, this expression is perfectly polite. It is perfectly polite by itself; indeed, Carrier has no word for "please".

give me, he/she is going to. *V* sghante'alh. [class: sdo-n] *[FA]* [0-'alh < 'ai$_1$]

give me (st), he/she will. *V* sghadutetilh. [class: lro] [abs: d] *[FA]* [0-tilh < tan$_1$]

glass, drinking. *N* betats'utnai.

glass, drinking. *N* toobets'utnai.

gloves. *N* butsuk hooni.

glutton, he/she is a. *V* buchahooncha. *[IA]* [0-cha < cha$_1$]

go around by boat, he/she is going to. *V* nutekelh. *[FA]* [0-kelh < ke$_2$]

go around by boat, he/she is not going to. *V* nuchateske. *[FN]* [0-ke < ke$_2$]

go around by boat, he/she is not going to. *V* nuchateskel. *[FN]* [0-kel < ke$_2$]

go around by boat, they are not going to. *V* nuchahateskel. *[FN]* [0-kel < ke$_2$]

go around by boat, they are not going to. *V* nucha:teske. *[FN]* [0-ke < ke$_2$]

go around by boat, they are not going to. *V* nucha:teskel. *[FN]* [0-kel < ke$_2$] [old-fashioned]

go around in a boat, I am not going to. *V* nuchatuzeske. *[FN]* [0-ke < ke$_2$]

go around in a boat, we (2) may. *V* noduke. *[OA]* [0-ke < ke$_2$]

go around in a boat, we (2) may not. *V* nuchasoduke. *[ON]* [0-ke < ke_2]

go by boat, I will. *V* teskelh. *[FA]* [0-kelh < ke_2]

go by boat, I will not. *V* chatuzeskel. *[FN]* [0-kel < ke_2]

go by boat, they will. *V* hutekelh. *[FA]* [0-kelh < ke_2]

go by boat, they will not. *V* chahuteskelh. *[FN]* [0-kelh < ke_2]

go by boat, we (2) will. *V* tadukelh. *[FA]* [0-kelh < ke_2]

go by boat, we (3+) will. *V* uztekelh. *[FA]* [0-kelh < ke_2]

go by boat, we (3+) will not. *V* chazteskel. *[FN]* [0-kel < ke_2]

go by plane, he/she will. *V* bulh 'utet'ah. *[FA]* [0-t'ah < $t'oh_2$]

go by plane, I will. *V* sulh 'utet'ah. *[FA]* [0-t'ah < $t'oh_2$]

go by plane, I will not. *V* sulh cha'test'ah. *[FN]* [0-t'ah < $t'oh_2$]

go by plane, they will. *V* hubhulh 'utet'ah. *[FA]* [0-t'ah < $t'oh_2$]

go by plane, we (2) will. *V* nawhulh 'utet'ah. *[FA]* [0-t'ah < $t'oh_2$]

go to the washroom, you (1) will. *V* 'az natanda. *[FA]* [0-da]

go to war, we (3+) may. *V* ts'utooba. *[OA]* [0-ba < ba_2]

goat. *N* dube.

God. *N* Yak'usda.

God. *N* Yoodughi.

going after (hunting, trapping), he/she is. *V* ka'uneh. *[IA]* [0-neh]

going after it, I am. *V* kwatesya. *[PA]* [0-ya]

going after it, they (3+) are. *V* hikatezdil. *[PA]* [0-dil < dil_1]

going after it, we (2) are. *V* kwatet'az. *[PA]* [0-'az < $'as_1$]

going after it, we (3+) are. *V* kwaztezdel. *[PA]* [0-del < dil_1]

going after it in a car, I am. *V* kwatesgoo. *[PA]* [0-goo < goo_1]

going after it in a car, we (2) are. *V* kwatedugoo. *[PA]* [0-goo < goo_1]

going after it in a car, we (3+) are. *V* kwaztezgoo. *[PA]* [0-goo < goo_1]

going around by boat, I am. *V* nuske. *[IA]* [0-ke < ke_2]

going around by sleigh, they are. *V* ne'hulooz. *[IA]* [0-looz < $looz_1$]

going around talking loudly, he/she is. *V* nuduldlut. *[IA]* [l-dlut]

going back by boat, he/she is. *V* nadukelh. *[IA–progressive]* [d-kelh < ke_2]

going back by boat, I am. *V* nasdukelh. *[IA–progressive]* [d-kelh < ke_2]

going back by boat, they are. V nahedukelh. *[IA–progressive]* [d-kelh < ke_2]

going back by boat, we (2) are. *V* naidukelh. *[IA–progressive]* [d-kelh < ke_2]

going back by boat, we (3+) are. *V* nats'edukelh. *[IA–progressive]* [d-kelh < ke_2]

going back by boat, you (1) are. *V* naindukelh. *[IA–progressive]* [d-kelh < ke_2]

going back by boat, you (2+) are. *V* nahdukelh. *[IA–progressive]* [d-kelh < ke_2]

going by boat, he/she is. *V* ukelh. *[IA–progressive]* [0-kelh < ke$_2$]

going by boat, I am. *V* uskelh. *[IA–progressive]* [0-kelh < ke$_2$]

going by boat, I am not going. *V* chaseskelh. *[IN–prog]* [0-kelh < ke$_2$]

going by boat, they are. *V* hekelh. *[IA–progressive]* [0-kelh < ke$_2$]

going by boat, they are not. *V* chaheskelh. *[IN–prog]* [0-kelh < ke$_2$]

going by boat, we (2) are. *V* idukelh. *[IA–progressive]* [0-kelh < ke$_2$]

going by boat, we (3+) are. *V* ts'ekelh. *[IA–prog]* [0-kelh < ke$_2$]

going by boat, we (3+) are not. *V* chats'eskelh. [*IN–prog]* [0-kelh < ke$_2$]

going by boat, you (1) are. *V* inkelh. *[IA–progressive]* [0-kelh < ke$_2$]

going by boat, you (2+) are. *V* uhkelh. *[IA–progressive]* [0-kelh < ke$_2$]

going by plane, he/she is. *V* bulh 'et'ah. *[IA–prog]* [0-t'ah < t'oh$_2$]

going by plane, I am. *V* sulh 'et'ah. *[IA–prog]* [0-t'ah < t'oh$_2$]

going by plane, I am not. *V* sulh cha'est'ah. *[IN–prog]* [0-t'ah < t'oh$_2$]

going by plane, they are. *V* hubulh 'et'ah. *[IA–prog]* [0-t'ah < t'oh$_2$]

going by plane, we (2) are. *V* nawhulhutet'ah. *[IA–prog]* [0-t'ah < t'oh$_2$]

Golden Eagle. [Aquila chrysaetos] *N* sjul.

gone by boat in a loop, I have. *V* nususki. *[PA]* [0-ki < ke$_2$]

gone in a loop by boat, they have. *V* nuhuzki. *[PA]* [0-ki < ke$_2$]

gone in a loop by boat, they have not. *V* nuchahikel. *[PN]* [0-kel < ke$_2$]

gone in a loop by boat, we (3+) have. *V* nuts'uzki. *[PA]* [0-ki < ke$_2$]

gone in a loop by plane, he/she has. *V* bulh ne'uzt'a. *[PA]* [0-t'a < t'oh$_2$]

gone in a loop by plane, I have. *V* sulh ne'uzt'a. *[PA]* [0-t'a < t'oh$_2$]

gone in a loop by plane, they have. V hubulh ne'uzt'a. *[PA]* [0-t'a < t'oh$_2$]

gone in a loop by plane, we (2) have. *V* nawhulh ne'uzt'a. *[PA]* [0-t'a < t'oh$_2$]

good, he/she is. *V* nzoo. *[IA]* [0-zoo < zoo$_1$]

good, I am. *V* uszoo. *[IA]* [0-zoo < zoo$_1$]

good, it (weather) is. *V* whudinzoo. [abs: wh] *[IA]* [0-zoo < zoo$_1$]

good, it is. *V* dinzoo. [abs: d] *[IA]* [0-zoo < zoo$_1$]

good, it is. *V* ninzoo. [abs: n] *[IA]* [0-zoo < zoo$_1$]

good, it is. *V* hoonzoo. [abs: wh] *[IA]* [0-zoo < zoo$_1$]

good, it is not. *V* chadizoo. [abs: d] *[IN]* [0-zoo < zoo$_1$]

good, it is not. *V* chahoozoo. [abs: wh] *[IN]* [0-zoo < zoo$_1$]

good, it is not. *V* chaizoo. [IN] [0-zoo < zoo$_1$]

good, it is not. *V* chanizoo. [abs: n] *[IN]* [0-zoo < zoo$_1$]

good, it tastes. *V* lhuk'i. *[IA]* [0-k'i]

good, they are. *V* hinzoo. *[IA]* [0-zoo < zoo_1]

good, they are not. *V* chahizoo. *[IN]* [0-zoo < zoo_1]

good, we (2) are. *V* idudzoo. *[IA]* [0-zoo < zoo_1]

good, we (3+) are. *V* ts'inzoo. *[IA]* [0-zoo < zoo_1]

good, you (1) are. *V* inzoo. *[IA]* [0-zoo < zoo_1]

good, you (2+). *V* uhzoo. *[IA]* [0-zoo < zoo_1]

good-natured, he/she is. *V* budzi nzoo. *[IA]* [0-zoo < zoo_1] Literally, "his heart is good".

goose. [Branta canadensis] *N* khoh.

goose, Canada. [Branta canadensis] *N* khah.

gooseberry, black. [Ribes oxycanthoides] *N* 'indawuz.

gosling. *N* khahyaz.

grade (st:road), he/she will. *V* nawtezo. [abs: wh] *[FA]* [0-zo]

grading (st:road), he/she is. *V* nawhuzo. [abs: wh] *[IA]* [0-zo]

grammar. *N* beyatuk.

grandchild. *N* -chai. Duoplural: -chaike

grandchild, his/her. *N* buchai.

grandchild, my. *N* schai.

grandchild, our. *N* nechai.

grandchild, their. *N* hubuchai.

grandchild, your (1). *N* nchai.

grandchild of first cousin. *N* -chai. Duoplural: -chaike

grandchild of sibling. *N* -chai. Duoplural: -chaike

grandchildren, his/her. *N* buchaike.

grandchildren, my. *N* schaike.

grandchildren, our. *N* nechaike.

grandchildren, their. *N* hubuchaike.

grandchildren, your (1). *N* nchaike.

grandfather. *N* 'utsiyan.

grandfather. *N* -tsiyan.

grandfather, his/her. *N* butsiyan.

grandfather, his/her younger. *N* butsiyanyaz.

grandfather, my. *N* stsiyan.

grandfather, my younger. *N* stsiyanyaz.

grandfather, our. *N* netsiyan.

grandfather, our younger. *N* netsiyanyaz.

grandfather, their. *N* hubutsiyan.

grandfather, their younger. *N* hubutsiyanyaz.

grandfather, younger. *N* -tsiyanyaz.

grandfather, your (1). *N* ntsiyan.

grandfather, your (1). *N* ntsiyanyaz.

grandmother, his/her maternal. *N* butsoo.

grandmother, his/her younger. *N* butsooyaz.

grandmother, maternal. *N* 'utsoo.

grandmother, maternal. *N* -tsoo.

grandmother, my maternal. *N* stsoo.

grandmother, my younger. *N* stsooyaz.

grandmother, our maternal. *N* netsoo.

grandmother, our younger. *N* netsooyaz.

grandmother, paternal. *N* 'uts'un.

grandmother, paternal. *N* -ts'un.

grandmother, their maternal. *N* hubutsoo.

grandmother, their younger. *N* hubutsooyaz.

grandmother, younger. *N* -tsooyaz.

grandmother, your (1) younger. *N* ntsooyaz.

grasping, he/she is. *V* whudzaih. *[IA]* [0-dzaih]

grass. *N* tl'o.

grasshopper. *N* tulk'us.

grave. *N* dune 'adilhti.

gravel. *N* tseyaz.

graveyard. *N* ts'unk'ut.

grease. *N* khe. Poss: ghe.

grease, black bear. *N* susghe.

grease, my. *N* sghe.

grease, oolichan. *N* sleghe.

greasing, he/she is. *V* ulhlhah. *[IA]* [lh-tlah]

greasy, it is. *V* khe unli. *[IA]* [0-li]

greater than you (1). *PPC* nyanus.

great-grandchild. *N* -chaicho.

great-grandchild. *N* -dulhchai.

great-grandchild, his/her. *N* buchaicho.

great-grandchild, my. *N* schaicho.

great-grandchild, our. *N* nechaicho.

great-grandchild, their. *N* hubuchaicho.

great-grandchild, your (1). *N* nchaicho.

great-grandfather. *N* -'lunatdultsiyan.

great-grandfather. *N* 'utsiyancho.

great-grandfather. *N* -tsiyancho.

great-grandfather, his/her. *N* be'lunatdultsiyan.

great-grandfather, his/her. *N* butsiyancho.

great-grandfather, my. *N* se'lunatdultsiyan.

great-grandfather, my. *N* stisyancho.

great-grandfather, our. *N* ne'lunatdultsiyan.

great-grandfather, our. *N* netsiyancho.

great-grandfather, their. *N* hube'lunatdultsiyan.

great-grandfather, their. *N* hubutsiyancho.

great-grandfather, your (1). *N* ntsiyancho.

great-grandfather, your (1). *N* nye'lunatdultsiyan.

great-grandmother, his/her maternal. *N* butsoocho.

great-grandmother, maternal. *N* -tsoocho.

great-grandmother, my maternal. *N* stsoocho.

great-grandmother, our maternal. *N* netsoocho.

great-grandmother, paternal. *N* 'uts'uncho.

great-grandmother, paternal. *N* -ts'uncho.

great-grandmother, their maternal. *N* hubutsoocho.

great-great-grandfather. *N* whulhtat-lunatdultsiyan.

great-great-grandfather, his/her. *N* whulhtatbulunatdultsiyan.

great-great-grandfather, my. *N* whulhtatselunatdultsiyan.

great-great-grandfather, our. *N* whulhtatnelunatdultsiyan.

great-great-grandfather, their. *N* whulhtathubalunatdultsiyan.

great-great-grandmother. *N* whulhtatlunatdultsoo.

great-great-grandmother, his/her. *N* whulhtatyubulunatdultsoo.

great-great-grandmother, my. *N* whulhtatselunatdultsoo.

great-great-grandmother, our. *N* whulhtatnelunatdultsoo.

great-great-grandmother, their. *N* whulhtathubelunatdultsoo.

Green Alder. [Alnus viridis] *N* k'us.

green, it is. *V* t'an dohot'en. [abs: wh] *[IA]* [d-'en < 'en1]

green, it is. *V* t'an dot'en. *[IA]* [d-'en < 'en1]

green, it is. *V* t'an doonat'en. [abs: n] *[IA]* [d-'en < 'en1]

green moss (on ground or tree trunks). *N* 'uyemba.

Greer Mountain. *N* Deduk Usk'en.

grey, it is. *V* dulgi. *[IA]* [l-gi < gi1]

grey, it is. *V* whudulgi. [abs: wh] *[IA]* [l-gi < gi1]

Grey Jay. [Perisoreus canadensis] *N* gwuzih.

grinding (st), he/she is. *V* dunulhdus. *[IA]* [lh-dus]

grindstone. *N* tsek'az.

grizzly bear. [Ursus arctos] *N* shas. Poss: shas.

grizzly bear, my. *N* sushas.

groin. *N* -tl'et.

groin. *N* -tl'it.

groin, his/her. *N* butl'ɛt.

groin, my. *N* stl'et.

groin, your (1). *N* ntl'et.

groins, our. *N* netl'et.

groins, their. *N* hubutl'et.

ground. *N* yun.

grouse, ruffed. [Bonasa umbellus] *N* 'utsut.

Grouse, Spruce. [Falcipennis canadensis] *N* nat'ah.

grizzly bear. [Ursus arctos] *N* shas.

growing, he/she (child) is. *V* nuyeh. *[IA]* [0-yeh]

growing, it (plant) is. *V* hanuyeh. *[IA]* [0-yeh]

growing (st:plant), he/she is. *V* hanulhyeh. *[IA]* [lh-yeh]

growing back, it (plant) is. *V* hananulyeh. *[IA]* [l-yɛh]

grows out of it, it. *V*

buts'ahainalhyi. *[IA]* [l-yi] E.g. a limb from the trunk of a tree.

guard, security. *N* huwunli.

guarding (s.o.), I am. *V* ghusli. *[IA]* [0-li]

guarding (s.o.), we (2) are. *V* ghidudli. *[IA]* [0-li]

guarding (s.o.), we (3+) are. *V* ghuts'inli. *[IA]* [0-li]

guarding him/her, he/she is. *V* yughunli. *[IA]* [0-li]

guarding him/her, they are. *V* highunli. *[IA]* [0-li]

guilty, he/she feels. *V* tinch'a'unt'en. *[IA]* [d-'en < 'en$_2$]

gum, chewing. *N* dzeh.

gums. *N* -ghootsun.

gums, his/her. *N* bughootsun.

gums, my. *N* sghootsun.

gums, our. *N* neghootsun.

gums, their. *N* hubughootsun.

gums, your (1). *N* nghootsun.

gushes, it. *V* tujulh. *[IA]* [d-yulh < yul$_3$]

guts. *N* -ts'ik.

guts, fish. *N* lhooktsik.

hailing, it is. *V* 'inlootsan nanukat. *[IA]* [0-kat < kat1]

hailstone. *N* 'inlootsan.

hair. *N* 'ugha.

hair, facial. *N* -dagha.

hair, his/her head. *N* butsigha.

hair, his/her own. *N* dugha.

hair, head. *N* 'utsigha.

hair, my head. *N* stsigha.

hair, our head. *N* netsigha.

hair, their head. *N* hubutsigha.

hair, your (1) head. *N* ntsigha.

hair braid. *N* -tsigha lhu'ool.

hair of body. *N* -gha.

hair of head. *N* -tsigha.

hair oil. *N* netsigha bets'ulha.

hairy, he/she is. *V* lhghai. *[IA]* [lh-ghai < ghai$_1$]

hairy, he/she is. *V* dughai. [abs: 0] *[IA]* [0-ghai < ghai$_1$]

hairy, it is. *V* dunughai. [abs: n] *[IA]* [0-ghai < ghai$_1$]

hairy, it is. *V* whudughai. [abs: wh] *[IA]* [0-ghai < ghai$_1$]

hairy, they are. *V* hudughai. *[IA]* [0-ghai < ghai$_1$]

hairy, we (2) are. *V* didughai. *[IA]* [0-ghai < ghai$_1$]

half. *ADV* lhulcho.

half full, it is. *V* lhulcho dezbun. *[IA]* [0-bun < bun$_1$]

half of it. *N* buk'uz.

halfway. *ADV* lhulcho.

hamburger soup. *N* 'utsun tazul.

hammer. *N* be'ul'uz.

hammering (u.o.), he/she is. *V* 'unulh'uz. *[IA]* [lh-'uz < tl'es$_1$]

hammering it, he/she is. *V* yunulh'uz. *[IA]* [lh-'uz < tl'es$_1$]

hand, back of. *N* -lat'ak.

hand, palm of. *N* -lak'et.

hand, the back of his/her. *N* bulat'ak.

hand, the back of my. *N* slat'ak.

hand, the back of your (1). *N* nlat'ak.

handkerchief. *N* dzezoh.

handle. *N* ti.

handle, its. *N* buti.

handling, he/she is. *V* nunulle. [class: mdo] [abs: n] *[IA]* [l-le]

hands, the backs of our. *N* nelat'ak.

hands, the backs of their. *N* hubulat'ak.

handsome, he/she is. *V* dunezoo'. *[IA]* [0-zoo'] Applicable only to men.

hangover, he/she has a. *V* nedotoo yundulhda. *[IA]* [lh-da] Literally, "liquor has made him sick".

happened, it. *V* 'uhooja. *[PA]* [0-ja]

happened, it has not. *V* nduchahoonil. *[PN]* [0-nil]

happened, something bad. *V* dzah 'uhooja. *[PA]* [0-ja]

happened to you (1), something. *V* 'inja. *[PA]* [0-ja]

happened to him/her, something. *V* 'uja. *[PA]* [0-ja]

happening, it is. *V* nduwhuneh. *[IA]* [0-neh]

happy, he/she is. *V* hoont'i. *[IA]* [0-t'i]

happy, I am. *V* hoonust'i. *[IA]* [0-t'i]

happy, they are. *V* huhoont'i. *[IA]* [0-t'i]

happy, we (3+) are. *V* ts'uhoont'i. *[IA]* [0-t'i]

hard, it became. *V* duts'un suli. [abs: 0] *[PA]* [0-li]

hard, it became. *V* whuduts'un suli. [abs: wh] *[PA]* [0-li]

hard, it is. *V* dunuts'un. [abs: n] *[IA]* [0-ts'un < $ts'un_1$]

hard, it is. *V* duts'un. [abs: 0] *[IA]* [0-ts'un < $ts'un_1$]

hard, it is. *V* huwawhulna. *[IA]* [l-na < na_2]

hard, it is. *V* whuduts'un. [abs: wh] *[IA]* [0-ts'un < $ts'un_1$]

hard for him/her, it is too. *V* yughulait'ah. *[IA]* [0-t'ah]

hardened, it. *V* duts'un suli. [abs: 0] *[PA]* [0-li]

hardened, it. *V* whuduts'un suli. [abs: wh] *[PA]* [0-li]

hare. [Lepus americanus pal.] *N* gah. The only species native to the territory is the Snowshoe Hare, also known as Varying Hare, Lepus americanus. The term is, however, applied to other varieties of rabbit and hare.

hare, snowshoe. [Lepus americanus pal.] *N* gah. The only species native to the territory is the Snowshoe Hare, also known as Varying Hare, Lepus americanus. The term is, however, applied to other varieties of rabbit and hare.

hare, varying. [Lepus americanus pal.] *N* gah. The only species native to the territory is the Snowshoe Hare, also known as Varying Hare, Lepus americanus. The term is, however, applied to other varieties of rabbit and hare.

harvest, I. *V* hanusle. *[IA]* [0-le]

harvest, we (3+). *V* hats'unule. *[IA]* [0-le]

harvest (u.o.), I. *V* ha'nusle. *[IA]* [0-le]

harvest (u.o.), they. *V* ha'hunule.

[IA] [0-le]

harvest (u.o.), we (2). *V* ha'nidulye. *[IA]* [0-le]

harvest (u.o.), we (3+). *V* ha'ts'unule. *[IA]* [0-le]

harvests (u.o.), he/she. *V* ha'nule. *[IA]* [0-le]

hat. *N* ts'ah.

hat. *N* ts'oh.

hat, cowboy. *N* ts'ahbul.

hat, his/her cowboy. *N* buts'ahbul.

hat, my cowboy. *N* sts'ahbul.

hat, our cowboy. *N* nets'ahbul.

hat, their cowboy. *N* hubuts'ahbul.

hat, wide-brimmed. *N* ts'ahbul.

hat, your (1) cowboy. *N* nts'ahbul.

have many, they have. *V* huldlai. *[IA]* [d-lai]

hawk. *N* dut'ai ooltas.

hawk. *N* ligok hoolht'as.

hay. *N* tl'o.

haystack. *N* lubuyos.

he. *PRO* 'en.

he. *PRO* didut.

head. *N* -tsi.

head, back of. *N* -tsit'ak.

head, back of. *N* -tsitl'ah.

head, back of the. *N* -tsint'ak.

head, fish. *N* lhooktsi.

head, his/her. *N* butsi.

head, my. *N* stsi.

head, the back of your (1). *N* ntsint'ak.

head, the back of your (1). *N* ntsitl'ah.

head, the back of his/her. *N* butsint'ak.

head, the back of his/her. *N* butsit'ak.

head, the back of his/her. *N* butsitl'ah.

head, the back of my. *N* stsint'ak.

head, the back of my. *N* stsit'ak.

head, the back of my. *N* stsitl'ah.

head, the back of your (1). *N* ntsit'ak.

head, the top of his/her. *N* butsidah.

head, the top of my. *N* stsidah.

head, the top of your (1). *N* ntsidah.

head, top of. *N* -tsidah.

head, your (1). *N* ntsi.

headache, I have a. *V* stsi nduda. *[IA]* [0-da < da_2] Literally, "my head is sore".

headache medicine. *N* netsiyoo.

headlight. *N* -na.

heads, our. *N* netsi.

heads, the backs of ours. *N* netsint'ak.

heads, the backs of ours. *N* netsit'ak.

heads, the backs of ours. *N* netsitl'ah.

heads, the backs of their. *N* hubutsint'ak.

heads, the backs of their. *N* hubutsit'ak.

heads, the backs of their. *N* hubutsitl'ah.

heads, the tops of our. *N* netsidah.

heads, the tops of their. *N* hubutsidah.

heads, their. *N* hubutsi.

healing, it is. *V* najih. *[IA]* [d-yih]

health centre. *N* yoobayah.

heart. *N* -dzi.

heart, his/her. *N* budzi.

heart, my. *N* sdzi.

heart, your (1). *N* ndzi.

hearth. *N* tseba.

hearts, our. *N* nedzi.

hearts, their. *N* hubudzi.

heat. *N* sazul.

heat waves. *N* sul.

heater. *N* stocho.

heater. *N* beoonezul.

heater, his/her. *N* bustocho.

heater, my. *N* sestocho.

heater, our. *N* nestocho.

heater, their big. *N* hubustocho.

heater, your (1). *N* nyestocho.

heats (st) up, he/she. *V* nulhghaz. *[IA]* [lh-ghaz < $ghaz_1$]

heats (st), he/she. *V* whunulhghaz. [abs: wh] *[IA]* [lh-ghaz < $ghaz_1$]

heaven. *N* yak'uz.

heavy, it is. *V* ndaz. *[IA]* [0-daz < daz_1]

heel. *N* -kelatsul.

heel. *N* -kentsul.

heel. *N* -ketsul.

heel, his/her. *N* bukelatsul.

heel, his/her. *N* bukentsul.

heel, my. *N* skelatsul.

heel, my. *N* skentsul.

heel, your (1). *N* nkelatsul.

heel, your (1). *N* nkentsul.

heels, his/her. *N* buketsul.

heels, my. *N* sketsul.

heels, our. *N* nekelatsul.

heels, our. *N* nekentsul.

heels, our. *N* neketsul.

heels, their. *N* hubukelatsul.

heels, their. *N* hubukentsul.

heels, their. *N* hubuketsul.

heels, your (1). *N* nketsul.

held, I am being. *V* soontun. *[IA]* [0-tun]

helicopter. *N* kw'ut nabadunelgih.

hell. *N* kwunchoyuk.

hello. *INT* hadih.

hello!. *INT* hadi'. An expression of surprise.

helped me, you (2+) have. *V* snachalhuya. *[PA]* [l-ya]

helped us, you (1) have. *V* nenachailya. *[PA]* [l-ya]

helping. *PP* la.

helping me. *PPC* sla.

helping them. *PPC* hubula.

her. *PRO* 'en.

her. *PRO* didut.

herbal medicine. *N* duchunyoo.

here. *ADV* njan.

here you are. *INT* nah. This is what you say to someone as you hand him or her something.

hereditary chief. *N* keyah whuduchun.

heron, blue. *N* tehhoongook.

herring eggs. *N* chilchunk'oon.

hers. *PRO* buch'i'.

hide. *N* -zuz.

hide. *N* 'uzuz.

hide, his/her. *N* buzuz.

hide, moose. *N* dunizuz.

hide, my. *N* szuz.

hide, our. *N* nezuz.

hide, their. *N* hubuzuz.

hide, your (1). *N* nzuz.

highway. *N* ti.

High-Bush Blueberries. [Vaccinium ovalifolium] *N* yalhtsul.

hillside. *N* whenun.

him. *PRO* 'en.

him. *PRO* didut.

himself. *PRO* dich'ah.

hipbone, his/her. *N* buk'its'un.

hipbone, my. *N* sk'its'un.

hipbone, your (1). *N* nk'its'un.

hipbones, our. *N* nek'its'un.

hipbones, their. *N* hubuk'its'un.

his. *PRO* buch'i'.

history. *N* 'udada.

hobo-ing around, he/she is. *V* nuduzas. *[IA]* [0-zas]

hold us, they. *V* newoontun. *[IA]* [0-tun]

holding me, they are. *V* suhoontun. *[IA]* [0-tun]

holds us, he/she. *V* nehoontun. *[IA]* [0-tun < tun_1]

hollering, he/she is. *V* dulghi. *[IA]* [l-ghi]

Holy Spirit. *N* Ndoni.

home. *N* yah.

home, at. *ADV* koonk'et.

home village. *N* kooz.

home village, your (1). *N* nkooz.

honest, he/she is. *V* ts'ihnus hoolh'i. *[IA]* [lh-'i]

honey. *N* hoolht'oghe.

honour (s.o.), I. *V* delhti. *[IA]* [lh-ti < ti_2]

honour (s.o.), we (2). *V* dedulti. *[IA]* [lh-ti < ti_2]

honour (st), we (3+). *V* uzdelhti. *[IA]* [lh-ti < ti_2]

honour him/her, they. *V* hidelhti. *[IA]* [lh-ti < ti_2]

honour them, we (2). *V* hubudedulti. *[IA]* [lh-ti < ti_2]

honour them, we (3+). *V* hubuzdelhti. *[IA]* [lh-ti < ti_2]

honours them, he/she. *V* hubudelhti. *[IA]* [lh-ti < ti_2]

hooked (fish), we (2). *V* seduljus. *[PA]* [lh-jus < jas_2]

hooked (fish), we (3+). *V* ts'ulhjus. *[PA]* [lh-jus < jas_2]

heron, blue. *N* tehhoongook.

hooked (st: fish), I. *V* sulhjus. *[PA]* [lh-jus < jas_2]

hooked it (fish), he/she. *V* yulhjus. *[PA]* [lh-jus < jas_2]

hooked it (fish), they. *V* huyulhjus. *[PA]* [lh-jus < jas_2]

hoop and holler so that the mountains resound, they. *V* tsehudil-hts'ai. *[IA]* [lh-ts'ai]

horizon. *N* yaba.

horn. *N* -de'.

hornet. *N* ts'ihna.

horns. *N* 'ude. Indefinitely possessed form.

horse. *N* yeztli.

horse, his/her. *N* buyeztli.

horse, his/her own. *N* duyeztli.

horse, my. *N* syeztli.

horse, our. *N* neyeztli.

horse, their. *N* hubuyeztli.

horse, your (1). *N* nyeztli.

horsey. *N* hoot'as.

hot, it has gotten. *V* nawnilghaz. *[PA]* [l-ghaz < $ghaz_1$]

hot, it is. *V* nilghaz. *[IA]* [l-ghaz < $ghaz_1$]

hot, it is. *V* nulwus. *[IA]* [l-wus]

hot, it is. *V* whunilghaz. *[IA]* [l-ghaz < $ghaz_1$]

hot, it will start to get. *V* nawntelwus. *[FA]* [l-wus]

hot, they are. *V* hunulwus. *[IA]* [l-wus]

hot, we (3+) are. *V* ts'unulwus. *[IA]* [l-wus]

hot, you (1) are. *V* nilwus. *[IA]* [l-wus]

hot and sunny, it is. *V* sa dendi. *[IA]* [0-di]

hotel. *N* sdutez.

house. *N* koo.

house. *N* yah.

house, his/her. *N* bukoo.

house, his/her. *N* buyah.

house, my. *N* skoo.

house, my. *N* syah.

house, our. *N* nekoo.

house, their. *N* hubukoo.

house, their. *N* hubuyah.

house, your (1). *N* nyah.

house, your (1). *N* nkoo.

how are they?. *V* dahint'ah. *[IA]* [0-t'ah < $t'oh_1$]

how are you (1)?. *V* daint'ah. *[IA]* [0-t'ah < $t'oh_1$]

how big is it?. *V* dalcho. *[IA]* [l-cho < cha_1]

how cold is it?. *V* dahoolk'uz. *[IA]* [l-k'uz]

how far?. *V* dahooldzoh. *[IA]* [l-dzoh]

how is he?. *V* dant'ah. *[IA]* [0-t'ah < $t'oh_1$]

how is it?. *V* dahot'e. *[OA]* [0-t'e < $t'oh_1$]

how is it?. *V* dahoont'ah. *[IA]* [0-t'ah < $t'oh_1$]

how long a time?. *V* dahooldzoh. *[IA]* [l-dzoh]

how long is it?. *V* dalyiz. *[IA]* [l-yiz]

how many (years of age). *QWH* nts'oh.

how many of them?. *V* dahuneltsuk. *[IA]* [l-tsuk]

how many of us?. *V* dazneltsuk. *[IA]* [l-tsuk]

how many of you?. V danelhutsuk. *[IA]* [l-tsuk]

how much?. *V* daltsuk. *[IA]* [l-tsuk]

how pretty is it?. *V* daldzoo. *[IA]* [l-d-zoo < zoo$_1$]

how?. *QWH* nts'ezun'a

huckleberries. [Vaccinium membranaceum] *N* tsulhcho.

humble, he/she is. *V* yuk 'ududildzun. *[IA]* [l-d-zun]

hummingbird. *N* ts'unulhduz.

hunchback. *N* 'uk'o'.

hunchback. *N* k'o'.

hunger. *N* dai.

hungry, he/she is. *V* oodaningi. *[IA]* [0-gi]

hungry, I am. *V* chawhuszum. *[IA]* [0-zum]

hungry, I am. *V* schahoonzum. *[IA]* [0-zum]

hungry, I am. *V* sdaningi. *[IA]* [0-gi]

hungry, I am. *V* sye'elts'ul. *[IA]* [l-ts'ul]

hungry, they are. *V* chahuhoonzum. *[IA]* [0-zum]

hungry, we (2) are. *V* chahoodudzum. *[IA]* [0-zum]

hungry, we (3+) are. *V* chats'uhoonzum. *[IA]* [0-zum]

hungry, you (1) are. *V* chahoonzum. *[IA]* [0-zum]

hungry, you (1) are. *V* nye'elts'ul. *[IA]* [l-ts'ul]

hunt (st), they will. *V* ka'hutet'ilh. *[FA]* [d-'ilh < 'en$_2$]

hunt (st), we (2) will. *V* ka'tadut'ilh. *[FA]* [d-'ilh < 'en$_2$]

hunt (st), we (3+) will. *V* ka'uztet'ilh. *[FA]* [d-'ilh < 'en$_2$]

hunt (st), you (1). *V* ka'tant'ilh. *[FA]* [d-'ilh < 'en$_2$]

hunt around (in the bush), I am going to. *V* nutelhudzulh. *[FA]* [l-dzulh]

hunted (st), they. *V* ka'het'en. *[PA]* [d-'en < 'en$_2$]

hunted (st), we (2). *V* ka'adut'en. *[PA]* [d-'en < 'en$_2$]

hunted (st), we (3+). *V* ka'ts'et'en. *[PA]* [d-'en < 'en$_2$]

hunted (st), you (1). *V* ka'ant'en. *[PA]* [d-'en < 'en$_2$]

hunting, I am going. *V* nutesdzulh. *[IA]* [l-dzulh]

hunting, I am going to go. *V* telhudzulh. *[FA]* [l-dzulh]

hunting (st), they are. *V* ka'hut'en. *[IA]* [d-'en < 'en$_2$]

hunting (st), we (2) are. *V* ka'idut'en. *[IA]* [d-'en < 'en$_2$]

hunting (st), we (3+) are. *V* ka'ts'ut'en. *[IA]* [d-'en < 'en$_2$]

hunting (st), you (1) are. *V* ka'int'en. *[IA]* [d-'en < 'en$_2$]

husband. *N* -kui.

husband, his/her. *N* bukui.

husband, my. *N* skui.

husbands. *N* -kuike.

husbands, our. *N* nekuike.

husbands, their. *N* hubukuike.

I. *PRO* si.

ice, a piece of. *N* lhum.

ice over a surface. *N* tun. This describes ice on a lake or road as opposed to pieces of ice.

icicle. *N* kw'uzghoo.

ignite, you (1). *V* nadilhk'aih. *[IA]* [lh-k'aih < k'aih$_1$]

ignite again, you (1). *V* nakwundilhk'aih. *[IA]* [lh-k'aih]

illness. *N* dada.

in aid of. *PP* la.

in aid of me. *PPC* sla.

index finger, my. *N* slasgek.

Indian Hellebore. [Veratrum viride] *N* whulhdulh. Locally known as "ginseng", but not at all the same plant.

Indian ice cream. [Shepherdia canadensis] *N* nawus.

Indian, non-Athabaskan. *N* 'utna. Especially, but not exclusively, those of the West Coast.

Indian Paint Brush. [Castilleja miniata] *N* ditnikwun.

Indigenous person. *N* yunkaot'en.

Indigenous person. *N* yunkawhut'en.

Indian popcorn. *N* k'ohch'az.

intestine, his/her large. *N* buts'ikcho.

intestine, large. *N* -ts'ikcho.

intestine, my large. *N* sts'ikcho.

intestine, your (1) large. *N* nts'ikcho.

intestines. *N* -ts'ik.

intestines, his/her. *N* buts'ik.

intestines, my. *N* sts'ik.

intestines, our. *N* nets'ik.

intestines, our large. *N* nets'ikcho.

intestines, their. *N* hubuts'ik.

intestines, their large. *N* hubuts'ikcho.

intestines, your (1). *N* nts'ik.

is, he/she. *V* unli. *[IA]* [0-li < li$_3$]

is, it. *V* 'int'ah. *[IA]* [0-t'ah]

is not, it. *V* chailah. *[IN]* [0-lah]

island. *N* noo.

its. *PRO* buch'i'.

jack pine. [Pinus contorta latifolia] *N* chundoo. Locally known as "**jack pine**", which standardly refers to Pinus banksiana.

jacket, skin. *N* 'uzuz dzoot.

jam. *N* mai be'ulha.

jam. *N* mai nesdliz.

jam. *N* mai nudutleh.

January. *N* Sachonun.

Jay, Canadian. [Perisoreus canadensis] *N* goozeh. Also known as Gray Jay and Canadian Jay.

Jay, Canadian. [Perisoreus canadensis] *N* gwuzih.

Jay, Gray. [Perisoreus canadensis] *N* goozeh. Also known as Gray Jay and Canadian Jay.

Jay, Grey. [Perisoreus canadensis] *N* gwuzih.

jealous, he/she is. *V* 'oolnih. *[IA]* [l-nih]

jeans, blue. N doso tl'asus.

jerry can. *N* tes hoongwun.

juice, fruit. *N* maitoo.

July. *N* Shen Whuniz.

jump rope. *N* ye'ubul.

jumping, it (fish) is. *V* halghok. *[IA]*

[l-ghok < ghok$_1$]

jumps out of, he/she. *V* hallhat. *[IA]* [l-lhat]

jumps out of, he/she. *V* hallhuk. *[IA–customary]* [l-lhuk]

June. *N* Duk'ai Ooza.

juniper. [Juniperus communis] *N* datsan'algut.

kerchief. *N* dzezoh.

Ketlo. *N* Ketloh.

kettle. *N* 'oosa'.

kettle. *N* betanizilh.

keyhole. *N* luglik'et.

kick (him) up the ass, you (1). *V* tsul'aintal. *[IA]* [0-tal < tulh$_1$]

kicked (d-class object), he/she. *V* deztulh. *[PA]* [0-tulh]

kicked a hole in it, he/she. *V* yuk'untal. *[PA]* [0-tal < tulh$_1$]

kicked a hole in it, I. *V* kw'unustal. *[PA]* [0-tal < tulh$_1$]

kicked a hole in it, they. *V* hik'untal. *[PA]* [0-tal < tulh$_1$]

kicked a hole in it, we (2). *V* kw'unidutal. *[PA]* [0-tal < tulh$_1$]

kicked him/her, they. *V* huyuztulh. *[PA]* [0-tulh]

kicked him/her in the face, he/she. *V* yuneztulh. *[PA]* [0-tulh]

kicked into a hole, he/she. *V* 'a'antal. *[PA]* [0-tal < tulh$_1$]

kicked it (the door) in, he/she. *V* daidental. *[PA]* [0-tal < tulh$_1$]

kicked me in the face, he/she. *V* suneztulh. *[PA]* [0-tulh]

kicked the door in, I. *V* danustal. *[PA]* [0-tal < tulh$_1$]

kicked the door in, they. *V* dahidental. *[PA]* [0-tal < tulh$_1$]

kicked the door in, we (2). *V* danidutal. *[PA]* [0-tal < tulh$_1$]

kicked the door in, we (3+). *V* dazdental. *[PA]* [0-tal < tulh$_1$]

kicking (st), I am. *V* oostalh. *[IA]* [0-talh < tulh$_1$]

kicking (st), we (2) are. *V* oodutalh. *[IA]* [0-talh < tulh$_1$]

kicking (st), we (3+) are. *V* ts'ootalh. *[IA]* [0-talh]

kicking it, he/she is. *V* yootalh. *[IA]* [0-talh]

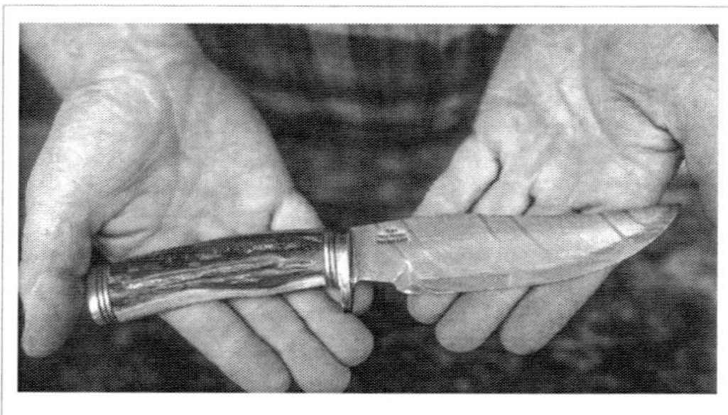

knife. *N* 'utes.

kicking it, they are. *V* huyootalh. *[IA]* [0-talh]

kidney. *N* -kw'uz.

kidneys, his/her. *N* bukw'uz.

kidneys, my. *N* skw'uz.

kidneys, our. *N* nekw'uz.

kidneys, their. *N* hubukw'uz.

kidneys, your (1). *N* nkw'uz.

kill each other, they (3+). *V* lhuhudughan. *[IA]* [d-ghan < ghan$_1$]

killed (st), I. *V* selhghi. *[PA]* [lh-ghi < ghi$_1$]

killed (st), we (2). *V* sedulghi. *[PA]* [lh-ghi < ghi$_1$]

killed by frost, it was. *V* ustun. *[PA]*

[0-tun < tun$_2$]

killed each other, they (3+). *V* lhuhedughan. *[PA]* [d-ghan < ghan$_1$]

killed it, he/she. *V* yuzelhghi. *[PA]* [lh-ghi < ghi$_1$]

killed it, they. *V* huyuzelhghi. [*PA*] [lh-ghi < ghi$_1$]

killed them (3+), he/she. *V* hubanghan. *[PA]* [0-ghan < ghan$_1$]

killed them (3+), I. *V* hubesghan. *[PA]* [0-ghan < ghan$_1$]

killed them (3+), they. *V* hubuhanghan. *[PA]* [0-ghan < ghan$_1$]

killed them (3+), we (2). *V* hubadughan. *[PA]* [0-ghan < ghan$_1$]

killed them (3+), we (3+). *V* hubuts'anghan. *[PA]* [0-ghan < ghan$_1$]

kinds, it is of (st). *V* dedah. *[IA]* [0-dah < dah$_1$]

king. *N* lelwe.

Kingfisher. *N* tehgwuzeh.

kinnickinnick leaves. *N* duniht'an.

kinnikinnick. [Arctostaphylos uva-ursi] *N* dunih.

kitchen. *N* ts'uyi ghutna-a.

kitchen, his/her. *N* buts'uyi ghutna-a.

kitchen, my. *N* sts'uyi ghutna-a.

kitchen, our. *N* nets'uyi ghutna-a.

kitchen, their. *N* hubuts'uyi ghutna-a.

kitchen, your (1). *N* nts'uyi ghutna-a.

Klez Lake. *N* Lhez Bunk'ut.

Knapp Creek. *N* Belhk'akoh.

kneading it, he/she is. *V* yunulhlhus. *[IA]* [lh-tlus]

knee. *N* -gwut.

knee, back of. *N* -gwutt'uk.

knee, his/her. *N* bugwut.

knee, my. *N* sgwut.

knee, the back of his/her. *N* bugwutt'uk.

knee, the back of my. *N* sgwutt'uk.

knee, the back of your (1). *N* ngwutt'uk.

knee, your (1). *N* ngwut.

kneecap. *N* -gwutduts'un.

kneecap. *N* -gwuttsi.

kneecap, his/her. *N* bugwutduts'un.

kneecap, my. *N* sgwuduts'un.

kneecap, my. *N* sgwuttsi.

kneecap, your (1). *N* ngwutduts'un.

kneecaps, our. *N* negwutduts'un.

kneecaps, their. *N* hubugwutduts'un.

knees, our. *N* negwut.

knees, the backs of our. *N* negwutt'uk.

knees, the backs of their. *N* hubugwutt'uk.

knees, their. *N* hubugwut.

knife. *N* 'utes.

knife. *N* tes.

knife, my. *N* se'utes.

knife, our. *N* netes.

knitting (u.o.), he/she is. *V* 'utl'oo. *[IA]* [0-tl'oo]

knitting (u.o.), I am. *V* 'ustl'oo. *[IA]* [0-tl'oo < ts'it$_1$]

knot. *N* lhuschuz.

knowledgable person. *N* moodih.

kokanee. [Oncorhynchus nerka] *N* gestl'ah.

Labrador Tea. [Rhododendron groenlandicum/Ledum groenlandicum] *N* ludi musjek.

Labrador Tea. [Rhododendron groenlandicum/Ledum groenlandicum] *N* yak'unulh'a.

ladle. *N* tazul benuduka.

lahal. *N* nehu'a.

lake. *N* bun.

lake trout. [Salvelinus namaycush] *N* bet.

Lambs Quarters. [Chenopodium album] *N* 'ut'antsi.

land in a boat, I. *V* yaskih. *[IA]* [0-kih < k$_2$]

land in a boat, they. *V* yahukih. *[IA]* [0-kih < k$_2$]

land in a boat, we (2). *V* yaidukih. *[IA]* [0-kih < k$_2$]

land in a boat, we (3+). *V* yats'ukeh. *[IA]* [0-keh < ke$_2$]

lands in a boat, he/she. *V* yakih. *[IA]* [0-kih < k$_2$]

language. *N* khunek.

language. *N* khunik. Poss: ghunik.

large, it is. *V* hooncha. [abs: wh] *[IA]* [0-cha < cha$_1$]

large, it is. *V* ncha. [abs: 0] *[IA]* [0-cha < cha$_1$]

large, it is. *V* nincha. [abs: n] *[IA]* [0-cha < cha$_1$]

large intestine. *N* -ts'ikcho.

larynges, our. *N* nezoolts'un.

larynges, their. *N* hubuzoolts'un.

larynx. *N* -yih k'ut nez'ai.

larynx. *N* -zoolts'un.

larynx, his/her. *N* buzoolts'un.

larynx, my. *N* szoolts'un.

larynx, your (1). *N* nzoolts'un.

last time. *ADV* yeda'.

last year. *N* ghada.

late, the. *ADJ* uzdani. Follows the head noun.

late (deceased). *V* inle. *[PA]* [0-le]

lawyer. *N* dune ba yalhduk.

laying eggs, it (fish) is. *V* tesjaz. *[PA]* [0-jaz]

lead (s.o.) to you (1), I will. *V* nts'untelhdlooh. *[FA]* [lh-dlooh]

leader, village, the first among equals of the clan heads. *NP* keyoh whuduchun.

leaf. *N* 'ut'an.

leaf. *N* -t'an.

learn, we (3+) will. *V* hoozdutel'eh. *[FA]* [l-'eh]

learn, you (1) will. *V* hoodutal'eh. *[FA]* [l-'eh]

learn, you (2+) will. *V* hoodutelhu'eh. *[FA]* [l-'eh]

learning, he/she is. *V* hodul'eh. *[IA]* [l-'eh < 'eh$_1$]

leather, fruit. *N* maitlus.

leech. *N* hoot'ukw. Also known as "blood sucker".

leech. *N* tehchunultsi.

left. *ADV* 'intl'us.

left by boat, they. *V* whehanki. *[PA]* [0-ki < ke_2]

left by boat, we (3+). *V* whets'anki. *[PA]* [0-ki < ke_2]

left side. *N* 'int'lus tsih.

leg. *N* -kechun.

leg, calf of. *N* -wuda.

leg, his/her. *N* bukechun.

Leg Lake. *N* Tsenabulh'a Bunk'ut. A crescent-shaped expansion of Smith Creek on the east side of Lily Lake Road, east of Chowsunkut Lake.

leg, my. *N* skechun.

leg, upper. *N* -wuz.

leg, your (1). *N* nkechun.

legend. *N* 'udada.

leggings. *N* ketsih.

legs. *N* 'ukechun.

legs, back of. *N* -kechunt'ak.

legs, our. *N* nekechun.

legs, the back of his/her. *N* bukechunt'ak.

legs, the back of my. *N* skechunt'ak.

legs, the back of our. *N* nekechunt'ak.

leaf. *N* 'ut'an.

legs, the back of their. *N* hubukechunt'ak.

legs, the back of your (1). *N* nkechunt'ak.

legs, their. *N* hubukechun.

Lejac. *N* Tseyaz.

less than. *PP* -k'elh'ih. This is the pivot for negative comparison. It attaches to the possessive prefixes.

less than me. *PPC* sk'elh'ih.

light. *N* kwun.

lightning strikes. *V* detnik oolhgis. *[IA]* [lh-gis]

like (st), I. *V* hoonust'i. [abs: wh] *[IA]* [0-t'i < $t'i_1$]

likes (st), he. *V* hoont'i. [abs: wh] *[IA]* [0-t'i < $t'i_1$]

ling cod. [Lota lota] *N* tsintel.

linoleum. *N* yun subal. Any floor covering.

lips. *N* -da.

lips, his/her. *N* buda.

lips, his/her own. *N* duda.

lips, my. *N* sda.

lips, our. *N* neda.

lips, their. *N* hubuda.

lips, your (1). *N* nda.

liquor. *N* nedotoo.

liquor, hard. *N* too ulhtus.

little. *SUFFIX* -yaz. Attaches to nouns and yields a new noun.

little bit, a. *N* ntsolyaz.

little finger. *N* -lasts'ah.

liver. *N* 'uzut.

liver. *N* -zut.

liver, his/her. *N* buzut.

liver, my. *N* szut.

liver, your (1). *N* nzut.

livers, our. *N* nezut.

livers, their. *N* hubuzut.

living room. *N* dune delts’i.

living room, his/her. *N* budune delts’i.

living room, my. *N* sdune delts’i.

living room, our. *N* nedune delts’i.

living room, their. *N* hubudune delts’i.

living room, your (1). *N* ndune delts’i.

load carried on back. *N* khelh. This refers to what one carries, not to the container.

located, he/she is. *V* usda. *[IA]* [0-da < da$_1$]

lodge, beaver. *N* tsaken.

lodge, muskrat. *N* tsek’etken.

lodgepole pine. [Pinus contorta latifolia] *N* chundoo. Locally known as “jack pine”, which standardly refers to Pinus banksiana.

long, it (generic) is. *V* nyez. *[IA]* [0-yez]

long, it is. *V* ninyez. [abs: n] *[IA]* [0-yez]

long, it is. *V* dinyiz. [abs: d] *[IA]* [0-yiz]

long, it is. *V* hoonyiz. [abs: wh] *[IA]* [0-yiz]

long ago. *ADV* ’uda.

long time, for a. *ADV* sa.

look at, you (1). *V* nilh’en. *[IA]* [lh-’en < ’en$_1$]

look at (st), we (2). *V* nidul’en. *[IA]* [l-’en]

look at it, they. *V* hinilh’en. *[IA]* [lh-’en]

look good for himself, he/she makes it. *V* dubanawhuldzit. *[IA]* [l-dzit] This could be said, for example, of someone who looks at himself in a mirror, sees that he looks bad, but tells himself he looks good. It could also be said of someone who covers up for a mistake.

looking at, I am. *V* nulh’en. *[IA]* [lh-’en < ’en$_1$]

looking at each other, they are. *V* lhahunul’en. *[IA]* [d-lh-’en < ’en$_1$]

looking at each other with admiration, they are. *V* lhahuntoolhe’en. *[IA]* [l-’en < ’en$_1$]

looking at me, he/she is. *V* sunilh’en. *[IA]* [lh-’en < ’en$_1$]

looking at me, they are. *V* sahunilh’en. *[IA]* [lh-’en < ’en$_1$]

looking at them, they are. *V* hubanilh’en. *[IA]* [lh-’en < ’en$_1$]

looking at you (1), he/she is. *V* nyunilh’en. *[IA]* [lh-’en]

looking at you (1), we (3+) are. *V* nyuznilh’en. *[IA]* [lh-’en]

looking for, you (1) are. *V* kuninta. *[IA]* [0-ta < ta$_1$]

looking for (st), I am. *V* hukwunusta. [abs: wh] *[IA]* [0-ta < ta$_1$]

looking for (st), you (1) are. *V* hukwuninta. [abs: wh] *[IA]* [0-ta < ta$_1$]

looking for him/her, he/she is. *V* yukunuta. *[IA]* [0-ta < ta$_1$]

looking for him/her, you (1) are. *V* bukuninta. *[IA]* [0-ta < ta_1]

looking for me, he/she is. *V* skunuta. *[IA]* [0-ta < ta_1]

looking for me, they are. *V* skanuta. *[IA]* [0-ta < ta_1]

looking for them, he/she is. *V* hubukanuta. *[IA]* [0-ta < ta_1]

looking for us, he/she is. *V* nekahunuta. *[IA]* [0-ta < ta_1]

looking for us (2), he/she is. *V* nahkunuta. *[IA]* [0-ta < ta_1]

looking for you (1), he/she is. *V* nkunuta. *[IA]* [0-ta < ta_1]

looking for you (2+), he/she is. *V* nahkunuta. *[IA]* [0-ta < ta_1]

loon, common. [Gavia immer] *N* dadzi.

louse. *N* ya.

love you (1), I. *V* nk'esi'. *[IA]* [0-tsi' < tsi'_1]

lower. *ADV* yusts'i.

low-bush blueberry. [Vaccinium myrtilloides?] *N* 'ilhtsul.

lungs. *N* -des.

lungs, his/her. *N* budes.

lungs, my. *N* sdes.

lungs, our. *N* nedes.

lungs, their. *N* hubudes.

lungs, your (1). *N* ndes.

lynx. *N* wasi.

mad, he/she got. *V* hunlhuch'e. *[PA]* [l-ch'e < $ch'eh_1$]

mad, he/she is. *V* hunilch'e. *[IA]* [l-ch'e < $ch'eh_1$]

maggot. *N* 'usgoo.

magician. *N* nedulhdehun.

make noise so as to announce myself, I. *V* nadewelhuts'ah. *[IA]* [l-ts'ah]

make noise so as to announce ourselves, we (3+). *V* nadewe'ts'elts'ah. *[IA]* [l-ts'ah]

make noise so as to announce themselves, they. *V* na'hududelhts'ah. *[IA]* [lh-ts'ah] This is what people might do if they were approaching someone's campsite at dusk.

makeup, they put on. *V* nehndunulhtsi. *[IA]* [lh-tsi]

makeup, we (3+) put on. *V* nehts'undunultsi. *[IA]* [l-tsi]

makeup, you (1) put on. *V* nehndunilhtsi. *[IA]* [lh-tsi]

male fish. *N* tl'uzni.

mallard duck. [Anas platyrhynchos] *N* t'acho.

man. *N* dune. Duoplurals: dunene, duneke.

man, old. *N* duneti. Duoplural: dunetike

man, older. *N* dunecho. Duoplural: dunechone.

manner, in that. *ADV* whuzun'a.

manually. *PPP* sla be.

many, there are. *V* whulai. [abs: wh] *[IA]* [0-lai]

many, they (generic) are. *V* lhai. *[IA]* [0-lhai]

many, they are. *V* hulan. *[IA]* [0-lan]

many, they are. *V* nulai. [abs: n] *[IA]* [0-lai < lai_1]

many (places). *Q* lhat-un.

many as, there are as. *V*

ndaneltsuk. [IA] [l-tsuk]

many as, there are as. *V* njaneltsuk. *[IA]* [0-tsuk]

many times. *Q* lhat.

many ways, in. *ADV* lhelhdun k'un'a.

many [abstract]. *Q* lhawh.

many [human]. *Q* lhane.

many [human]. *Q* lhanun.

many [multiplicative]. *Q* lhat.

Maple, Dwarf. [Acer glabrum] *N* khasdzoon.

March. *N* Tsintel Ooza.

marijuana. *N* beni hoolah 'ut'ot.

marmot. [Marmota caligata] *N* dutni.

married, he is. *V* 'at ut'i. *[IA]* [0-t'i < $t'i_2$] Literally,"he/she has a wife".

married, she is. *V* ki ut'i. *[IA]* [0-t'i < $t'i_2$] Literally,"she has a husband".

married to him/her, she is. *V* yughusda. *[IA]* [0-da < da_1]

marten. [Martes americana] *N* chunih.

Mary Lake. *N* Basghelh Bunk'ut. The small lake behind (North) of Basghelh, the rock bluff right behind Nadleh village.

Masked Shrew. [Sorex cinereus] *N* dlim dats'olh.

Masked Shrew. [Sorex cinereus] *N* khunaichoya.

massaging it, he/she is. *V* yulhguk. *[IA]* [lh-guk < guk_1]

mastodon. *N* khunaicho.

matches. *N* kwunyoo.

mattress. *N* 'ustes.

mattress. *N* madus.

mattress, his/her. *N* be'ustes.

mattress, my. *N* se'ustes.

mattress, our. *N* ne'ustes.

mattress, their. *N* hube'ustes.

mattress, your (1). *N* nye'ustes.

mattress. *N* kw'ut ts'uzti.

May. *N* Dugoos Ooza.

maybe. *ADV* tusih.

me. *PRO* si.

meadow. *N* tl'otelk'ut.

meadow. *N* tl'ok'et.

meadow. *N* tl'ok'ut.

meat. *N* 'utsun.

meat, bird. *N* dut'aitsun.

meat, black bear. *N* sustsun.

meat, deer. *N* yests'etsun.

meat, dried. *N* 'utsungui.

meat, dried. *N* 'utsun sugi.

meat, dried beaver. *N* tsaikaih.

meat, dried beaver. *N* tsatsun sugi.

meat, fried. *N* 'utsun sut'e.

meat, ground. *N* 'utsun duneldus.

meat, his/her. *N* be'utsun.

meat, my. *N* se'utsun.

meat, rabbit. *N* gahtsun.

meat, raw. *N* t'eh.

meat soup. *N* 'utsun tazul.

medicine. *N* yoo.

medicine, herbal. *N* duchunyoo.

medicine cabinet. *N* yoo besula.

medicine man. *N* duyun.

meek, he/she is. V cha'dudilti. *[IN]*

[l-ti < ti$_2$]

melt, it has started to. *V* nawhenilghen. *[PA]* [l-ghen < ghe$_2$]

melt, it will. *V* natelghen. *[FA]* [l-ghen < ghe$_2$]

melted, it has. *V* nalhughen. *[PA]* [l-ghen < ghe$_2$]

melting, it is. *V* nalgheh. *[IA]* [l-gheh < ghe$_2$]

melting (st), I am. *V* nalhgheh. *[IA]* [lh-gheh < ghe$_2$]

melting (u.o.), he/she is. *V* na'ulhgheh. *[IA]* [lh-gheh < ghe$_2$]

melting it, he/she is. *V* naiyulhgheh. *[IA]* [lh-gheh < ghe$_2$]

men. *N* dunene.

merciful, he/she is. *V* te'ninzun. *[IA]* [0-zun < zun$_1$]

merciless, he/she is. *V* budzi hoolah. [IA] [0-lah] Literally, "his heart is absent".

mercury. *N* nudubun.

mercury. *N* too dulk'un. This refers to the "mercury" in a thermometer, that is, to the red-colored alcohol. By extension, it might be used in reference to actual mercury in a mercury thermometer. It does not refer to the metal mercury in general.

message. N khunek.

messenger. *N* khunek nu'a.

messenger. *N* nawhulnuk.

messing around, he/she is. *V* nuduzas. [IA] [0-zas]

meteorite shower. *N* kwuncho nainkat.

meteorite shower. *N* sumcho nainkat.

meteorites. *N* sum nainkat.

meteorites. *N* sum nakat.

microwave oven. *N* 'a be'ut'es.

midnight. *N* tuzniz.

midstream. P*PP* 'ukoh whuniz.

midsummer. *N* dayun.

migrating, he/she is. *V* nukedudeh. *[IA]* [0-deh]

migrating, they are. *V* nukehududeh. *[IA]* [0-deh]

milfoil. [Achillea millefolium] *N* lacholbai.

milk. *N* lilet.

milk, I am going to. *V* tesjas. *[FA]* [lh-jas < jas$_1$]

milk, powdered. *N* lilet sugi.

milking it, he/she is. *V* yulhjas. *[IA]* [lh-jas < jas$_1$]

milt. *N* lhooktl'uz.

milt. *N* -tl'uz.

mind. *N* -ni. Poss: -eni.

mind, his/her. *N* beni.

mind, my. *N* seni.

mind, our. *N* neni.

mind, your (1). *N* nyeni.

mine. *PRO* sch'e'.

mine. *PRO* sch'i'.

mine, it is. *V* se'ilhdzun. *[IA]* [lh-dzun]

mink. *N* telhjoos.

mirror. *N* tooznal'en.

misses you (1), he/she. *V* nk'enentai. *[IA]* [0-tai]

mistake, he/she is making a. *V*

nanut'a. *[IA]* [d-'a]

mistake, you (1) made a. *V* nananda. *[PA]* [0-da]

mittens. *N* bat.

mittens, their. *N* hububat.

mitts, his/her. *N* bubat.

mitts, my. *N* sbat.

mitts, your (1). *N* mbat.

moccasins. *N* kesgwut.

moccasins, old, worn-out. *N* kesgwutch'ul.

Monday. *N* Dimos k'elh'az.

Monday. *N* Landidzen.

money. *N* chigamin.

money. *N* sooniya.

money. *N* soonuya.

monster, prehistoric. *N* khunaicho.

month. *N* sanun.

moose. [Alces alces andersoni] *N* duni.

moose, bull. *N* jenyo.

moose, cow. *N* duni'at.

moose, dry cow. *N* yats'it.

moose hide. *N* dunizuz.

moose meat. *N* dunitsun.

morning. *N* bundada.

morning, this. *N* bunda.

morning, this. *N* k'an bundada.

morning, tomorrow. *N* bunde bundada.

morning, yesterday. *N* lhda bundada.

morning star. *N* sumcho.

Morris Creek. *N* Tats'utnai.

moss, green, on ground or tree trunks. *N* 'uyemba.

moss, tree. [Bryoria lanestris] *N* dahgha.

most. *ADV* 'uk'enus.

most of them. *V* k'us hulan. *[IA]* [0-lan]

moth. *N* musdziyaz.

moss, green, on ground or tree trunks. *N* 'uyemba.

mother. *N* 'uloo.

mother. *N* -loo. Duoplural: -loo.

mother, his/her. *N* buloo.

mother, my. *N* sloo.

mother, our. *N* neloo.

mother, their. *N* hubuloo.

mother, your (1). *N* nloo.

mother's sister. *N* -ak'i.

mother's sister, his/her. *N* bak'u.

mother's sister, my. *N* sak'i.

mother's sister, someone's. *N* 'ak'i.

mother's sister, their. *N* hubak'i.

mother's sister, your (1). *N* nyak'i.

mother-in-law. *N* -bez.

mother-in-law, his/her. *N* bubez.

mother-in-law, my. *N* sbez.

mother-in-law, our. *N* nebez.

mother-in-law, someone's. *N* 'ubez.

mother-in-law, their. *N* hububez.

mother-in-law, your (1). *N* mbez.

mother-of-pearl. *N* mulla.

mountain. *N* dzulh.

mountain, big. *N* dzulhcho.

Mount Greer. *N* Deduk Usk'en.

Mountain Ash. [Sorbus sitchensis] *N* ch'ok.

Mouse Mountain. *N* Lhkw'etsilhchola.

Mouse Mountain. *N* Neyilasts'ah.

mouse. *N* dats'ooz.

moustache. *N* -dagha.

moustache, his/her. *N* budagha.

moustache, my. *N* sdagha.

moustache, your (1). *N* ndagha.

moustaches, our. *N* nedagha.

moustaches, their. *N* hubudagha.

mouth. *N* -zek.

mouth, his/her. *N* buzek.

mouth, my. *N* szek.

mouth, river. *N* tachek.

mouth, roof of. *N* -zek dusts'ai.

mouth, roof of. *N* -zek onduk.

mouth, your (1). *N* unzek.

mouths, our. *N* nezek.

mouths, their. *N* hubuzek.

mowing, I am. *V* ust'as. *[IA]* [0-t'as]

mud. *N* lhetl'us.

mukluks. *N* kesgwutcho.

muscle. *N* nut'uk.

muscle, his/her. *N* bunut'uk.

muscle, my. *N* snut'uk.

muscle, your (1). *N* unnut'uk.

muscles, our. *N* nenut'uk.

muscles, their. *N* hubunut'uk.

muskrat. [Ondatra zibethicus] *N* tsek'et.

muskrat lodge. *N* tsek'etken.

mustard. *N* skitsan.

myself. *ADV* sich'ah.

Mouse Mountain. *N* Lhkw'etsilhchola.
Mouse Mountain. *N* Neyilasts'ah.

nail. *N* lugloo.

name. *N* 'oozi.

name. *N* -oozi.

name, his/her. *N* boozi.

name, your (1). *N* nyoozi.

name of trickster culture hero. *N* 'Usdas.

nape of neck, the. *N* buts'ilchunt'ak.

nape of my neck, the. *N* sts'ilchunt'ak.

nape of the neck. *N* -ts'ilchunt'ak.

nape of your (1) neck, the. *N* nts'ilchunt'ak.

napes of our necks, the. *N* nets'ilchunt'ak.

napes of their necks, the. *N* hubuts'ilchunt'ak.

Nautley. *N* Nadleh.

Nautley River. *N* Nadlehkoh. Drains Fraser Lake into the Nechako River.

navel. *N* -tsakelhgwus.

navel, his/her. *N* butsakelhgwus.

navel, my. *N* stsakelhgwus.

navel, your (1). *N* ntsakelhgwus.

navels, our. *N* netsakelhgwus.

navels, their. *N* hubutsakelhgwus.

near, it is. *V* nilhdukw. *[IA]* [lh-dukw < $dukw_1$]

neck. *N* -ts'ilchun.

neck, his/her. *N* buts'ilchun.

neck, my. *N* sts'ilchun.

neck, nape of. *N* -ts'ilchunt'ak.

neck, your (1). *N* nts'ilchun.

necks, our. *N* nets'ilchun.

necks, their. *N* hubuts'ilchun.

needle. *N* yabatsoh.

nephew. *N* -gwaz.

nephew, his/her. *N* bugwaz.

nephew, my. *N* ngwaz.

nephew, my. *N* sgwaz.

nephew, our. *N* negwaz.

nephew, their. *N* hubugwaz.

net, fish. *N* lhombilh.

next time. *ADV* 'on'at.

next to it. *PPC* bughah.

next year. *N* 'on'at yusk'ut.

next year. *N* nade.

niece. *N* -gwaz.

niece, my. *N* ngwaz.

niece, my. *N* sgwaz.

niece, our. *N* negwaz.

niece, their. *N* hubugwaz.

niece (younger than ego). *N* -gwastl'oh.

night. *N* lh'ek.

nine [abstract]. *NUM* 'ilhohoolawh.

nine [generic]. *NUM* 'ilhah hoolah.

nine [human]. *NUM* 'ilhah hoolanun.

nine [locative]. *NUM* 'ilho hooladun.

nine [multiplicative]. *NUM* 'ilha hoolat.

ninety [generic]. *NUM* 'ilhah hoolah lanezi.

nipple. *N* -ts'oola.

nipples, his/her. *N* buts'oola.

nipples, my. *N* sts'oola.

nipples, our. *N* nets'oola.

nipples, their. *N* hubuts'oola.

nipples, your (1). *N* nts'oola.

no. *INT* 'awundooh.

noble in clan system, female. *N* ts'ekeza'.

noble in the clan system, of either sex. *N* 'uza'. Duoplural: 'uza'ne.

noble in the clan sytsem, male. *N* duneza'. Duoplural: duneza'ne.

noisy, it (wh-class) is. *V* taoodetnuk. [IA] [d-nuk]

noisy, it is. *V* tawdetnuk. *[IA–wh]* [d-nuk]

noon. *N* dzetniz.

north, from the. *ADV* nizde.

north wind. *N* yootsiz nilhts’i.

northern lights. *N* yawhudoos.

Northern Black Currant. [Ribes hudsonianum] *N* tsasdlimai.

Northern Pikeminnow. *N* khusih.

Northwest Dace. [Mylocheilus caurinus] *N* daltsi’. Also known as Northwest Dace.

northward. *ADV* ni.

nostril. *N* -nik.

not. *PART* ’aw.

nothing. *N* soo gak.

November. *N* Banghan Nuts’uke.

now. *ADV* k’an.

now. *ADV* k’andit.

numerous, they are. *V* hulan. *[IA]* [0-lan]

numerous, they are. *V* dulai. [abs: d] *[IA]* [0-lai < lai$_1$]

numerous, they are. *V* whulai. [abs: wh] *[IA]* [0-lai]

nurse. *N* ndudaneghunli.

oar. *N* chos.

oatmeal. *N* mus.

obedient, they are. *V* ’uk’une’hut’en. *[IA]* [d-’en < ’en$_2$]

obese, he/she has become again. *V* nanilk’a. *[PA]* [l-k’a]

obese, he/she has not become. *V* nachanlhuk’a. *[PN]* [l-k’a]

obese, he/she is. *V* nelhuk’a. *[IA]* [l-k’a]

obese, he/she is not. *V* chanilk’ah. *[IN]* [l-k’ah]

obey him/her, I. *V* buk’une’ust’en. *[IA]* [d-’en < ’en$_2$]

obey him/her, we (2). *V* buk’une’idut’en. *[IA]* [d-’en < ’en$_2$]

obey him/her, we (3+). V buk’une’ts’ut’en. *[IA]* [d-’en < ’en$_2$]

obeys him/her, he/she. *V* yuk’une’ut’en. *[IA]* [d-’en < ’en$_2$]

obsidian. *N* tselgai.

occasionally. *ADV* whulutah.

ocean. *N* yatoo.

October. *N* Bet Ooza.

offset, they (teeth) are. *V* dzohtsulyaz tink’us unli. *[IA]* [0-li]

off-white, it is. *V* dulba. *[IA]* [l-ba < ba$_1$] This includes shades such as cream, tan, and blond.

oil. *N* khe. Poss: ghe.

oil, hair. *N* netsigha bets’ulha.

ointment. *N* lhootyoo

okay. *INT* a’ah.

old, he/she is. *V* nesjan. *[PA]* [d-yan < yan$_1$]

old, I am. *V* nesjan. *[PA]* [d-yan < yan$_1$]

old, it (wh-class) is. *V* whujut. *[IA]* [0-jut]

old age. *N* jan.

older than them. *PPC* hubutsah.

Old Fort Fraser. *N* K’atsizchun. The area at the base of Fraser Mountain. The site of the original Northwest Company fort and of the grave of the prophet Boba, now Beaumont Provincial Park.

omen. *N* me.

on. *PP* -k’ut.

on behalf of. *PP* ba.

on earth. *PPP* yun k'ut.

on it. *PPC* buk'ut.

on me. *PPC* sk'ut.

on the snow. *PPP* yusk'ut.

on top of each other. *PPC* lhk'ut.

once. *NUM* 'ilhah.

one hundred [generic]. *NUM* lanezi lanezi.

one side of it. *N* buk'uz.

one [abstract]. *NUM* 'ilhowh.

one [generic]. *NUM* 'ilhuk'i.

one [human]. *NUM* 'ilhunun.

one [locative]. *NUM* 'ilhuk'i.

one [multiplicative]. *NUM* 'ilhah.

onion. *N* tl'otsun.

only. *PART* ze.

only if. *CONJ* 'et ze.

Oona Lake. *N* Yoonoo Bunk'ut.

oops!. *INT* hadi'. An expression of surprise.

opinated, he/she is. *V* dubat'eoonaoonudzun. *[IA]* [d-zun]

or. *CONJ* k'us.

orange, it is. *V* kwun dohot'en. [abs: wh] *[IA]* [d-'en < 'en$_1$]

orange, it is. *V* kwun doonat'en. [abs: n] *[IA]* [d-'en < 'en$_1$]

oranges. *N* ooyutalhai.

orates, he/she. *V* duldlut. *[IA]* [l-dlut]

ordeal, he/she endured an. *V* dzah nusuzut. *[PA]* [0-zut]

Ormond Creek. *N* Noozkoh.

Ormond Lake. *N* Choostl'o.

Ormond Lake River. *N* Choostl'okoh.

orphan. *N* tsunah.

other side, on the. *ADV* 'onyaz ts'i.

otter. [Lontra canadensis] *N* tsis.

ouch!. *INT* 'ayah.

ouch!. *INT* 'ayoh ho.

ours. *PRO* nech'i'.

ourselves. *PRO* whenich'ah.

outhouse. *N* tsanbayah.

outhouse, his/her. *N* butsanbayah.

outhouse, my. *N* stsanbayah.

outhouse, our. *N* netsanbayah.

outhouse, their. *N* hubutsanbayah.

outhouse, your (1). *N* ntsanbayah.

outlet, electrical. *N* 'ana'dutsih.

outside. *ADV* 'az.

oven. *N* be'ut'es.

oven, microwave. *N* 'a be'ut'es.

over it. *PPC* butus.

over there. *ADV* whuz.

overcast and stormy, it (the sky) becomes. *V* whudintsuk. *[IA–customary]* [0-tsuk]

overcast and stormy, it (the sky) is. *V* whudintsi'. *[IA]* [0-tsi' < tsi'$_1$]

overflowed, it. *V* dadembun. *[PA]* [0-bun < bun$_1$]

overseas. *N* yaba.

overslept, he/she. *V* whunandunesti. [PA] [0-ti < ti$_4$]

overwhelmed, he/she is. *V* 'oot'e ghainil. *[IA]* [0-nil]

over-ripe, it is. *V* 'on'un hanit'ai. [IA] [0-t'ai]

owes him (st), he/she. *V*

yuts'uwhulh'ai. *[IA]* [lh-'ai]

owl. *N* musdzi.

pack. *N* khelh. This refers to what one carries, not to the container.

packed (st) back home, I. *V* whusanasasdughi. *[PA]* [d-ghi < ghe$_1$]

packed it back home, he/she. *V* whusanayedughi. *[PA]* [d-ghi < ghe$_1$]

packing, he/she is. *V* eghilh. *[IA–progressive]* [0-ghilh < ghe1]

packing, we (2) are. *V* edughilh. *[IA–progressive]* [0-ghilh < ghe$_1$]

packing (u.o.), he/she is. *V* 'eghilh. *[IA–prog]* [0-ghilh < ghe$_1$]

packing (u.o.), I am. *V* 'esghilh. *[IA–prog]* [0-ghilh < ghe$_1$]

packing (u.o.), they are. *V* 'uheghilh. *[IA–prog]* [0-ghilh < ghe$_1$]

packing around (u.o.), he/she is. *V* ne'ughe. *[IA]* [0-ghe < ghe$_1$]

packing around (u.o.), I am. *V* ne'usghe. *[IA]* [0-ghe < ghe$_1$]

owl. *N* musdzi.

packing back, he/she is. *V* na'eghilh. *[IA–progressive]* [0-ghilh < ghe$_1$]

packing back, he/she is. *V* natesdughi. *[IA]* [d-ghi < ghe$_1$]

packrat. [Neotoma cinerea] *N* dlimcho.

Pack Rat. [Neotoma cinerea] *N* dlooncho.

paddle (canoe). *N* ches.

pail. *N* 'oosa'.

pain, he/she has. *V* bulh ults'ulh. [abs: 0] *[IA]* [l-ts'ulh]

pain, he/she has severe. *V* ootsas. *[IA]* [0-tsas]

pain, I cause (s.o.). *V* ndulhda. *[IA]* [lh-da]

paint, I. *V* 'uk'une'ulhtsi. *[IA]* [lh-tsi]

paint (st), I. *V* dustl'us. [abs: 0] *[IA]* [0-tl'us]

paint (st), I. *V* whudustl'us. [abs: wh] *[IA]* [0-tl'us]

pajamas. *N* bulh sdutez.

pale, it is. *V* t'alint'ah. *[IA]* [0-t'ah]

palm, his/her. *N* bulak'et.

palm, my. *N* slak'et.

palm, your (1). *N* nlak'et

palm of hand. *N* -lak'et.

palms, our. *N* nelak'et.

palms, their. *N* hubulak'et.

pan, frying. *N* be'ut'es.

pancreas. *N* -zumcho.

pancreas, his/her. *N* buzumcho.

pancreas, my. *N* szumcho.

pancreas, your (1). *N* nzumcho.

pancreases, our. *N* nezumcho.

pancreases, their. *N* hubuzumcho.

panicked, he/she. *V* khunantesja. *[PA]* [0-ja]

panties. *N* ts'ekoo tl'asus.

pants. *N* dune tl'asus.

pants. *N* tl'asus.

pants, his/her. *N* butl'asus.

pants, my. *N* stl'asus.

pants, our. *N* netl'asus.

pants, their. *N* hubutl'asus.

pants, your (1). *N* ntl'asus.

paper. *N* dustl'us.

paper, our (3+). *N* nedustl'us.

paper, your (2+). *N* nahdustl'us.

paper bag. *N* dustl'uszuz.

passed him in a boat, I. *V* beghoosuski. *[PA]* [0-ki < ke_2]

passed them driving, I. *V* hubeghoosusgoo. *[PA]* [0-goo < goo_1]

passing, (time) is. *V* whudezulh. *[IA–prog]* [0-zulh]

pasture. *N* tl'ok'ut.

patches it, he/she. *V* naiyulhkut. *[IA]* [lh-kut]

paw, dog's. *N* lhike.

peach-fuzz on his face, he/she has. *V* dunuyo. [abs: n] *[IA]* [0-yo]

peak of Greer Mountain. *N* Skenicho.

peaks, mountain. *N* tah. [archaic]

Peamouth Chub. [Mylocheilus caurinus] *N* daltsi'. Also known as Northwest Dace.

peavine, creamy. [Lathyrus ochroleucus] *N* chunalhduz.

pee. *N* lhuz. Poss: -luz.

pee, I am going to. *V* natesdudluz. *[FA]* [d-luz < luz_1]

pee, I am going to. *V* natesdluz. *[FA]* [d-luz < luz_1]

pee, they will. *V* natedudluz. *[FA]* [d-luz]

pee, we (3+) will. *V* naztedudluz. *[FA]* [d-luz]

pee, you (1) will. *V* natandudluz. *[FA]* [d-luz]

pelvic bone. *N* -k'itsun'.

pelvic region, his/her. *N* buk'i.

pelvic region, my. *N* sk'i.

pelvic region, your (1). *N* nk'i.

pelvic region. *N* -k'i.

pelvic regions, our. *N* nek'i.

pelvic regions, their. *N* hubuk'i.

peninsula. *N* sdughak'ut.

penis. *N* 'utsukw.

penis. *N* lait'i.

penis. *N* -tsukw.

penis, boy's. *N* kw'ulh. Poss: kw'ulh.

penis, his. *N* butsukw.

penis, his (boy's). *N* bukw'ulh.

penis, my. *N* stsukw.

penis, someone's. *N* 'utsukw.

penis, your (1). *N* ntsukw.

penises, our. *N* netsukw.

penises, their. *N* hubutsukw.

penniless, he/she is. *V* k'untuk. *[IA]* [0-tuk < tuk_2]

Pentecost. *N* Ndonidzen.

people. *N* dunene.

pepper. *N* sulhts'i.

perch. *N* lhooz.

Peregrine Falcon. *N* hoolht'as.

peritoneum. *N* 'uchanyoo. The fatty lining of the abdominal cavity.

person. *N* dune. Duoplurals: dunene, duneke

perspiration. *N* tsentoo.

Peta Lake. *N* Nanguz Bunk'ut.

Peterson's Beach Forestry Campsite. *N* Sainaditi. A sheltered bay with a sandy beach. A fishing area for netting char and whitefish.

Peterson's beach and point. *N* Nuswhutisdi.

pick it up, you (1). *V* dudi'a. [class: sdo-gen] *[IA]* [0-'a < 'ai$_1$]

picking (berries), he/she is. *V* oonuyin. *[IA]* [0-yin]

picking (berries), I am. *V* oonusyin. *[IA]* [0-yin]

picking (berries), we (2). *V* oonidujin. *[IA]* [0-yin]

pie. *N* bedutleh.

pig. *N* gugoos.

Pigweed. [Chenopodium album] *N* 'ut'antsi.

pill. *N* ts'untulwis.

pillow. *N* tsi'alh.

pillow, his/her. *N* butsi'alh.

pillow, my. *N* stsi'alh.

pillow, our. *N* netsi'alh.

pillow, their. *N* hubutsi'alh.

pillow, your (1). *N* ntsi'alh.

pimple. *N* netsa uneltsut.

pin. *N* 'unibek. This includes both straight pins and sewing pins.

pin, straight. *N* 'unubek.

pine, jack. [Pinus contorta latifolia] *N* chundoo. Locally known as "jack pine", which standardly refers to Pinus banksiana.

pine, lodgepole. [Pinus contorta latifolia] *N* chundoo. Locally known as "jack pine', which standardly refers to Pinus banksiana.

pine cone. *N* 'angwul.

pink, it is. *V* 'inchooh isdak dohot'en. [abs: wh] *[IA]* [d-'en < 'en$_1$]

pink, it is. *V* 'inchooh isdak dot'en. *[IA]* [d-'en < 'en$_1$]

pink, it is. *V* 'inchooh isdak doonat'en. [abs: n] *[IA]* [d-'en < 'en$_1$]

pinkie. *N* -lasts'ah.

pinkie, his/her. *N* bulasts'ah.

pinkie, my. *N* slasts'ah.

pinkie, your (1). *N* nlasts'ah.

pinkies, our. *N* nelasts'ah.

pinkies, their. *N* hubulasts'ah.

pitch. *N* dzeh.

Pitka Mountain. N Buzdaiyus.

pity him/her, I. *V* wate'nuszun. *[IA]* [0-zun < zun$_1$]

place (st), you (1). *V* nininle. [class: mdo-c] *[IA]* [0-le < lɛ$_1$]

planing, he/she is. *V* dughas. *[IA]* [0-ghas < ghaz$_2$]

planing (u.o.), we (2) are. *V* 'udidughas. *[IA]* [0-ghas < ghaz$_2$]

plantain. [Plantago major] *N* khuzbat'an.

plate. *N* lasyet.

plate. *N* tsets'ai.

plow, he/she will. *V* nawtelhch’ul. [abs: wh] *[FA]* [lh-ch’ul < ch’ul$_1$]

plow (st:road), he/she will. *V* nawtezo. [abs: wh] *[FA]* [0-zo]

plowed, he/she. *V* nawhalhch’ul. [abs: wh] *[PA]* [lh-ch’ul < ch’ul$_1$]

plowing (st:road), he/she is. *V* nawhuzo. [abs: wh] *[IA]* [0-zo]

plucked it, he/she. *V* hayanyuz. *[PA]* [0-yuz]

plucking (st), I am. *V* hasyis. *[IA]* [0-yis]

plucking it, he/she is. *V* hayuyis. *[IA]* [0-yis]

plus. *CONJ* ’on’at.

pocket. *N* lubos.

point of land on Fraser Lake below Mouse Mountain. *N* Lhkw’etsilhchola.

pointed, it is. *V* dujooz. *[IA]* [0-jooz < jooz$_1$]

poison. *N* lubezo.

poker, fire. *N* daskwun.

poker, stove. *N* koonangus.

pole, fishing. *N* juschun.

police officer. *N* neilhchuk.

pond. *N* tadez’ai.

poop. *N* tsan.

poor, he is. *V* tel’en. *[IA]* [l-’en]

poor, I am. *V* telhu’en. *[IA]* [l-’en]

poor, they are. *V* tehul’en. *[IA]* [l-’en]

poor, we (2) are. *V* teidul’en. *[IA]* [l-’en]

pop. *N* too lhuk’i.

Poplar. [Populus tremuloides] *N* t’ughus. Locally known as Poplar.

porch. *N* buntsilh.

porch. *N* dadeltel.

porch, his/her. *N* budadeltel.

porch, my. *N* sdadeltel.

porch, our. *N* nedadeltel.

porch, their. *N* hubudadeltel.

porch, your (1). *N* ndadeltel.

porcupine. [Erethizon dorsatum] *N* ’oojoonih.

porcupine. [Erethizon dorsatum] *N* ’ujoohnih.

porcupine. [Erethizon dorsatum] *N* duch’ukw.

porcupine. *N* duneza’.

porcupine quills used for embroidery. *N* lhbai.

pork. *N* gugoostsun.

porridge. *N* mus.

possessor, first person dual. *PREFIX* nah-. Attaches to class 1 nouns.

possessor, first person plural. *PREFIX* ne-. Attaches to class 1 nouns.

possessor, first person singular. *PREFIX* s-. Attaches to class 1 nouns.

possessor, second person duo-plural. *PREFIX* nah-.

possessor, second person singular. *PREFIX* n-. Attaches to class 1 nouns.

possessor, third person duo-plural, for class 1 nouns. *PREFIX* hubu-.

possessor, third person reflexive, for class 1 nouns. *PREFIX* du-.

possessor, third person singular. *PREFIX* bu-. Attaches to class 1 nouns.

post office. *N* dustl'us dakat.

pot. *N* 'oosa'.

pot, coffee. *N* lugafi be'udlez.

pot, tea. *N* ludibot.

potato. *N* lubudak.

potato chips. *N* lubudakt'ooz.

potlatch. *N* balhats.

pouch. *N* zus.

pouch, bullet. *N* k'azus.

pouring down, rain is. *V* chan nainli. *[IA]* [0-li < li$_1$]

pouring it back into (st), he/she is. *V* denayulhdzeh. *[IA]* [lh-dzeh < dzeh$_1$]

pouring it into (st), he/she is. *V* deyulhdzeh. *[IA]* [lh-dzeh < dzeh$_1$]

pouring it into it, he/she is. *V* yeyulhdzeh. *[IA]* [lh-dzeh < dzeh$_1$]

pouring it into it, I am. *V* beyulhdzeh. *[IA]* [lh-dzeh < dzeh$_1$]

pouting, he/she is. *V* nulch'e'. *[IA]* [l-ch'e' < ch'eh$_1$]

powerful, it is. *V* beoonujut. *[IA]* [0-jut < joot$_1$]

pray, I. *V* tenadusdli. *[IA]* [0-dli < dli$_2$]

pray, we (3+). *V* tenazdudli. *[IA]* [0-dli < dli$_2$]

pray, we (3+) may. *V* tenazdoodli. *[OA]* [0-dli < dli$_2$]

pray, you (1). *V* tenadindli. *[IA]* [0-dli]

pray, you (1) may. *V* tenadondlih. *[OA]* [0-dlih < dli$_2$]

pray, you (2+). *V* tenadahdli. *[IA]* [0-dli]

prays, he/she. *V* tenadudli. *[IA]* [0-dli < dli$_2$]

preceding. *PP* tsah.

pregnant, I am. *V* ulhuchan. *[IA]* [l-chan < chan$_1$]

pregnant, she is. *V* ulchan. *[IA]* [l-chan < chan$_1$]

pregnant, you (2+) are. *V* ulhuchan. *[IA]* [l-chan < chan$_1$]

premonition, he/she had a. *V* nus whan'en. *[PA]* [0-'en < 'en$_1$]

pretty, it is. *V* nzoo. *[IA]* [0-zoo < zoo$_1$]

priest. *N* nawhulnuk.

prisoner. *N* tsak'etdin'ai.

probably. *ADV* tusih.

procrastinating, he/she is. *V* nus whe'alh. *[IA]* [0-'alh]

properly. *ADV* soocho.

property. *N* -ch'e'. With possessive prefixes forms the possessive pronouns mine, yours, etc.

Protestant. *N* luminis.

proud, he/she is not. *V* cha'dudilti. *[IN]* [l-ti < ti$_2$]

ptarmigan, willow. [Lagopus lagopus] *N* k'azba.

pterodactyl. *N* dune oolhdzai.

pubic hair. *N* -tl'egha.

pubic hair, his/her. *N* butl'egha.

pubic hair, my. *N* stl'egha.

pubic hair, our. *N* netl'egha.

pubic hair, their. *N* hubutl'egha.

pubic hair, your (1). *N* ntl'egha.

puddle. *N* tadez'ai.

puppy. *N* lhiyaz. Duoplural: lhiyazke

purple, it is. *V* 'ilhtsultoo dohot'en. [abs: wh] *[IA]* [d-'en < 'en$_1$]

purple, it is. *V* 'ilhtsultoo dot'en. *[IA]* [d-'en < 'en$_1$]

purple, it is. *V* 'ilhtsultoo doonat'en. [abs: n] *[IA]* [d-'en < 'en$_1$]

purple, it is. *V* maitoo dot'en. *[IA]* [d-'en < 'en$_1$] Literally, "it is the colour of berry juice".

puppy. *N* lhiyaz. Duoplural: lhiyazke

purse. *N* 'ezdlai.

purse. *N* soonuy uti zuz.

purse. *N* soonuyazuz.

pus. *N* khuz.

pussy willow. *N* k'edlits'ilhba.

put his/her foot inside, he/she. *V* dadel'ez. *[PA]* [l-'ez < 'es$_1$]

put on (st), he/she. *V* ghunadulhu'oo'. *[PA]* [l-'oo' < tl'oo$_1$] Something like a belt or sash that one "ties" on.

put on (st), I. *V* ghunadulhu'oo'. *[PA]* [l-'oo' < tl'oo$_1$] Something like a belt or sash that one "ties" on.

put on your seatbelt, you (1). *V* denawhudinges. *[IA]* [0-ges]

put them on each other, he/she. *V* lhk'einla. [class: mdo-c] *[PA]* [0-la] Among other things, this can mean that he crossed his feet while walking or running and tripped.

put your shoes on, you (1). *V* kenaindutsi. *[IA]* [d-tsi]

quarrelsome, to be. *V* -zek taoodetnuk. *[IA]* [d-nuk] This is literally "X's mouth is noisy". The "subject" is therefore indicated by possessive prefixation of -zek "mouth", e.g. buzek taoodetnuk "his mouth is noisy", i.e. "he/she is quarrelsome".

quarter. *N* gwada'. That is, a twenty-five cent coin.

question marker. *PART* lak.

question marker. *PART* ih.

quick to anger, he/she is. *V* 'a buzkai tujulh. *[IA]* [0-julh]

quickly. *ADV* 'a.

quickly. *ADV* 'acho.

quickly. *ADV* soo 'a.

quiet, he/she became. *V* t'edinil. *[PA]* [0-nil]

quiet, I am. *V* t'edusnih. *[IA]* [0-nih]

quiet, it is. *V* t'ewdusnih. [abs: wh] *[IA]* [0-nih]

quietly. *ADV* nagoostliyaz.

quilt. *N* lhtanalkat.

rabbit. [Lepus americanus pal.] *N* gah. The only species native to the territory is the Snowshoe Hare, also known as Varying Hare, Lepus americanus. The term is, however, applied to other varieties of rabbit and hare.

rabbit meat, dried. *N* gahgi.

rabbit trail. *N* gahti.

rabbit snare. *N* gambilh.

rabbit tracks. *N* gahk'ah.

rack for drying food. *N* nin'ai.

raft. *N* khinyus.

rain. *N* chan.

rain coat. *N* toodzoot.

rainbow, there is a. *V* detnik na'nanguz. *[IA]* [0-guz]

raining into water, it is. *V* tawhulhti. [abs: wh] *[IA]* [lh-ti < ti$_3$]

ran in a loop, I. *V* nusulhugai. *[PA]* [l-gai < gaih$_1$]

ran in a loop, they (2). *V* nuhulhugai. *[PA]* [l-gai < gaih$_1$]

ran in a loop, we (2). *V* nusidulgai. *[PA]* [l-gai < gaih$_1$]

ran in a loop, we (3+). *V* nuts'ulhughaz. *[PA]* [l-ghaz]

ran in a loop, you (1). *V* nusilgai. *[PA]* [l-gai < gaih$_1$]

raspberries. [Rubus strigosus] *N* bindak khastl'ah.

raven, common. [Corvus corax] *N* datsancho.

RCMP officer. *N* neilhchuk.

really. *ADV* soo.

recliner. *N* kw'usduda t'az whebulh.

reclining chair. *N* duneti kw'uts'uda.

rectum. *N* -boo'. [babytalk]

rectum. *N* -tsul.

rectum, his/her. *N* buboo'.

rectum, his/her. *N* butsul.

rectum, my. *N* sboo'.

rectum, my. *N* stsul.

rectum, your (1). *N* mboo'.

rectum, your (1). *N* ntsul.

rectums, our. *N* neboo'.

rectums, our. *N* netsul.

rectums, their. *N* hububoo.

rectums, their. *N* hubutsul.

Red Columbine. [Aquilegia formosa] *N* whulecho.

red, it is. *V* dulk'un. *[IA]* [l-k'un < k'un$_1$]

red, it is. *V* dunulk'un. [abs: n] *[IA]* [l-k'un < k'un$_1$]

red, it is. *V* whudulk'un. [abs: wh] *[IA]* [l-k'un < k'un$_1$]

Red-Osier Dogwood. [Cornus stolonifera] *N* k'en dulk'un. Locally known as Red Willow.

red willow. [Cornus stolonifera] *N* k'en dulk'un. Locally known as Red Willow.

Red-winged Blackbird. [Agelaius phoeniceus] *N* ch'uk.

reef. *N* tehnoo.

refrigerator. *N* be'nezk'uz.

relative. *N* t'eke.

reside, you (1). *V* hoont'i. *[IA]* [0-t'i]

respect them, you (2+). *V* hubudulhti. *[IA]* [lh-ti < ti$_2$]

rest, you (1). *V* nailyis. *[IA]* [l-yis]

rest, you (1). *V* naolyis. *[OA]* [l-yis]

rest, you (1) will. *V* natalyis. *[FA]* [l-yis]

resting, they are. *V* nahulyis. *[IA]* [l-yis]

resting, we (3+) are. *V* nats'ulyis. *[IA]* [l-yis]

rib. *N* -chak.

ribs, his/her. *N* buchak.

ribs, my. *N* schak.

ribs, our. *N* nechak.

ribs, their. *N* hubuchak.

ribs, your (1). *N* nchak.

rice sack. *N* rice 'uzus.

rifle. *N* 'ulhti.

rifle, big. *N* 'ulhticho. E.g. .306 or .303.

rifle shell. *N* k'a.

right. *ADV* nailhni.

right now. *ADV* k'an 'awet.

right side. *ADV* nalhni.

rifle. *N* lhti.

rim of a basket. *N* bal'ah.

river. *N* 'ukoh.

river. *N* koh.

river. *N* took'oh.

river bank. *N* toobus.

river bank, underpart of overhanging. *N* toobust'ah.

river mouth. *N* tachek.

road. *N* ti.

robin, American. [Turdus migratorius] *N* sooh.

rock. *N* tse.

rocking chair. *N* kw'usduda nubalh.

roe, fish. *N* 'uk'oon.

roll (st) to you (1), I will. *V* nts'untelhbus. *[FA]* [lh-bus < bas$_1$]

roll down, it will. *V* nts'ununtelbus. *[FA]* [l-bus]

rolled over, it. *V* nadezghuz. *[PA]* [0-ghuz < ghis$_1$]

rolled over (in vehicle), I. *V* nadesghuz. *[PA]* [0-ghuz < ghis$_1$]

roof. *N* bun.

roof, his/her. *N* bubun.

roof, my. *N* sbun.

roof, our. *N* nebun.

roof, peaked. *N* ch'usghak.

roof, their. *N* hububun.

roof, your (1). *N* mbun.

roof of the mouth. *N* -zek dusts'ai.

room, dining. *N* 'uts'uyi ba whuz'ai.

rope. *N* tl'oolh.

rope, my. *N* stl'oolh.

rots out, it. *V* hawhulhjuk. *[IA–customary]* [lh-juk < jut$_1$]

rough, it is. *V* dutsiz. *[IA]* [0-tsiz]

rubbing it, he/she is. *V* yulhguk. *[IA]* [lh-guk < guk$_1$]

Ruffed Grouse. [Bonasa umbellus] *N* 'utsut.

rug. N yunlhutel.

run, I will. *V* telhugih. *[FA]* [l-gih < gaih$_1$]

run, they (2) will. *V* hutelgih. *[FA]* [l-gih < gaih$_1$]

run, they (3+) will. *V* hutelwus. *[FA]* [l-wus]

run, we (2) will. *V* tadulgih. *[FA]* [l-gih < gaih$_1$]

run, we (3+) will. *V* uztelwus. *[FA]* [l-wus]

run, you (1) will. *V* talgih. *[FA]* [l-gih]

run, you (1) will. *V* talgih. *[FA]* [l-gih < gaih$_1$]

running, he/she is. *V* ulgih. *[IA–*

progressive] [l-gih < gaih$_1$]

running, I am. *V* ulhugih. *[IA–prog]* [l-gih < gaih$_1$]

running, we (2) are. *V* idulgih. *[IA–prog]* [l-gih < gaih$_1$]

running, we (3+) are. *V* ts'elwus. *[IA–prog]* [l-wus]

running, you (1) are. *V* ilgih. *[IA–prog]* [l-gih < gaih$_1$]

running after it, I am. *V* kwatelhugai. *[PA]* [l-gai < gaih$_1$]

running after it, they (3+) are. *V* hikatelhughaz. *[PA]* [l-ghaz]

running after it, we (2) are. *V* kwatedulgai. [PA] [l-gai < gaih$_1$]

running after it, we (3+) are. *V* kwaztelhughaz. *[PA]* [l-ghaz]

running around, he/she is. *V* nulgaih. *[IA]* [l-gaih < gaih$_1$]

runs, he/she. *V* nulgaih. *[IA]* [l-gaih < gaih$_1$]

salmon. *N* talook.

Saddle Mountain. *N* Buzdaiyus.

safety pin. *N* dughul.

sagebrush, Arctic. [Artemisia frigida] *N* tse'ul.

salamander. [Ambystoma macrodactylum] *N* chunlai. The only species of salamander found in the region is the Long-toed Salamander Ambystoma macrodactylum. No lizards are found in the region. However, this term is applied to other varieties of salamander and to lizards, such as the geckos sold as pets.

saliva. *N* -zekw.

saliva, my. *N* suzekw.

salmon. *N* talook.

salmon, dried. *N* talook sugi.

salmon, Spring. *N* ges.

salmon, steelhead. *N* sdle.

salt. *N* lusel.

sand. *N* sai.

sandpiper. *N* wedlew.

sang, I. *V* esjun. *[PA]* [d-yun < yun$_1$]

Sandhill Crane. [Grus canadensis/ Antigone canadensis] *N* dilh.

sasquatch. *N* ts'eslol. Not to be confused with neyi, sasquatch are different from human beings. Their bodies are covered with long hair. They are said to eat plants and fish, not meat, and are not considered dangerous to people.

Saskatoon berries. [Amelanchier alnifolia] *N* k'emai.

Saturday. *N* Dzenwhuyaz.

Saturday. *N* Sumdi.

saw. *N* be'dudughut.

sawing (st), he/she is. *V* dughut. *[IA]* [0-ghut < ghut$_1$]

sawing (st), we (2) are. *V* didughut. *[IA]* [0-ghut < ghut$_1$]

sawing (u.o.), we (3+) are. *V* 'uzdughut. *[IA]* [0-ghut < ghut$_1$]

says, he/she. *V* ni. *[IA]* [0-ni]

scarf. *N* tsinezdelya.

scattered clouds, there are. *V* ninawhunde. *[IA]* [0-de]

scissors. *N* lheyih.

scoop, he/she tries to. *V* ookaih. [class: coc] *[IA–customary]* [0-kaih]

Scouring Rush. [Equisetum hyemale] *N* khahdai.

scraper, bone skin. *N* be'ulje.

screw in, you (1). *V* 'ailhghis. *[IA]* [lh-ghis < ghis$_1$]

scrotum. *N* -ghezzus.

scrotum, his/her. *N* bughezzus.

scrotum, my. *N* sghezzus.

scrotum, your (1). *N* nghezzus.

scrotums, our. *N* neghezzus.

scrotums, their. *N* hubughezzus.

scurry around, we (3+). *V* nukats'ulgas. *[IA]* [l-gas]

seagull. *N* besk'i.

seaweed, dried Red Laver. [Porphyra abbottae Krishnamurthy] *N* lhaga'as. When fresh this is reddish, purple, or greenish and has the consistency of cellophane. When dried it is black and brittle. It grows on rocks in the lower intertidal zone. It was and to some extent still is traded in from the coast; it does not grow in Carrier territory.

second person duo-plural possessor. *PREFIX* nah-.

second person singular possessor. *PREFIX* n-. Attaches to class 1 nouns.

secretary. *N* 'uk'une'uguz.

see, he/she cannot. *V* chawes'en. *[IN]* [0-'en < 'en$_1$]

see, I cannot. *V* chawzes'en. *[IN]* [0-'en < 'en$_1$]

see you (1) again, I may. *V* nanyoost'en. *[OA]* [d-'en < 'en$_1$]

seeder. *N* behooyeznuleh. A machine for planting seeds.

seeds, its (plant's). *N* bumai.

seep. *N* whuzainli. A place where water seeps out of the ground.

sees, he/she. *V* in'en. *[IA]* [0-'en < 'en$_1$]

sees (wh-class object), he/she. *V* hon'en. *[IA]* [0-'en < 'en$_1$]

semen, fish. *N* -tl'uz.

seniors. *N* netsawhudelhzulhne.

seniors. *N* netsowhuzulhne.

September. *N* Gestlah Ooza.

servant. *N* 'ustlen.

servant, his/her. *N* be'ustlen.

servant, my. *N* se'ustlen.

servant, our. *N* ne'ustlen.

servant, their. *N* hube'ustlen.

servant, your (1). *N* nye'ustlen.

set net, he/she. *V* te'ninla. *[PA]* [0-la]

set net, I. *V* te'nusla. *[PA]* [0-la]

set net, I. *V* te'usdle. *[IA]* [0-le]

set net, I will. *V* te'teslilh. *[FA]* [0-lilh]

set off by boat, I. *V* wheski. *[PA]* [0-ki < ke$_2$]

set off by boat, I have not. *V* whechaskel. *[PN]* [0-kel < ke$_2$]

set off by car, I. *V* whesgoo. *[PA]* [0-goo < goo$_1$]

set off by plane, I have not. *V* sulh

whechait’ah. *[PN]* [0-t’ah < t’oh$_2$]

set off driving, they did not. *V* whechahigooh. *[PN]* [0-gooh < goo$_1$]

set off on foot, they (3+) have. *V* whehandil. *[PA]* [0-dil < dil$_1$]

set off on foot, we (3+). *V* whets’andil. *[PA]* [0-dil < dil$_1$]

sets, the sun. *V* sa k’ah naidutut’ih. *[IA–customary]* [d-’ih < ’en$_1$]

setting, the sun is. *V* sa k’ah naidutat’en. *[IA]* [d-’en < ’en$_1$]

setting, the sun is. *V* yuk ’utez’ai. *[IA]* [0-’ai < ’a$_3$]

seven [abstract]. *NUM* lhtak’alt’iwh.

seven [generic]. *NUM* lhtak’alt’i.

seven [human]. *NUM* lhtak’alt’inun.

seven [locative]. *NUM* lhtak’alt’idun.

seven [multiplicative]. *NUM* lhtak’alt’it.

seventy [generic]. *NUM* lhtak’alt’it lanezi.

shady spot. *N* ’uscheh.

shaggy, he/she is. *V* duyo. [abs: 0] *[IA]* [0-yo]

shaggy, it is. *V* dunuyo. [abs: n] *[IA]* [0-yo]

shaman. *N* duyun.

shaman, female. *N* duyun ts’eke.

sharp, it is. *V* desgwut. *[IA]* [0-gwut]

sharp, it is. *V* dujooz. *[IA]* [0-jooz < jooz$_1$]

shat, I have. V nasusdutsan. [PA] [d-tsan < tsan$_1$]

shawl. *N* lasal.

she. *PRO* ’en.

she. *PRO* didut.

sheet. *N* lilik’usulhchooz.

sheet, his/her. *N* bulilik’usulhchooz.

sheet, my. *N* slilik’usulhchooz.

sheet, our. *N* nelilik’usulhchooz.

sheet, their. *N* hubulilik’usulhchooz.

sheet, your (1). *N* nlilik’usulhchooz.

shin. *N* -gwutyuschun.

shins, his/her. *N* bugwutyuschun.

shins, my. *N* sgwutyuschun.

shins, our. *N* negwutyuschun.

shins, their. *N* hubugwutyuschun.

shins, your (1). *N* ngwutyuschun.

shirt. *N* dzoozt’an.

shirt, his/her. *N* budzoozt’an.

shirt, my. *N* sdzoozt’an.

shirt, our. *N* nedzoozt’an.

shirt, their. *N* hubudzoozt’an.

shirt, your (1). *N* ndzoozt’an.

shit. *N* tsan.

shit, I am going to. *V* natesdutsun. *[FA]* [d-tsun]

shit, we (3+) will. *V* naztedutsun. *[FA]* [d-tsun]

shit, you (1) will. *V* natandutsun. *[FA]* [d-tsun]

shit, you (2+) will. *V* natehdutsun. *[FA]* [d-tsun]

shoe. *N* kegon.

shoe laces. *N* ketl’oolh.

shooting stars. *N* sum nainkat.

shooting stars. *N* sum nakat.

shore. *N* taba.

shore of river. *N* kohtaba.

short, he/she is. *V* ndukw. *[IA]* [0-dukw < dukw$_1$]

short, it is. *V* dindukw. [abs: d] *[IA]* [-dukw < dukw$_1$]

short, it is. *V* hoondukw. [abs: wh] *[IA]* [0-dukw < dukw$_1$]

shortly after. *ADV* k'adliyaz.

shorts. *N* tl'abesu'ai.

shoulder. *N* -wus.

shoulders, his/her. *N* buwus.

shoulders, my. *N* swus.

shoulders, our. *N* newus.

shoulders, their. *N* hubuwus.

shoulders, your (1). *N* nwus.

show how, I. *V* whunulhtun. *[IA]* [lh-tun < tun$_3$]

show them how, I. *V* whuhubunulhtun. *[IA]* [lh-tun < tun$_3$]

shows me how, he/she. *V* whusunulhtun. *[IA]* [lh-tun < tun$_3$]

shrew, masked. [Sorex cinereus] *N* dlim dats'olh.

shrew, masked. [Sorex cinereus] *N* khunaichoya.

sibling's spouse. *N* -ghe.

sick, he/she is. *V* nduda. *[IA]* [0-da < da$_2$]

sick, they are. *V* hunduda. *[IA]* [0-da]

sick, we (3+) are. *V* ts'unduda. *[IA]* [0-da]

sick, you (1) are. *V* ndinda. *[IA]* [0-da]

sinew. *N* 'uts'eh.

sing, he/she will. *V* 'utejun. *[FA]* [d-yun]

sing, he/she will. *V* tejun. *[FA]* [d-yun < yun$_1$]

sing, I. *V* usjun. *[IA]* [d-yun < yun$_1$]

sing, I did not. *V* chasjun. *[PN]* [d-yun < yun$_1$]

sing, I do not. *V* chasusjun. *[IN]* [d-yun < yun$_1$]

sing, I will. *V* 'utesjun. *[FA]* [d-yun]

sing, I will. *V* tesjun. *[FA]* [d-yun]

sing, they are going to. *V* hutejun. *[FA]* [d-yun]

sing, we (2) will. *V* 'utadujun. *[FA]* [d-yun]

sing, we (3+) will. *V* 'uztejun. *[FA]* [d-yun]

sing, we (3+) will not. *V* ts'utesjun. *[FN]* [d-yun]

sing, you (1) will. *V* 'utanjun. *[FA]* [d-yun]

sing, you (2+) will. *V* 'utehjun. *[FA]* [d-yun]

sing, you (2+) will. *V* tehjun. *[FA]* [d-yun < yun$_1$]

single, he/she is. *V* bu'at hoolah. *[IA]* [0-lah] Literally, "he/she has no wife".

sinus cavities, our. *N* neninchustah.

sinus cavities, their. *N* hubuninchustah.

sinus cavity. *N* -ninchustah.

sinus cavity, his/her. *N* buninchustah.

sinus cavity, my. *N* sninchustah.

sinus cavity, your (1). *N* nninchustah.

sister. *N* -ulhtus.

sister, his/her. *N* bulhtus.

sister, his/her. *N* yulhtus.

sister, his/her older. *N* buyat.

sister, his/her own. *N* dulhtus.

sister, his/her younger. *N* budesyaz.

sister, my. *N* sulhtus.

sister, my older. *N* syat.

sister, my younger. *N* sdesyaz.

sister, older. *N* -yat.

sister, our. *N* nelhtus.

sister, our. *N* neyulhtus.

sister, our older. *N* neyat.

sister, our younger. *N* nedesyaz.

sister, their. *N* hubulhtus.

sister, their older. *N* hubuyat.

sister, their younger. *N* hubudesyaz.

sister, younger. *N* -destl'oh.

sister, younger. *N* -desyaz.

sister, your (1). *N* nyulhtus.

sister, your (1) older. *N* nyat.

sister, your (1) younger. *N* ndesyaz.

sister, your (2+). *N* nawhulhtus.

sisters. *N* lhulhtuske.

sisters. *N* -ulhtuske. This is the plural of -ulhtus.

sister-in-law, his/her. *N* bughe.

sister-in-law, my. *N* sghe.

sister-in-law, our. *N* neghe.

sister-in-law, their. *N* hubughe.

sister-in-law, your (1). *N* nghe.

Sitka Mountain Ash. [Sorbus sitchensis] *N* ch'ok.

sitting, he/she is. *V* usda. *[IA]* [0-da < da_1]

sitting, I am. *V* susda. *[IA]* [0-da < da_1]

sitting, they (2) are. *V* huzke. *[IA]* [0-ke < ke_1]

sitting, they (3+) are. *V* hudelhts'i. *[IA]* [l-ts'i < $ts'i_1$]

sitting, we (2) are. *V* siduke. *[IA]* [0-ke < ke_1]

sitting, we (3+) are. *V* ts'udelhts'i. *[IA]* [l-ts'i < $ts'i_1$]

sitting, we (3+) are. *V* uzdelhts'i. *[IA]* [lh-ts'i < $ts'i_1$]

sitting, you (1) are. *V* sinda. *[IA]* [0-da < da_1]

sitting, you (2) are. *V* sahke. *[IA]* [0-ke < ke_1]

sitting, you (3+) are. *V* delhuts'i. *[IA]* [l-ts'i < $ts'i_1$]

six [abstract]. *NUM* lhk'utakiwh.

six [generic]. *NUM* lhk'utak'i.

six [human]. *NUM* lhk'utanun.

six [locative]. *NUM* lhk'utadun.

six [multiplicative]. *NUM* lhk'utat.

sixty [generic]. *NUM* lhk'utat lanezi.

skating around, he/she is. *V* nuzoot. *[IA]* [0-zoot < $zoot_1$]

skating around for pleasure, we (2) are. *V* nunidudzoot. *[IA]* [0-zoot < $zoot_1$]

skewer. *N* daskwun.

skidoo. *N* yusk'ut nugoo.

skin. *N* -zuz.

skin. *N* 'uzuz.

skin, deer. N yests'ezuz.

skin jacket. *N* 'uzuz dzoot.

skinny, he/she is. *V* budaningi. *[IA]* [0-gi]

skinny, he/she is. *V* oodaningi. *[IA]* [0-gi]

skinny, I am. *V* sdaningi. *[IA]* [0-gi]

skirt. *N* ts'itukdukw.

skull. *N* -tsints'un.

skull. *N* tsits'un.

skull, his/her. *N* butsints'un.

skull, my. *N* stsints'un.

skull, your (1). *N* ntsints'un.

skulls, our. *N* netsints'un.

skulls, their. *N* hubutsints'un.

skunk. [Mephitis mephitis] *N* hoonliz.

sky. *N* yat.

sky, clear. *N* yazai.

slapped it thoroughly, he. *V* yayankat. *[PA]* [0-kat < kat_3]

sled. *N* nugoo.

sleep, I will go to. *V* nantestelh. *[FA]* [0-telh < ti_4]

sleep, they will. *V* hutetez. *[FA]* [0-tez]

sleep, we (2). *V* odutez. *[OA]* [0-tez]

sleep, we (3+) will. *V* nants'untetus. *[FA]* [0-tus]

sleep, we (3+) will. *V* uztetez. *[FA]* [0-tez]

sleep, you (1). *V* sinti. *[IA]* [0-ti < ti_4]

sleep, you (1) are going to. *V* nantantelh. *[FA]* [0-telh]

sleep, you (1) may. *V* onte. *[OA]* [0-te < ti_4]

sleeping, I am. *V* susti. *[IA]* [0-ti < ti_4]

sleeping, they are. *V* huztez. *[IA]* [0-tez]

sleeping, we (2) are. *V* sidutez. *[IA]* [0-tez]

sleeping, we (3+) are. *V* ts'uztez. *[IA]* [0-tez]

sleepy, we are. *V* bulh nenideltsut. *[IA]* [l-tsut]

sleepy, you (1) are. *V* bulh nideltsut. *[IA]* [l-tsut]

sleepy, you (2+) are. *V* bulh nahnideltsut. *[IA]* [l-tsut]

slept, we (2). *V* idutez. *[PA]* [0-tez]

slept, we (3+). *V* ts'antez. *[PA]* [0-tez]

slept, you (1). *V* inte. *[PA]* [0-te]

slime, fish. *N* -tl'us.

slip. *N* deyoh tl'asus.

slippery it is. *V* lhkut. *[IA]* [lh-kut]

slowly. *ADV* sa.

slug. *N* talukwya.

slug, a variety of. *N* gesya.

small. *SUFFIX* -tsol.

small. *SUFFIX* -yaz. Attaches to nouns and yields a new noun.

small, it is. *V* ntsol. *[IA]* [0-tsol < $tsol_1$]

small it is!, how. *V* lhe'ultsolyaz. *[IA]* [l-tsol < $tsol_1$]

smallpox. *N* natsauneltsut.

smart, he/she is. *V* beni hooni. *[IA]* [0-ni] Literally, "his mind exists".

smashed, it is easily. *V* k'unuduwul. *[IA]* [0-wul] E.g. glass.

smashed into many pieces, it is. *V* yadanwul. *[PA]* [0-wul]

smeared, it has been. *V* talhutloh. *[PA]* [l-tloh]

smearing, he/she is. *V* talhlhoh. *[IA]* [lh-tloh]

smearing, we (2) are. *V* idullhah. *[IA]* [lh-tlah]

smearing it, he/she is. *V* yulhlhah. *[IA]* [lh-tlah]

smells good, it. *V* sooltsun. *[IA]* [l-tsun]

Smilacina amplexicaulis. [Maianthemum racemosum] *N* lhiluzchun. This species is also known by common names Canada mayflower, sugarberry, False Solomon's Seal.

smiling, I am. *V* dlonuszun. *[IA]* [0-zun]

smiling, they are. *V* dlohuninzun. *[IA]* [0-zun]

smiling, we (3+) are. *V* dlots'uninzun. *[IA]* [0-zun]

smoke. *N* lhut.

smoke, I. *V* 'ust'ot. *[IA]* [0-t'ot < $t'ot_1$]

smoke house. *N* slenyah.

smoking marijuana, he/she will be. *V* t'antet'ot. *[FA]* [0-t'ot < $t'ot_1$]

smooth, it is. *V* lhkut. *[IA]* [lh-kut]

smoulders, it. *V* diyooh. *[IA]* [0-yooh < $yooh_1$]

snail. *N* talook ts'it.

snake. *N* tl'ughus.

snare. *N* bilh.

snow. *N* yus.

snow, it is blowing. *V* yus bulh nilhts'i. *[IA]* [lh-ts'i < $ts'i_2$]

snow machine. *N* yusk'ut nugoo.

snow up high. *N* dahyus.

snowbird. *N* kwuz oonde

snowing, it is. *V* najas. *[IA]* [d-yas]

snowing, it is. *V* nawhujas. [abs: wh] *[IA]* [d-yas]

snowshoe hare. [Lepus americanus pal.] *N* gah. The only species native to the territory is the Snowshoe Hare, also known as Varying Hare, Lepus americanus. The term is, however, applied to other varieties of rabbit and hare.

soap. *N* latl'us.

soapberry. [Shephercia canadensis] *N* nawus.

socks. *N* ketul.

soft, it is. *V* dutleh. *[IA]* [0-tleh < tle_2]

soldiers. *N* bahne.

soldiers. *N* nubah.

sole, his/her. *N* buketl'ah.

sole, my. *N* sketl'ah.

sole, your (1). *N* nketl'ah.

sole of foot. *N* -ketl'ah.

soles, our. *N* neketl'ah.

soles, their. *N* hubuketl'ah.

sombrero. *N* ts'ahbal.

some more. *N* 'uyoocha.

son, her. *N* buyaz.

son, his. *N* buye'.

son, man's. *N* -ye'.

son, my. *N* sye'.

son, our. *N* neye'.

son, their. *N* hubuye'.

son, woman's. *N* -yaz.

son, your (1). *N* nye'.

song. *N* shun. Poss: yun.

song, clan. *N* dedohneyun.

song, clan. *N* dedohshun.

song, face-saving. *N* kayanahudlishun.

song, he/she broke into. *V* shun hadan'ai. *[PA]* [0-'ai]

song, longing. *N* k'ehududlishun. A song expressing longing for someone.

song, my. *N* syun.

song, our. *N* neyun.

song, personal. *N* duneyun.

song, public. *N* netsiyun.

song, ritual. *N* duyun-ishun. Ritual songs are sung only by 'uza'.

song, shaming. *N* lhuhoonultsasshun.

song, sung during the distribution of fruit at a balhats. *N* maiyun.

song, sung to embarass someone. *N* 'udulhtsiyun.

song, sweetheart. *N* 'indumanukshun. A song sung when couples dance into a balhats.

son-in-law. *N* -ghundan.

son-in-law, his/her. *N* bughundan.

son-in-law, my. *N* sghundan.

son-in-law, our. *N* neghundan.

son-in-law, their. *N* hubughundan.

son-in-law, your (1). *N* nghundan.

sore, it is. *V* nduda. *[IA]* [0-da < da_2]

sore, it is. *V* whunduda. [abs: wh] *[IA]* [0-da < da_2]

sore feeling. *N* kwenduda.

soup. *N* tazul.

soup, fish. *N* lhook tazul.

soup, hamburger. *N* 'utsun tazul.

soup, meat. *N* 'utsun tazul.

sour, it is. *V* dunink'ooz. *[IA]* [0-k'ooz]

south, from the. *ADV* ntsizde.

south wind. *N* yoo'az nilhts'i.

southward. *N* nun.

spade. *N* lubel.

spanked it thoroughly, he/she. *V* yayankat. *[PA]* [0-kat < kat_3]

speak, I. *V* yalhduk. *[IA]* [lh-duk < duk_1]

speak, I will. *V* yatelhduk. *[FA]* [lh-duk < duk_1]

speak, they will. *V* yahutelhduk. *[FA]* [lh-duk < duk_1]

speak, we (2). *V* yaiduldu k. *[IA]* [lh-duk < duk_1]

speak, we (2) will. *V* yataduldu k. *[FA]* [l-duk < duk_1]

speak, we (3+). *V* yats'ulhduk. *[IA]* [lh-duk < duk_1]

speak, we (3+) will. *V* yaztelhduk. *[FA]* [lh-duk < duk_1]

speak, you (1). *V* yailhduk. *[IA]* [lh-duk < duk_1]

speak, you (1) will. *V* yatalhduk. *[FA]* [lh-duk < duk_1]

speaking, they are. *V* yahulhduk. *[IA]* [lh-duk < duk_1]

speaks, he/she. *V* yalhtuk. *[IA]* [lh-tuk]

speaks (language), he/she. *V*

k'uyalhtuk. *[IA]* [lh-tuk]

speaks broken Carrier, he/she. *V* k'uneduyus. *[IA]* [0-yus]

speaks very loudly, he. *V* duldlut. *[IA]* [l-dlut]

spear. *N* ts'oh.

spear, fish. *N* dagwut.

sperm, fish. *N* lhooktl'uz.

Sphagnum moss, Common Red. [Sphagnum capillaceum] *N* ts'al. Locally known as Diaper Moss.

spider. *N* tsootsi.

spider. *N* whutsootsi.

spider web. *N* tsootsibilh.

spine. *N* -yunts'un.

spine. *N* -t'akts'un.

spine. *N* -yun.

spiritual traveler. *N* duyun.

split (st) into many pieces lengthwise, he/she. *V* lhtadelnat. *[PA]* [l-nat]

split it in two lengthwise, he/she. *V* lhts'eidalnat. *[PA]* [l-nat]

splitting (st), he/she is. *V* dulhnat. *[IA]* [lh-nat]

spoke, they. *V* yahalhduk. *[PA]* [lh-duk < duk$_1$]

spoke, we (3+). *V* yats'alhduk. *[PA]* [lh-duk < duk$_1$]

spoon. *N* tsunts'alh.

spouse's sibling. *N* -ghe.

spring. *N* 'olulh.

Spring Salmon. *N* ges.

Spruce. *N* ts'oo.

spruce, black. [Picea mariana] *N* ts'oobez.

spruce bark. *N* ts'ooladel.

spruce roots. *N* khi.

spurts, it. *V* tujulh. *[IA]* [d-yulh < yul$_3$]

squash. *N* khusih.

squirrel. [Tamiasciurus hudsonicus col.] *N* tsaluk.

squirrel, flying. [Glaucomys sabrinus alp.] *N* ts'unulhbuz.

stable for cattle. *N* musdusbayah.

stable for horses. *N* yeztlibayah.

stand amongst them, I. *V* hubutususyin. *[IA]* [0-yin < yin$_1$]

ag Lake. *N* Yests'e Bunk'ut.

star. *N* sum.

stars, shooting. *N* kwuncho nainkat.

stars, shooting. *N* sum nainkat.

stars, shooting. *N* sum nakat .

stars, shooting. *N* suncho nainkat.

started on foot, we (3+) have. *V* uztezdil. *[PA]* [0-dil < dil$_1$]

Stellakoh River. *N* Stellakoh.

Stellakoh village. *N* Stellakoh.

step-father. *N* -tai.

step-father, his/her. *N* butai.

step-father, my. *N* stai.

step-father, our. *N* netai.

step-father, their. *N* hubutai.

step-father, your (1). *N* ntai.

step-mother. *N* -k'i.

step-son. *N* -ulhye'.

step-son, my. *N* sulhye'.

step-son, your (1). *N* nyulhye'.

stereo. *N* bets'ujun.

sternum. *N* -t'ayukts'un.

stew. *N* tazul.

stewing (st), I am. *V* uslez. *[IA]* [0-lez]

stick. *N* duchun.

still. *ADV* 'awhuz.

Stinging Nettle. [Urtica dioica] *N* hoolhts'ik.

stomach. *N* -but.

stomach, bottom of the. *N* -buttl'ah.

stomach, his/her. *N* bubut.

stomach, my. *N* sbut.

stomach, my. *N* sbut.

stomach, the bottom of his/her. *N* bubuttl'ah.

stomach, the bottom of my. *N* sbuttl'ah.

stomach, the bottom of your (1). *N* mbuttl'ah.

stomach, your (1). *N* mbut.

stomachs, our. *N* nebut.

stomachs, the bottoms of our. *N* nebuttl'ah.

stomachs, the bottoms of their. *N* hububuttl'ah.

stomachs, their. *N* hububut.

stone. *N* tse.

story. *N* 'udada.

stove. *N* sto.

stove, your (1) big. *N* nyestocho.

stove, air-tight. *N* chalhyalsto.

stove, big. *N* stocho.

stove, cook. *N* 'ook'et'as.

stove, his/her big. *N* bustocho.

stove, his/her cook. *N* be'ook'et'as.

stove, my big. *N* sestocho.

stove, my cook. *N* se'ook'et'as.

stove, our big. *N* nestocho.

stove, our cook. *N* ne'ook'et'as.

stove, their big. *N* hubustocho.

stove, their cook. *N* hube'ook'et'as.

stove, your (1) cook. *N* nye'ook'et'as.

stove poker. *N* koonangus.

strawberry. [Fragaria species] *N* 'indzi.

Strawberry Blite. [Chenopodium capitatum] *N* koonk'etsih.

stretch, we (3+) may. *V* ts'oobuz. *[OA]* [0-buz < buz_1]

stretch, you (1) may. *V* ombuz. *[OA]* [0-buz < buz_1]

stretch (st), I may. *V* noosbuz. [abs: n] *[OA]* [0-buz < buz_1]

stretch (st), I may. *V* wusbuz. [abs: gen] *[OA]* [0-buz < buz_1]

stretched it, they. *V* huyuzbuz. *[PA]* [0-buz]

stretching frame. *N* busdi.

stretching it, he/she is. *V* yubuz. *[IA]* [0-buz < buz_1]

stroller. *N* 'uski benunulbas.

strong, it is. *V* ulhtus. *[IA]* [lh-tus < tus_2] This can describe either physical strenghth or the strength of a beverage.

stubborn, to be. *V* -tsi dunuts'un. *[IA]* [0-ts'un] This literally means "X's head is hard", so it is conjugated for "subject" by possessive prefixation of tsi "head".

sturgeon. [Acipenser transmontanus] *N* lhkw'encho.

Subalpine Fir. [Abies lasiocarpa] *N* ts'ootsun. Locally known as Balsam Fir.

Subalpine Fir bark. *N* ts'ootsunt'ooz.

subject, his/her. *N* be'ustlen.

subject, my. *N* se'ustlen.

subject, our. *N* ne'ustlen.

subject, their. *N* hube'ustlen.

subject, your (1). *N* nye'ustlen.

subject (of ruler). *N* 'ustlen.

sucked on me, it has. *V* suzt'ukw. *[PA]* [0-t'ukw < $t'ukw_1$]

sucker species. *N* dugoos.

sucker species. *N* dulgiyaz.

sucker species. *N* goosibai.

sucking, he/she is. *V* ulht'ukw. *[IA]* [lh-t'ukw]

suffered greatly, he/she. *V* dzah nusuzut. *[PA]* [0-zut]

suffering, I am. *V* dzah nuszut. *[IA]* [0-zut]

sugar. *N* lusook.

sugar. *N* soogah.

sugar, my. *N* sulusook.

Sugarberry. [Maianthemum racemosum] *N* lhiluzchun. This species is also known by the older scientific name Smilacina amplexicaulis.

sugar sack. *N* soogah 'uzus.

summer. *N* shen.

summit. *N* tl'adak.

sun. *N* sa.

Sunday. *N* Dimosdzen.

sundogs. *N* sadzak'uilke.

sunny, it is. *V* hasadanat *[IA]* [0-nat]

sunny and hot, it is. *V* sa dendi. *[IA]* [0-di]

sunrise, red. *N* sa ilk'un han'ai.

sunset, red. *N* lhk'un nasain'aih.

sunset, red. *N* nasain'ai hoolk'un.

Sutherland River. *N* Natl'alikoh. Runs in a northwesterly direction into the southeast tip of Babine Lake.

strawberry. [Fragaria species] *N* 'indzi.

Sutherland River. *N* Natl'ilikoh.

swallow. *N* schas.

swamp. *N* whulhtsultah.

swamp. *N* yak'ut.

swan. *N* ts'uncho.

swather (machine for mowing hay). *N* tl'o bedudut'as.

sweat. *N* tsentoo.

sweat lodge. *N* tsezul ba yah.

sweetie. *INT* goolh. Endearment addressed children and younger brothers and sisters.

sweet, it (generic) is. *V* lhuki. *[IA]* [l-ki]

sweet, it is. *V* lhuk'i. *[IA]* [0-k'i]

sweet, it is. *V* nulk'i. [abs: n] *[IA]* [l-k'i]

sweetheart song. *N* 'indumanukshun. A song sung when couples dance into a balhats.

switch (for whipping). *N* tut'oh.

table. *N* ludab.

table. *N* ook'utudai.

table, his/her. *N* be'ook'utudai.

table, his/her. *N* buludab.

table, my. *N* se'ook'utudai.

table, my. *N* sludab.

table, our. *N* ne'ook'utudai.

table, our. *N* neludab.

table, their. *N* hube'ook'utudai.

table, their. *N* hubuludab.

table, your (1). *N* nludab.

table, your (1). *N* nye'ook'utudai.

table cloth. *N* ludab k'usulhchooz.

Tahltan person. *N* Talht'en.

tail, fish. *N* lhookchetl'a.

tail lights of vehicle. *N* butsul kwun dilhk'ai.

tailbone. *N* -chets'un.

tailbone, his/her. *N* buchets'un.

tailbone, my. *N* schets'un.

tailbone, your (1). *N* nchets'un.

tailbones, their. *N* hubuchets'un.

take from you (1), I am going to. *V* nghuteschulh. *[FA]* [0-chulh < $choot_1$]

take it from you (1), they will. *V* nghahitelhchulh. *[FA]* [lh-chulh < $choot_1$]

taken aback, I am. *V* sba hooncha. *[IA]* [0-cha < cha_1]

taking a deliberate look at each other, they are. *V* hubuhunilh'en. *[IA]* [lh-'en < $'en_1$]

Takla person. *N* Nadot'en.

talk dirty, to. *V* -zek hoontsi'. [IA] [0-tsi' < tsi'_1] This is literally "X's mouth is dirty". The "subject" is therefore indicated by possessive prefixation of -zek "mouth", e.g. buzek hoontsi' "his mouth is dirty", i.e. "he/she talks dirty".

talking loudly, he/she is going in a loop. *V* nuduldlut. *[IA]* [l-dlut]

talking to you (1), I am. *V* nts'uyasduk. *[IA]* [lh-duk]

tall, he/she is. *V* nyez. *[IA]* [0-yez]

tall, he/she is. *V* nyiz. *[IA]* [0-yiz]

tall, he/she is comparatively. *V* 'ilyez. *[IA]* [l-yez < yez_1]

tall, I am. *V* usyez. *[IA]* [0-yez]

tall, I am. *V* usyiz. *[IA]* [0-yiz]

tall, it (generic) is. *V* nyez. *[IA]* [0-yez]

tall, it is. *V* hoonyiz. [abs: wh] *[IA]* [0-yiz]

tall, they are. *V* hinyez. *[IA]* [0-yez]

tall, we (2) are. *V* idujez. *[IA]* [0-yez]

tall, we (3+) are. *V* ts'inyez. *[IA]* [0-yez]

tall, you (1) are. *V* inyez. *[IA]* [0-yez]

tall, you (2+) are. *V* uhyez. *[IA]* [0-yez]

tall man. *N* duneyez.

tall person. *N* dunecho. Duoplural: dunechone.

tall woman. *N* ts'ekeyez. Duoplural: ts'ekooyez.

tamarack. [Larix laricina] *N* netsi'ul.

tamarack. *N* kasdzoon.

tan, I got a. *V* nisdutsil. *[PA]* [d-tsil]

tap-dancing, he/she is. *V* duke be 'ulhghalh. *[IA]* [lh-ghalh < ghalh$_1$]

tastes good, it. *V* nulk'i. [abs: n] *[IA]* [l-k'i]

Tatin Lake. *N* Tanyiz Bun.

tea. *N* ludi.

tea, my. *N* sludi.

tea, their. *N* hubuludi.

tea pot. *N* ludibot.

tea pot. *N* ludi 'oosa'.

teacher. *N* dune hodulh'eh.

teaching, he/she is. *V* hodulh'eh. *[IA]* [lh-'eh < 'eh$_1$]

teaching him/her, he/she is. *V* whuyodulh'eh. *[IA]* [lh-'eh < 'eh$_1$]

teaching me, he/she is. *V* whusodulh'eh. *[IA]* [lh-'eh < 'eh$_1$]

teaching them, he/she is. V hobudulh'eh. *[IA]* [lh-'eh < 'eh$_1$]

teaching you (1), he/she is. *V* whunyodulh'eh. *[IA]* [lh-'eh < 'eh$_1$]

teeth. *N* -ghoo.

teeth, his/her. *N* bughoo.

teeth, his/her own. *N* dughoo.

teeth, my. *N* sghoo.

teeth, our. *N* neghoo.

teeth, their. *N* hubughoo.

teeth, your (1). *N* nghoo.

telephone. *N* nilhdza beyaztuk.

telephone. *N* beyatuk.

television. *N* nul'en-i.

television, his/her. *N* bunul'en-i.

television, my. *N* snul'en-i.

television, our. *N* nenul'en-i.

television, their. *N* hubunul'en-i.

television, your (1). *N* nyenul'en-i.

ten [abstract]. *NUM* laneziwh.

ten [generic]. *NUM* lanezi.

ten [human]. *NUM* lanezinun.

ten [locative]. *NUM* lanezidun.

ten [multiplicative]. *NUM* lanezit.

tentatively. *ADV* khuntusinilhholhdulh.

territory. *N* keyah.

testicles. *N* 'ughez.

testicles. *N* -ghez.

testicles, his/her. *N* bughez.

testicles, my. *N* sghez.

testicles, our. *N* neghez.

testicles, their. *N* hubughez.

testicles, your (1). *N* nghez.

than. *PP* -anus.

thank you (one person to more than one). *VP* slahja. *[PA]*

thank you (one person to one person). *V* snachailya. *[PA]* [l-ya]

thank you (one person to one person). *VP* sla inja. *[PA]*

thank you (1), we (3+). *V* nenachailya. *[PA]* [l-ya]

thank you (2+), I. *V* snachalhuya. *[PA]* [l-ya]

that's all, the end. *INT* 'awetze.

theirs. *PRO* hubuch'i'.

them. *PRO* 'enne.

themselves. *PRO* hudich'ah.

themselves, by. *PRO* hudich'ah.

therapist. *N* duyun.

there. *ADV* 'et.

there is not. *V* hoolah. *[IA]* [0-lah]

thermometer. *N* kw'uzbenul'en.

thermos. *N* be'nezul.

they. *PRO* 'enne.

thigh. *N* -wuz.

thigh bone. *N* -wuzts'un.

thighs, his/her. *N* buwuz.

thighs, my. *N* swuz.

thighs, our. *N* newuz.

thighs, their. *N* hubuwuz.

thighs, your (1). *N* nwuz.

thimbleberry. [Rubus parviflorus] *N* dukdunenkai.

things we (3+) use. *N* bene'ts'ut'en.

think, I. *V* nuninuszut. *[IA]* [0-zut < zit$_4$] This describes thinking as a mental activity, not the mere holding of an opinion.

think, I will. *V* nuninteszut. *[FA]* [0-zut < zit$_4$]

think, they. *V* nuninuhinzut. *[IA]* [0-zut < zit$_4$]

think, they will. *V* nunihuntezut. *[FA]* [0-zut < zit$_4$]

think, we (2). *V* nuninidudzut. *[IA]* [0-zut < zit$_4$]

think, we (2) will. *V* nunintadudzut. *[FA]* [0-zut < zit$_4$]

think, you (1). *V* nunininzut. *[IA]* [0-zut < zit$_4$]

think, you (1) will. *V* nunintanzut. *[FA]* [0-zut < zit$_4$]

think about it, I will. *V* hukw'unintelhdzut. *[FA]* [lh-zut]

think about it, they will. *V* hukw'unihuntelhdzut. *[FA]* [lh-zut]

think about it, we (2) will. *V* hukw'unintaduldzut. *[FA]* [lh-zut]

think about it, we (3+) will. *V* hukw'units'untelhdzut. *[FA]* [lh-zut]

think about it, you (1). *V* hukwunilhdzut. *[IA]* [d-zut]

think about it, you (1) will. *V* hukw'unintalhdzut. *[FA]* [lh-zut]

think that, we (3+). *V* ts'uninzun. *[IA]* [0-zun < zun$_1$]

thinking about it, I am. *V* hukw'uninulhdzut. *[IA]* [lh-zut]

thinking about it, they are. *V* hukw'unihunilhdzut. *[IA]* [lh-zut]

thinking about it, we (2) are. *V* hukw'uniniduldzut. *[IA]* [lh-zut]

thinking about it, we (3+) are. *V* hukw'uniznilhdzut. *[IA]* [lh-zut]

thinking about it, you (1) are. *V* hukw'uninilhdzut. *[IA]* [lh-zut]

third person singular possessor. *PREFIX* bu-. Attaches to class 1 nouns.

thirsty, I am. *V* taoosde. *[IA]* [0-de]

thirsty, they are. *V* tahoosde. *[IA]* [0-de]

thirsty, you (1) are. *V* taoonde. *[IA]* [0-de]

thirty [generic]. *NUM* tat lanezi.

thought about it, I. *V* hukw'uninelhdzut. *[PA]* [lh-zut]

thought about it, we (3+). *V* hukw'uniznalhdzut. *[PA]* [lh-zut]

thought about it, you (1). *V* hukw'uninalhdzut. *[PA]* [lh-zut]

thought that, we (3+). *V* uznanzin. *[PA]* [0-zin]

thought that, you (1). *V* nanzin. *[PA]* [0-zin]

thread. *N* yabatseh.

three [abstract]. *NUM* tawh.

three [generic]. *NUM* tak'i.

three [human]. *NUM* tanun.

three [locative]. *NUM* tadun.

three [multiplicative]. *NUM* tat.

thrice. *NUM* tat.

throat, his/her. *N* buzesdak.

throat, my. *N* szesdak.

throat, your (1). *N* nzesdak.

throat (interior). *N* -zesdak.

throats, our. *N* nezesdak.

throats, their. *N* hubuzesdak.

thrush, varied. *N* 'ints'ilh.

thumb. *N* -ninchuz.

thumb, his/her. *N* buninchuz.

thumb, his/her own. *N* duninchuz.

thumb, my. *N* sninchuz.

thumb, your (1). *N* unninchuz.

thumbs, our. *N* neninchuz.

thumbs, their. *N* hubuninchuz.

thunders, it. *V* detnik 'utni. *[IA]* [d-ni]

Thursday. *N* Whulhditdzen.

thus. *ADV* whuzun'a.

tie yourself down, you (1). *V* denawhudinges. *[IA]* [0-ges]

times, many. *Q* lhat.

tires. *N* kesgwut.

to. *PP* ts'i.

to. *PP* ts'un.

to me. *PPC* sts'un.

tobacco. *N* dek'a.

tobacco. *N* ts'ut'ot.

tobacco, my. *N* sdek'a.

tobacco pipe. *N* dek'atse.

toboggan. *N* 'usdloos.

today. *N* k'anditdzen.

today. *N* k'andzen.

toe. *N* -kelamai.

toe, big. *N* -kelamaicho.

toe, big. *N* -kelascho. This means the big toe, not just any large toe, that is, the rightmost toe on the left foot or the leftmost toe on the right foot.

toe, fourth. *N* -kelamaiyazts'unts'e. The toe next to the little toe.

toe, his/her big. *N* bukelamaicho.

toe, his/her big. *N* bukelascho.

toe, his/her fourth. *N* bukelamaiyazts'unts'e.

toe, his/her little. *N* bukelamaiyaz.

toe, his/her long. *N* bukelamaiyez.

toe, his/her middle. *N* bukelamaiyezts'unts'e.

toe, little. *N* -kelamaiyaz. This refers to the outermost toe, the leftmost toe on the left foot, the rightmost toe on the right foot, not just any small toe.

toe, long. *N* -kelamaiyez. This names the toe next to the big toe, not just any long toe.

toe, middle. *N* -kelamaiyezts'unts'e.

toe, my big. *N* skelamaicho.

toe, my big. *N* skelascho.

toe, my fourth. *N* skelamaiyazts'unts'e.

toe, my little. *N* skelamaiyaz.

toe, my long. *N* skelamaiyez.

toe, my middle. *N* skelamaiyezts'unts'e.

toe, your (1) big. *N* nkelamaicho.

toe, your (1) big. *N* nkelascho.

toe, your (1) fourth. *N* nkelamaiyazts'unts'e.

toe, your (1) little. *N* nkelamaiyaz.

toe, your (1) long. *N* nkelamaiyez.

toe, your (1) middle. *N* nkelamaiyezts'unts'e.

toe spaces, his/her. *N* buketl'i.

toe spaces, our. *N* neketl'i.

toe spaces, their. *N* hubuketl'i.

toe spaces, your (1). *N* nketl'i.

toenail. *N* -kelagi.

toenail. *N* -kengi.

toenail, his/her. *N* bukelagi.

toenail, his/her. *N* bukengi.

toenail, my. *N* skelagi.

toenail, my. *N* skengi.

toenail, our. *N* nekengi.

toenail, your (1). *N* nkelagi.

toenail, your (1). *N* nkengi.

toenails, our. *N* nekelagi.

toenails, their. *N* hubukelagi.

toenails, their. *N* hubukengi.

toes, his/her. *N* bukelamai.

toes, my. *N* skelamai.

toes, our. *N* nekelamai.

toes, our big. *N* nekelamaicho.

toes, our big. *N* nekelascho.

toes, our fourth. *N* nekelamaiyazts'unts'e.

toes, our little. *N* nekelamaiyaz.

toes, our long. *N* nekelamaiyez.

toes, our middle. *N* nekelamaiyezts'unts'e.

toes, spaces between. *N* -ketl'i.

toes, their. *N* hubukelamai.

toes, their big. *N* hubukelamaicho.

toes, their big. *N* hubukelascho.

toes, their fourth. *N* hubukelamaiyazts'unts'e.

toes, their little. *N* hubukelamaiyaz.

toes, their long. *N* hubukelamaiyez.

toes, their middle. *N* hubukelamaiyezts'unts'e.

toes, your (1). *N* nkelamai.

toes spaces, my. *N* sketl'i.

together with each other. *PPC* lhulh.

toilet. *N* tsan bayah.

told him/her, he/she. *V* yudani. *[PA]* [0-ni]

told me, he/she. *V* sudani. *[PA]* [0-ni]

told them, he/she. *V* hubu dani. *[PA]* [0-ni]

told us, he/she. *V* ne dani. *[PA]* [0-ni]

told us (2), he/she. *V* nawhudani. *[PA]* [0-ni]

told you (1), he/she. *V* nyudani. *[PA]* [0-ni]

told you (2+), he/she. *V* nahdani. *[PA]* [0-ni]

tomorrow. *N* bunde.

tomorrow morning. *N* bunde bundada.

tongue. *N* -tsoola.

tongue, your (1). *N* ntsoola.

tongue, his/her. *N* butsoola.

tongue, my. *N* stsoola.

tongues, our. *N* netsoola.

tongues, their. *N* hubutsoola.

tonsils. *N* -zesdazum.

tonsils, his/her. *N* buzesdazum.

tonsils, my. *N* szesdazum.

tonsils, our. *N* nezesdazum.

tonsils, their. *N* hubuzesdazum.

tonsils, your (1). *N* nzesdazum.

too. *PART* cha.

tooth. *N* -ghoo.

tooth, wisdom. *N* -naghoo.

Top Lake. *N* Hoolhtan Bunk'ut. A small lake west of Peta Lake, at the head of Duncan Creek.

toppled, it. *V* nananghiz. *[PA]* [0-ghiz < ghis$_1$]

toque. *N* jook.

toque, his/her. *N* bujook.

toque, my. *N* sjook.

toque, our. *N* nejook.

toque, their. *N* hubujook.

toque, your (1). *N* njook.

tornado. *N* nilhts'idus.

towel. *N* lusooma.

trachea. *N* -zool.

tracks. *N* 'uk'ah.

tracks. *N* k'ah.

tracks, horse. *N* yeztlik'ah.

trail. *N* ti.

tramp. *N* kegudsatalh.

transporting me by boat, he/she is. *V* selhkelh. *[IA–progressive]* [lh-kelh < ke$_2$]

trap, deadfall. *N* gooh.

trap, fish. *N* 'uk'oondzai.

trapline. *N* keyah.

trapline. *N* yuntah.

trawling line. *N* nubalhtl'ool.

tree. *N* duchun.

tree from which rabbit snare is hung. *N* nache'untut'ah.

Trembling Aspen. [Populus tremuloides] *N* t'ughus. Locally known as Poplar.

trigger for rabbit snare. *N* haltoo'.

tripe. *N* 'utsutle.

tripe. *N* chalhzus.

triplets. *N* lhulhcho.

tripped and fell, he/she. *V* dughuntezghal. *[PA]* [0-ghal < ghal$_1$]

trout, lake. [Salvelinus namaycush] *N* bet.

Trout Lake. *N* Duk'ai Hooni Bunk'ut.

trout, rainbow. *N* duk'ai.

trunk. *N* duchun khelh.

Tuesday. *N* Whulhnatdzen.

turtle. *N* kw'uyukunenkai.

turn (st), you (1). *V* dilhghis. *[IA]* [lh-ghis < ghis$_1$]

turned around, he/she was. *V* ni lhe'nintananesja. *[PA]* [d-ya]

turnips. *N* lusoosam.

twenty [generic]. *NUM* nat lanezi.

twice. *NUM* nat.

Twinberry. [Lonicera involucrata] *N* susmai.

twine used for stringing nets, making dipnets, and so forth. *N* 'ubalt'i. This term is used in reference to the twine when it is not part of the net.

twine used to string net. *N* balt'i. This form is used in reference to the twine when on the net.

twins. *N* lhtudu'alts'e.

two, they are. *V* nanehult'ah. *[IA]* [l-t'ah < t'oh$_1$]

two (locative). *NUM* nadun.

two [abstract]. *NUM* nawh.

two [generic]. *NUM* nankah.

two [generic]. *NUM* nankah.

two [human]. *NUM* nanun.

two [multiplicative]. *NUM* nat.

ugly, he/she is. *V* dunetsi'. *[IA]* [0-tsi' < tsi'$_1$] Applicable only to men.

ugly, she is. *V* ts'eketsi'. *[IA]* [0-tsi' < tsi'$_1$]

uncle. *N* 'utai. father's brother

uncle, someone's. *N* 'uz'e.

uncle (father's brother, father's sister's husband, mother's sister's husband). *N* -tai.

uncle (mother's brother). *N* -z'e.

uncle (mother's brother), his/her. *N* buz'e.

uncle (mother's brother), my. *N* suz'e.

uncle (mother's brother), our. *N* nez'e.

uncle (mother's brother), their. *N* hubuz'e.

uncle (mother's brother), your (1). *N* nyuz'e.

uncle other than mother's brother, his/her. *N* butai.

uncle other than mother's brother, our. *N* netai.

uncle other than mother's brother, their. *N* hubutai.

uncle other than mother's brother, your (1). *N* ntai.

uncle other than mother's brother, my. *N* stai.

under. *PP* yah.

underwear. *N* deyoh naih.

unnamed hill. *N* Chundooyus. The hill on the sooutheст side of Ormond Lake.

unscrew, you (1). *V* hadilhghis. *[IA]* [lh-ghis < ghis$_1$]

unyielding, it is. *V* dunuts'un. [abs: n] *[IA]* [0-ts'un < ts'un$_1$]

upper. *ADV* dusts'i.

urinary tract. *N* -lhuzkah.

urinary tract. *N* lhuzk'et.

urinary tract, his/her. *N* buluzkah.

urinary tract, my. *N* sluzkah.

urinary tract, their. *N* hubuluzkah.

urinary tract, your (1). *N* nluzkah.

urinary tracts, our. *N* neluzkah.

urine. *N* lhuz. Poss: -luz.

urine, his/her. *N* buluz.

urine, my. *N* sluz.

urine, our. *N* neluz.

urine, their. *N* hubuluz.

urine, your (1). *N* nluz.

us. *PRO* wheni.

use, things we (3+). *N* be'ts'ut'en.

used to. *V* inle. *[IA]* [0-le]

uvula. *N* -zesdak'al.

uvula. *N* -zesdalos.

uvula. *N* -zesdamai.

uvula, his/her. *N* buzesdak'al.

toque, his/her. *N* bujook.

uvula, his/her. *N* buzesdalos.

uvula, his/her. *N* buzesdamai.

uvula, my. *N* szesdak'al.

uvula, my. *N* szesdalos.

uvula, my. *N* szesdamai.

uvula, your (1). *N* nzesdak'al.

uvula, your (1). *N* nzesdalos.

uvula, your (1). *N* nzesdamai.

uvulas, our. *N* nezesdak'al.

uvulas, our. *N* nezesdalos.

uvulas, our. *N* nezesdamai.

uvulas, their. *N* hubuzesdak'al.

uvulas, their. *N* hubuzesdalos.

uvulas, their. *N* hubuzesdamai.

Vanderhoof. *N* Tehk'eilchuk.

Varied Thrush. *N* 'ints'ilh.

varying hare. [Lepus americanus pal.] *N* gah. The only species native to the territory is the Snowshoe Hare, also known as Varying Hare, Lepus americanus. The term is, however, applied to other varieties of rabbit and hare.

vase. *N* t'an bedezdla.

vehicle. *N* benugoo-i.

vehicle. *N* benuts'ugoo.

vehicle. *N* hiyeneyulhgoo.

vehicle, my. *N* benusgoo-i.

vehicle, your (1). *N* beningoo-i.

vein. *N* -zkaich'ooz.

veins, his/her. *N* buzkaich'ooz.

veins, my. *N* suzkaich'ooz.

veins, our. *N* nezkaich'ooz.

veins, their. *N* hubuzkaich'ooz.

veins, your (1). *N* nyuzkaich'ooz.

venison. *N* yests'etsun.

vest. *N* dzootdukw.

village. *N* keyah.

wagon. *N* luwagun.

waiting for me, he/she is. *V* sbalh'i. *[IA]* [lh-'i < $'i_1$]

waiting for you (1), I am. *V* mbalh'i. *[IA]* [lh-'i < $'i_1$]

walk, he/she is going to. *V* teyalh. *[FA]* [0-yalh < ya_1]

walk, he/she will not. *V* chatesyalh. *[FN]* [0-yalh < ya_1]

walk, I will. *V* tesyalh. *[FA]* [0-yalh < ya_1]

walk, they (3+) will. *V* hutedulh. *[FA]* [0-dulh < dil_1]

walk, they (3+) will not. *V* chahutesdulh. *[FN]* [0-dulh < dil$_1$]

walk, we (2) will. *V* tat'us. *[FA]* [0-'us < 'as$_1$]

walk, we (2) will not. *V* chatuzat'us. *[FN]* [0-'us < 'as$_1$]

walk, we (3+) will. *V* uztedulh. *[FA]* [0-dulh < dil$_1$]

walk, we (3+) will not. *V* chaztesdulh. *[FN]* [0-dulh < dil$_1$]

walk, you (1) will. *V* tanyalh. *[FA]* [0-yalh]

walk, you (3+) will. *V* tehdulh. *[FA]* [0-dulh]

walk in a loop, he/she will. *V* nuteya. *[FA]* [0-ya < ya$_1$]

walk in a loop, he/she will not. *V* nuchatesya. *[FN]* [0-ya < ya$_1$]

walk in a loop, I will. *V* nutesya. *[FA]* [0-ya < ya$_1$]

walk in a loop, I will not. *V* nuchatuzesyal. *[FN]* [0-yal < ya$_1$]

walk in a loop, we (2) will. *V* nutat'us. *[FA]* [0-'us < 'as$_1$]

walk in a loop, we (2) will not. *V* nuchatuzat'us. *[FN]* [0-'us < 'as$_1$]

walk in a loop, we (3+) will. *V* nuztedulh. *[FA]* [0-dulh < dil$_1$]

walk in a loop, we (3+) will not. *V* nuchaztesdil. *[FN]* [0-dil < dil$_1$]

walk in a loop, you (1) may. *V* nonya. *[OA]* [0-ya < ya$_1$]

walk in a loop, you (1) will. *V* nutanya. *[FA]* [0-ya < ya$_1$]

walk in a loop, you (1) will not. *V* nuchatuzanya. *[FN]* [0-ya < ya$_1$]

walked in a loop, he/she. *V* nusuya. *[PA]* [0-ya < ya$_1$]

walked in a loop, he/she has not. *V* nuchaiyal. *[PN]* [0-yal < ya$_1$]

walked in a loop, I. *V* nususya. *[PA]* [0-ya < ya$_1$]

walked in a loop, I have not. *V* nuchasyal. *[PN]* [0-yal < ya$_1$]

walked in a loop, they (3+) have. *V* nuhuzdil. *[PA]* [0-dil < dil$_1$]

walked in a loop, they (3+) have not. *V* nuchahidulh. *[PN]* [0-dulh < dil$_1$]

walked in a loop, we (2). *V* nusit'az. *[PA]* [0-'az < 'as$_1$]

walked in a loop, we (2) have not. *V* nuchasit'az. *[PN]* [0-'az < 'as$_1$]

walked in a loop, we (3+) have. *V* nuts'uzdil. *[PA]* [0-dil < dil$_1$]

walked in a loop, we (3+) have not. *V* nuchats'idulh. *[PN]* [0-dulh < dil$_1$]

walked in a loop, you (1). *V* nusinya. *[PA]* [0-ya < ya$_1$]

walked in a loop, you (1) have not. *V* nuchainyal. *[PN]* [0-yal < ya$_1$]

walked in a loop, you (2) have. *V* nusah'az. *[PA]* [0-'az < 'as$_1$]

walking, he/she did not set off. *V* whechaiyal. *[PN]* [0-yal < ya$_1$]

walking, he/she is. *V* tezya. *[PA]* [0-ya] This means that he has already begun to walk. It contrasts with teyalh, which means that at some point in the future he is going to walk somewhere.

walking, he/she is. *V* uyalh. *[IA–progressive]* [0-yalh < ya$_1$]

walking, he/she is not. *V* chasyalh. *[IN–prog]* [0-yalh]

walking, he/she set off. *V* wheinya.

[PA] [0-ya < ya_1]

walking, I am. *V* usyalh. *[IA–progressive]* [0-yalh < ya_1]

walking, they (3+) are. *V* hedulh. *[IA–prog]* [0-dulh < dil_1]

walking, we (2) are. *V* it'us. *[IA–prog]* [0-'us < $'as_1$]

walking, we (2) are not. *V* chasit'us. *[IN–prog]* [0-'us < $'as_1$]

walking, we (2) set off. *V* wheit'az. *[PA]* [0-'az < $'as_1$]

walking, we (3+) are. *V* ts'edulh. *[IA–prog]* [0-dulh < dil_1]

walking forward, we (3+) are. *V* nus ts'edulh. *[IA–prog]* [0-dulh < dil_1]

walking in a loop, he/she is. *V* nuya. *[IA]* [0-ya]

walking in a loop, he/she is not. *V* nuchasyah. *[IN]* [0-yah < ya_1]

walking in a loop, I am. *V* nusya. *[IA]* [0-ya < ya_1]

walking in a loop, I am not. *V* nuchasusyah. *[IN]* [0-yah < ya_1]

walking in a loop, we (2) are. *V* nit'as. *[IA]* [0-'as < $'as_1$]

walking in a loop, we (2) are not. *V* nuchasit'as. *[IN]* [0-'as < $'as_1$]

walking in a loop, we (3+) are. *V* nuts'udilh. *[IA]* [0-dilh < dil_1]

walking in a loop, we (3+) are not. *V* nuchats'usdil. *[IN]* [0-dil < dil_1]

walking in a loop, you (1) are. *V* ninya. *[IA]* [0-ya < ya_1]

walking in a loop, you (1) are not. *V* nuchasinyah. *[IN]* [0-yah < ya_1]

walking in a slow, stately manner, he/she is. *V* del'us. *[IA–prog]* [l-'us]

walking quickly, he/she is. *V* gal be 'ut'en. *[IA]* [d-'en < $'en_2$]

walking quickly, he/she is. *VP* gal be nuye. *[IA]*

wall. *N* sih.

wall, his/her. *N* busih.

wall, my. *N* susih.

wall, our. *N* nesih.

wall, their. *N* busih.

wall, your (1). *N* nsih.

wallet. *N* sooniyazuz.

want (st), I. *V* hukwa'nuszun. [abs: wh] *[IA]* [0-zun < zun_1]

want (st), I. *V* ka'nuszun. *[IA]* [0-zun < zun_1]

want (st), they. *V* ka'huninzun. *[IA]* [0-zun < zun_1]

want (st), we (2). *V* ka'nidudzun. *[IA]* [0-zun < zun_1]

want (st), we (3+). *V* ka'uzninzun. *[IA]* [0-zun < zun_1]

want (st), you (1). *V* ka'ninzun. *[IA]* [0-zun < zun_1]

want (st), you (2+). *V* ka'nahzun. *[IA]* [0-zun < zun_1]

want it, I. *V* ooka'nuszun. *[IA]* [0-zun]

want it, they. *V* hika'ninzun. *[IA]* [0-zun]

want it, we (2). *V* ooka'nidudzun. *[IA]* [0-zun]

want it, we (3+). *V* ooka'ts'uninzun. *[IA]* [0-zun]

wants (st), he/she. *V* ka'ninzun. *[IA]* [0-zun < zun_1]

was, he/she. *V* inle. *[IA]* [0-le]

was, he/she. *V* inle. *[PA]* [0-le]

wash, you (1). *V* toonaingus. *[IA]* [0-gus]

wash basin. *N* bela'ts'uldeh.

washer, clothing. *N* be'tunadugus.

washing, he/she is. *V* nawhugus. [abs: wh] *[IA]* [0-gus]

washing for himself, he/she is. *V* nawhudugus. [abs: wh] *[IA]* [d-gus]

wasp. *N* stlesicho.

waste basket. *N* bena'ulkuk.

watch. *N* tsadzi.

watch what you say. *V* neghunik whuts'inli. *[IA]* [0-li]

watchmen. *N* huwunline.

water. *N* too.

water, cold. *N* too nezk'uz.

water, his/her. *N* butoo.

water, his/her. *N* yutoo.

water, his/her own. *N* dutoo.

water, my. *N* sutoo.

water, our. *N* netoo.

water, their. *N* hubutoo.

water, your (1). *N* nyutoo.

water, your (2+). *N* nahtoo.

water hole. *N* took'et.

water lily. [Nuphar lutea] *N* khelht'az.

waterfall. *N* duyuk.

waterspout. *N* nilhts'i tainya.

waves on water. *N* tatsi.

way, this. *ADV* ndezuna.

we. *PRO* wheni.

weak, it is. *V* lait'ah. *[IA]* [0-t'ah] This may mean physically weak or may apply to flavour.

weasel. [Martes erminea/Musela rixosa] *N* nahbai. In Carrier territory the weasel usually encountered is the Short-tailed Weasel (Mustela erminea richardsoni). However this term appears to be applicable as well to the Least Weasel (Mustela rixosa) which is more rarely encountered.

web, spider. *N* tsootsibilh.

Wednesday. *N* Whulhtatdzen.

weeds in water. *N* dlat.

week. *N* yanilhghel.

weight at end of fish net. *N* tse toos'ai.

well. *ADV* soo.

well. *INT* adih.

went in a loop by boat, we (2). *V* nusiduki. *[PA]* [0-ki < ke$_2$]

west, from the. *ADV* nusde.

west, to the. *ADV* nuk.

wet, they are. *V* nalhutsul. *[IA]* [l-tsul]

wet, we (3+) are. *V* naznelhtsul. *[IA]* [l-tsul]

wet, you are. *V* naneltsul. *[IA]* [l-tsul]

whale. *N* lhook'icho.

what happened?. *V* dahooja. *[PA]* [0-ja]

what will happen?. *V* dawtenilh. *[FA]* [0-nilh]

what?. *QWH* dant'i.

what?. *QWH* di.

what!. *INT* hadi'. An expression of surprise.

whence?. *QWH* nts'ez.

where. *QWH* nts'e.

where?, from. *QWH* nts'ez.

whips around, he/she. *V* nunutsus. *[IA]* [0-tsus] he/she runs around very fast, frequently changing direction

whirlpool. *N* 'ok'et.

Whiskey Jack. [Perisoreus canadensis] *N* goozeh. Also known as Gray Jay and Canadian Jay.

Whiskey Jack. [Perisoreus canadensis] *N* gwuzih.

whistling. *N* yooyooz.

White Goosefoot. [Chenopodium album] *N* 'ut'antsi.

white, it is. *V* nulyul. [abs: n] *[IA]* [l-yul < yul$_1$]

white, it (generic) is. *V* lhuyul. *[IA]* [l-yul < yul$_1$]

white, it (d-class) is. *V* dulyul. [abs: d] *[IA]* [l-yul < yul$_1$]

white, it is. *V* whulyul. [abs: wh] *[IA]* [l-yul < yul$_1$]

white person. *N* bosdun.

white person. *N* meljah.

white person. *N* nasdlez.

whitefish. *N* lhooh.

who?. *QWH* mbe.

Whooping Crane. [Grus americana] *N* dilhtsul.

why?. *QWH* dika.

wide, it is. *V* dintel. [abs: d] *[IA]* [0-tel]

wide, it is. *V* hoontel. [abs: wh] *[IA]* [0-tel < tel$_1$]

wide, it is. *V* ntel. [abs: 0] *[IA]* [0-tel < tel$_1$]

wide as this, it is. *V* njahooltel. [abs: wh] *[IA]* [l-tel < tel$_1$] The standard of comparison is something designated by the speaker using a gesture.

widow. *N* tsandelh.

widower. *N* tsandelh.

wife. *N* 'at. Duoplural: 'atke.

wife, his/her. *N* bu'at.

wife, my. *N* s'at.

Willow Ptarmigan. [Lagopus lagopus] *N* k'azba.

willow, pussy. *N* k'edlits'ilhba.

waves on water. *N* tatsi.

willow, red. [Cornus stolonifera] *N* k'en dulk'un. Locally known as Red Willow.

wind. *N* nilhts'i.

wind, east. *N* ndazde nilhts'i.

wind pipe. *N* -zool.

wind pipe, his/her. *N* buzool.

wind pipe, my. *N* szool.

wind pipe, your (1). *N* nzool.

wind pipes, our. *N* nezool.

wind pipes, their. *N* hubuzool.

window. *N* dadent’az.

window, his/her. *N* budadent’az.

window, my. *N* sdadent’az.

window, our. *N* nedadent’az.

window, their. *N* hubudadent’az.

window, your (1). *N* ndadent’az.

windy, it is. *V* nilhts’i. *[IA]* [lh-ts’i < ts’i$_2$]

wine. *N* ts’ekootoo.

winter. *N* khit.

wisdom tooth. *N* -naghoo.

with, together. *PP* -lh.

with (by means of). *PP* be.

with him/her. *PPC* bulh.

with me. *PPC* sulh.

with them. *PPC* hubulh.

with us. *PPC* nelh.

within the house. *ADV* kooz.

without it. *PPC* bu’et.

wives. *N* ’atke.

wives, our. *N* ne’atke.

wives, their. *N* hubu’atke.

wolf. [Canis lupus] *N* yus.

wolverine. [Gulo gulo] *N* noostel.

woman. *N* ts’eke. Duoplural: ts’ekoo

woman, large. *N* ts’ekecho.

woman, respected. *N* ts’ekecho.

woman, young. *N* noh.

women. *N* ts’ekoo. This is the irregular plural of ts’eke.

wonder, I. *PART* subah.

wood chuck. [Marmota monax] *N* k’ani.

woodpecker. *N* chundulht’a.

woodpecker. *N* kwalhcho.

word. *N* khunek.

word. *N* khunik. Poss: ghunik.

work, he/she will. *V* ’utet’ilh. *[FA]* [d-’ilh]

work, we (3+) will. *V* ’uztet’ilh. *[FA]* [d-’ilh]

work, you (1) will. *V* ’utant’ilh. *[FA]* [d-’ilh]

working, he/she is. *V* ’ut’en. *[IA]* [d-’en < ’en$_2$]

working, they are. *V* ’uhut’en. *[IA]* [d-’en < ’en$_2$]

working hard, we are. *V* usnedut’en. *[IA]* [d-’en]

working hard, you (1) are. *V* undendut’en. *[IA]* [d-’en]

worried, I am. *V* ni usli. *[IA]* [0-li]

wretched, he/she is. *V* tel’en. *[IA]* [l-’en]

wretched, I am. *V* telhu’en. *[IA]* [l-’en]

wretched, they are. *V* tehul’en. *[IA]* [l-’en]

wretched, we (2) are. *V* teidul’en. *[IA]* [l-’en]

wrist. *N* -lachunnah.

wrist. *N* -lat’ah.

wrist, his/her. *N* bulachunah.

wrist, his/her. *N* bulat’ah.

wrist, my. *N* slachunah.

wrist, my. *N* slat'ah.

wrist, your (1). *N* nlachunah.

wrist, your (1). *N* nlat'ah.

wrists, our. *N* nelachunah.

wrists, our. *N* nelat'ah.

wrists, their. *N* hubulachunah.

wrists, their. *N* hubulat'ah.

write, you (1). *V* 'uk'une'inguz. *[IA]* [0-guz]

writing, they are. *V* 'uk'une'huguz. *[IA]* [0-guz]

writing, we (3+) are. *V* 'uk'unets'uguz. *[IA]* [0-guz]

yarrow plant. [Achillea millefolium] *N* lacholbai.

year. *N* yusk'ut.

year, last. *N* ghada.

year, next. *N* 'on'at yusk'ut.

year, next. *N* nade.

years, your (1). *N* nyeyusk'ut.

Yensischuck. *N* 'Indzik'et. A berry-picking area and hay-cutting meadow at I.R. 3.

yellow, it is. *V* dultl'uz. *[IA]* [l-tl'uz < tl'uz$_1$]

yellow, it is. *V* dunultl'uz. [abs: n] *[IA]* [l-tl'uz < tl'uz$_1$]

yellow, it is. *V* sa dot'en. *[IA]* [d-'en < 'en$_1$] Literally, "it is the colour of the sun".

yes. *INT* a.

yesterday. *N* hulhda.

you (2+). *PRO* nawheni.

young man. *N* chilh. Duoplurals: chilhuka, chilhke.

young men. *N* chilhuke.

young woman. *N* t'et. Duoplurals: t'edukoo, t'eduke.

wolf. [Canis lupus] *N* yus.

GLOSSARY LISTS

1. Animals

animal khunai
Bushy-tailed Wood Rat dlimcho
Bushy-tailed Wood Rat dlooncho
bat t'az
bear, black sus
bear, grizzly shas
bear, my grizzly sushas
beaver tsa
beaver, big tsacho
beaver, large tsati
bitch lhits'e
black bear, little susyaz
buffalo tl'o musdus
bull musdusdune
Canadian River Otter tsis
calf musdusyaz
cariboo whudzih
cat boos
cat, little boosyaz
cat, my sboos
cat, our neboos
cat, their hububoos
cat, your (1) mboos
cat, his/her buboos
chipmunk ts'uwhuljos
colt yeztliyaz
cougar booscho
cow musdus
cow, female musdus ts'eke
cow, milk musdus uljas
cow, my smusdus
cow, our nemusdus
cow, their hubumusdus
cow, his/her bumusdus
coyote chuntulhi
coyote chuntunuye
coyote tintulhi
deer yests'e
dentalium shell dulhbai
dinosaur dondeti
dog lhi
dog, big lhicho
dog, female lhits'e
dog, little lhitsol
dog, my slik
dog, my big slikcho
dog, my little sliktsol
dog, old lhijut
dog, our nelik
dog, our big nelikcho
dog, their hubulik
dog, their big hubulikcho
dog, worthless lhich'ul
dog, your (1) nlik

dog, your (1) big nlikcho
dog, his/her bulik
dog, his/her big bulikcho
dogs lhike
dry cow moose yats'it
elephant khunaicho
elk yusi
ermine nahbaicho
Flying Squirrel ts'unulhbuz
fisher chunihcho
fox nanguz
goat dube
grizzly bear shas
grizzly bear, my sushas
horse yeztli
horse, my syeztli
horse, our neyeztli
horse, their hubuyeztli
horse, your (1) nyeztli
horse, his/her buyeztli
horse, his/her own duyeztli
lynx wasi
Masked Shrew dlim dats'olh
Masked Shrew khunaichoya
marmot dutni
marten chunih
mastodon khunaicho
mink telhjoos
monster, prehistoric khunaicho
moose duni
moose, bull jenyo
moose, cow duni'at
moose, dry cow yats'it
mouse dats'ooz
muskrat tsek'et
otter tsis
Pack Rat dlooncho
packrat dlimcho
pig gugoos
porcupine 'oojoonih
porcupine 'ujoohnih
porcupine duch'ukw
porcupine duneza'
puppy lhiyaz
rabbit gah
shrew, masked dlim dats'olh
shrew, masked khunaichoya
skunk hoonliz
slug talukwya
slug, a variety of gesya
squirrel tsaluk
squirrel, flying ts'unulhbuz
weasel nahbai
wolf yus
wolverine noostel
wood chuck k'ani

2. Fish and Shellfish

burbot tsintel
char bet
clam dulkw'ah besk'um
Dolly Varden tsabai
eel tselk'eh
female fish k'ooni
fish lhook
fish, female k'ooni
fish, male tl'uzni
herring eggs chilchunk oon
kokanee gestl'ah
lake trout bet

ling cod tsintel
male fish tl'uzni
Northern Pikeminnow khusih
Northwest Dace daltsi'
Peamouth Chub daltsi'
perch lhooz
Spring Salmon ges
salmon talook
salmon, Spring ges
salmon, steelhead sdle
squawsh khusih
sturgeon lhkw'encho
sucker species dugoos
sucker species dulgiyaz
sucker species goosibai
trout, lake bet
trout, rainbow duk'ai
whale lhook'icho
whitefish lhooh

3. Amphibians and Reptiles

dinosaur nondeti
frog dulkw'ah
salamander chunlai
turtle kw'uyukunenkai

4. Birds

American Crow datsan
American Robin sooh
Bald Eagle sbalyan
Blue Grouse dihcho
Blue Heron tehhoongook
bald eagle tsebalyan
bird dut'ai
blackbird, red-winged ch'uk
bluejay teh gwuzeh
Canada goose khoh
Common Loon dadzi
Common Raven datsancho
Crane, Whooping dilhtsul
chickadee ts'usgak
chicken ligok
crane, sandhill dilh
crow datsan
duck dut'ai
Eagle, Bald sbalyan
eagle, bald tsebalyan
eagle, golden sjul
Fool's Hen nat'ah
falcon, peregrine hoolht'as
Golden Eagle sjul
Grey Jay gwuzih
Grouse, Spruce nat'ah
giant bird, extinct dune oolhdzai
goose khoh
goose, Canada khah
gosling khahyaz
grouse, ruffed 'utsut
hawk dut'ai ooltas
hawk ligok hoolht'as
heron, blue tehhoongook
hummingbird ts'unulhduz
Jay, Canadian goozeh
Jay, Canadian gwuzih
Jay, Gray goozeh
Jay, Grey gwuzih
Kingfisher tehgwuzeh
loon, common dadzi
mallard duck t'acho

owl musdzi
Peregrine Falcon hoolht'as
ptarmigan, willow k'azba
pterodactyl dune oolhdzai
Red-winged Blackbird ch'uk
Ruffed Grouse 'utsut
raven, common datsancho
robin, american sooh
Sandhill Crane dilh
sandpiper wedlew
seagull besk'i
snowbird kwuz oonde
swallow schas
swan ts'uncho
thrush, varied 'ints'ilh
Varied Thrush 'ints'ilh
Whiskey Jack goozeh
Whiskey Jack gwuzih
Whooping Crane dilhtsul
Willow Ptarmigan k'azba
woodpecker chundulht'a
woodpecker kwalhcho

5. Bugs

ant 'andih
bee hoolht'o
beetle hoolhkw'ul
blood sucker hoot'ukw
bug yun nulh'as
bumblebee hoolht'ocho
butterfly tsagwulht'ai
caterpillar t'angoo
caterpillar ts'anikwa
Dialect: Nadleh specific
caterpillar, thick black susya
cobweb tsootsibilh
dragonfly nuduk'ui
earthworm 'usgoo
eggs, ant 'andihghez
eggs, fly sts'uzts'uz
flea lhiya
fly, black dahjolh
fly, house sts'uz
grasshopper tulk'us
hornet ts'ihna
horsefly hoot'as
leech hoot'ukw
leech tehchunultsi
louse ya
maggot 'usgoo
moth musdziyaz
snail talook ts'it
snake tl'ughus
spider tsootsi
spider whutsootsi
spider web tsootsibilh
wasp stlesicho
web, spider tsootsibilh

6. Tracks

footprints 'uk'ah
footprints k'ah
rabbit tracks gahk'ah
tracks 'uk'ah
tracks k'ah
tracks, horse yeztlik'ah

7. Plants

VArctic sagebrush tse'ul

Black Spruce ts'oobez

Brittle Horsehair Lichen dahgha

beaver's ears whuledzo

Canada mayflower lhiluzchun

Columbine, Red whulecho

Common Cattail tl'ok'uzih

Common Red Sphagnum moss ts'al

Cow Parsnip goos

Creamy Peavine chunalhduz

carrot, wild tsacheschun

cattail plant tl'ok'uzih

clover nedochunalhduz

currant, northern black tsasdlimai

Devil's Club hoolhghulh

dandelion ditnikwun

dandelion salat t'an

diaper moss ts'al

False Solomon's Seal lhiluzchun

Fir, Douglas tsuntsi

Fireweed khast'an

ginseng whulhdulh

grass tl'o

hay tl'o

Indian Hellebore whulhdulh

Indian Paint Brush ditnikwun

juniper datsan'algut

kinnikinnick dunih

Labrador Tea ludi musjek

Labrador Tea yak'unulh'a

Lambs Quarters 'ut'antsi

marijuana beni hoolah 'ut'ot

milfoil lacholbai

moss, tree dahgha

Northern Black Currant tsasdlimai

Pigweed 'ut'antsi

peavine, creamy chunalhduz

plantain khuzbat'an

Red Columbine whulecho

Red-osier Dogwood k'en dulk'un

red willow k'en dulk'un

Scouring Rush khahdai

Smilacina amplexicaulis lhiluzchun

Sphagnum moss, Common Red ts'al

Stinging Nettle hoolhts'ik

Strawberry Blite koonk'etsih

Sugarberry lhiluzchun

sagebrush, Arctic tse'ul

seaweed, dried Red Laver lhaga'as

spruce, black ts'oobez

thimbleberry dukdunenkai

White Goosefoot 'ut'antsi

water lily khelht'az

weeds in water dlat

willow, red k'en dulk'un

yarrow plant lacholbai

7.2. Berries

Bearberry susmai

Black Gooseberry 'indawuz

Black Twinberry susmai

Blueberries, High-Bush yalhtsul

berries mai

cranberries, bog yak'umai

cranberries, high-bush tsalhtse

cranberries, low-bush yak'umai

gooseberry, black 'indawuz

High-Bush Blueberries yalhtsul

huckleberries tsulhcho

Indian ice cream nawus

low-bush blueberry 'ilhtsul

Saskatoon berries k'emai

soapberry nawus

strawberry 'indzi

Twinberry susmai

7.3. Domestic Plants

cabbage 'utancho

onion tl'otsun

7.4. Other

green moss (on ground or tree trunks) 'uyemba

moss, green (on ground or tree trunks) 'uyemba

7.5. Products

gum, chewing dzeh

pitch dzeh

7.6. Plant Parts

acorn 'angwul

Balsam Fir bark ts'ootsunt'ooz

bark 'ula

bark 'ut'ooz

bark of Balsam Fir ts'ootsunt'ooz

bark of Subalpine Fir ts'ootsunt'ooz

birch conk diyooh

branch of conifer 'ul

branch of deciduous tree 'uyooschum

cambium k'unih

cedar bark hat'al

conk, birch diyooh

flower 'indak

flower -indak

flower, my se'indak

flower, our ne'indak

flower, their hube'indak

flower, your (1) xnyxe'indak

flower, your (2+) nawhe'indak

flower, his/her be'indak

flower, his/her ye'indak

flower, his/her own de'indak

flowers 'indakya

flowers, its bindak

kinnickinnick leaves duniht'an

leaf 'ut'an

leaf -t'an

pine cone 'angwul

Subalpine Fir bark ts'ootsunt'ooz

seeds, its (plant's) bumai

spruce bark ts'ooladel

spruce roots khi

7.7. Trees and Shrubs

Alder, Green k'us

Ash, Mountain ch'ok

Balsam Fir ts'ootsun

Birch, Black nach'ulh

Birch, Water nach'ulh

Birch, Western nach'ulh

Black Cottonwood landooz

Cherry, Bitter dulgoosmai

Chokecherry dulgoosmai

cedar chunzool

cottonwood landooz

Dwarf Maple khasdzoon

Green Alder k'us

jack pine chundoo

lodgepole pine chundoo

Maple, Dwarf khasdzoon

Mountain Ash ch'ok

Poplar t'ughus

pine, jack chundoo

pine, lodgepole chundoo

pussy willow k'edlits'ilhba

Sitka Mountain Ash ch'ok

Spruce ts'oo

Subalpine Fir ts'ootsun

Trembling Aspen t'ughus

tamarack netsi'ul

tamarack kasdzoon

willow, pussy k'edlits'ilhba

7.8. Tree Related

board dzihtel

stick duchun

tree duchun

8. Fabulous Beings

cannibal neyi

elephant khunaicho

giant bird, extinct dune oolhdzai

mastodon khunaicho

monster, prehistoric khunaicho

pterodactyl dune oolhdzai

sasquatch ts'eslol

9. Relatives

ancestors dusneke

ancestors, our nedusneke

ancestors, their hubudusneke

ancestors, your (1) xnyxudusneke

ancestors, his/her budusneke

aunt, my paternal sbezan

aunt, paternal -bez

aunt, their paternal hububezan

aunt, his/her paternal bubezan

aunt (father's sister) 'ubezan

aunt (father's sister) -bezan

aunt (mother's sister, father's brother's wife, mother's brother's wife) -k'i

aunt (mother's sister) -ak'i

aunt (mother's sister), my sak'i

aunt (mother's sister), someone's 'ak'i

aunt (mother's sister), their hubak'i

aunt (mother's sister), your (1) xnyxak'i

aunt (mother's sister), his/her bak'i

brother -ulhutsin

brother, my sulhutsin

brother, my elder soona

brother, my younger schul

brother, older -oona

brother, our nelhutsin

brother, our older neyoona

brother, their hubulhutsin

brother, their older huboona

brother, younger -chist'loh

brother, younger -chul

brother, your (1) older xnyxoona

brother, his/her bulhutsin

brother, his/her older boona

brother-in-law, my sghe

brother-in-law, my sghe

brother-in-law, our neghe

brother-in-law, their hubughe

brother-in-law, your (1) nghe

brother-in-law, his/her bughe

cousin -nadun

cousin, each other's lhoonde

cousin, father's female first -bezan

cousin, father's male first -tai

cousin, female cross -zet

cousin, female first -ulhtus

cousin, female second older than ego -dezcho

cousin, female second younger than ego -dezyaz

cousin, male first -ulhutsin

cousin, male first older than ego -oona

cousin, male second -oonde

cousin, mother's female first -k'i

cousin, mother's male first -z'e

cousin, my snadun

cousin, younger male -chul

cousins, female first lhulhtuske

cousins, female first -ulhtuske

co-wife -ulhtus

daughter, favorite t'etyaz

daughter, her buts'e

daughter, man's -tse'

daughter, my syats'e

daughter, my (man's) stse'

daughter, our netse'

daughter, our neyats'e

daughter, their hubutse'

daughter, their hubuyats'e

daughter, unfavoured t'etch'ul

daughter, woman's -yats'e

daughter, your (1) ntse'

daughter, your (1) nyats'e

daughter, his/her butse'

daughter-in-law -yas'at

daughter-in-law, my syas'at

daughter-in-law, our neyas'at

daughter-in-law, their hubuyas'at

daughter-in-law, your (1) nyas'at

daughter-in-law, his/her buyas'at

father 'uba

father -ba

father, my sba

father, our neba

father, their hububa

father, your (1) mba

father, his/her buba

father-in-law -zaz

father-in-law, my szaz

father-in-law, our nezaz

father-in-law, their hubuzaz

father-in-law, your (1) nzaz

father-in-law, his/her buzaz

grandchild -chai

grandchild, my schai

grandchild, our nechai

grandchild, their hubuchai

grandchild, your (1) nchai

grandchild, his/her buchai

grandchild of first cousin -chai

grandchild of sibling -chai

grandchildren, my schaike

grandchildren, our nechaike

grandchildren, their hubuchaike

grandchildren, your (1) nchaike

grandchildren, his/her buchaike

grandfather 'utsiyan

grandfather -tsiyan

grandfather, my stsiyan

grandfather, my younger stsiyanyaz

grandfather, our netsiyan

grandfather, our younger netsiyanyaz

grandfather, their hubutsiyan

grandfather, their younger hubutsiyanyaz

grandfather, younger -tsiyanyaz

grandfather, your (1) ntsiyan

grandfather, your (1) ntsiyanyaz

grandfather, his/her butsiyan

grandfather, his/her younger butsiyanyaz

grandmother, maternal 'utsoo

grandmother, maternal -tsoo

grandmother, my maternal stsoo

grandmother, my younger stsooyaz

grandmother, our maternal netsoo

grandmother, our younger netsooyaz

grandmother, paternal 'uts'un

grandmother, paternal -ts'un

grandmother, their maternal hubutsoo

grandmother, their younger hubutsooyaz

grandmother, younger -tsooyaz

grandmother, your (1) younger ntsooyaz

grandmother, his/her maternal butsoo

grandmother, his/her younger butsooyaz

great-grandchild -chaicho

great-grandchild -dulhchai

great-grandchild, my schaicho

great-grandchild, our nechaicho

great-grandchild, their hubuchaicho

great-grandchild, your (1) nchaicho

great-grandchild, his/her buchaicho

great-grandfather -'lunatdultsiyan

great-grandfather 'utsiyancho

great-grandfather -tsiyancho

great-grandfather, my se'lunatdultsiyan

great-grandfather, my stisyancho

great-grandfather, our ne'lunatdultsiyan

great-grandfather, our netsiyancho

great-grandfather, their hube'lunatdultsiyan

great-grandfather, their hubutsiyancho

great-grandfather, your (1) ntsiyancho

great-grandfather, your (1) xnyxe'lunatdultsiyan

great-grandfather, his/her be'lunatdultsiyan

great-grandfather, his/her butsiyancho

great-grandmother, maternal -tsoocho

great-grandmother, my maternal stsoocho

great-grandmother, our maternal netsoocho

great-grandmother, paternal 'uts'uncho

great-grandmother, paternal -ts'uncho

great-grandmother, their maternal hubutsoocho

great-grandmother, his/her maternal butsoocho

great-great-grandfather whulhtat-lunatdultsiyan

great-great-grandfather, my whulhtatselunatdultsiyan

great-great-grandfather, our whulhtatnelunatdultsiyan

great-great-grandfather, their whulhtathubalunatdultsiyan

great-great-grandfather, his/her whulhtatbulunatdultsiyan

great-great-grandmother whulhtatlunatdultsoo

great-great-grandmother, my whulhtatselunatdultsoo

great-great-grandmother, our whulhtatnelunatdultsoo

great-great-grandmother, their whulhtathubelunatdultsoo

great-great-grandmother, his/her whulhtatyubulunatdultsoo

husband -kui

husband, my skui

husband, his/her bukui

husbands -kuike

husbands, our nekuike

husbands, their hubukuike

mother 'uloo

mother -loo

mother, my sloo

mother, our neloo

mother, their hubuloo

mother, your (1) nloo

mother, his/her buloo

mother's sister -ak'i

mother's sister, my sak'i

mother's sister, someone's 'ak'i

mother's sister, their hubak'i

mother's sister, your (1) xnyxak'i
mother's sister, his/her bak'i
mother-in-law -bez
mother-in-law, my sbez
mother-in-law, our nebez
mother-in-law, someone's 'ubez
mother-in-law, their hububez
mother-in-law, your (1) mbez
mother-in-law, his/her bubez
nephew -gwaz
nephew, my ngwaz
nephew, my sgwaz
nephew, our negwaz
nephew, their hubugwaz
nephew, his/her bugwaz
niece -gwaz
niece, my ngwaz
niece, my sgwaz
niece, our negwaz
niece, their hubugwaz
niece (younger than ego) -gwastl'oh
sibling's spouse -ghe
sister -ulhtus
sister, my sulhtus
sister, my older syat
sister, my younger sdesyaz
sister, older -yat
sister, our nelhtus
sister, our neyulhtus
sister, our older neyat
sister, our younger nedesyaz
sister, their hubulhtus
sister, their older hubuyat
sister, their younger hubudesyaz
sister, younger -destl'oh
sister, younger -desyaz
sister, your (1) xnyxulhtus
sister, your (1) older nyat
sister, your (1) younger ndesyaz
sister, your (2+) nawhulhtus
sister, his/her bulhtus
sister, his/her yulhtus
sister, his/her older buyat
sister, his/her own dulhtus
sister, his/her younger budesyaz
sisters lhulhtuske
sisters -ulhtuske
sister-in-law, my sghe
sister-in-law, my sghe
sister-in-law, our neghe
sister-in-law, their hubughe
sister-in-law, your (1) nghe
sister-in-law, his/her bughe
son, her buyaz
son, his buye'
son, man's -ye'
son, my sye'
son, our neye'
son, their hubuye'
son, woman's -yaz
son, your (1) nye'
son-in-law -ghundan
son-in-law, my sghundan
son-in-law, our neghundan
son-in-law, their hubughundan
son-in-law, your (1) nghundan
son-in-law, his/her bughundan

spouse's sibling -ghe

step-father -tai

step-father, my stai

step-father, our netai

step-father, their hubutai

step-father, your (1) ntai

step-father, his/her butai

step-mother -k'i

step-son -ulhye'

step-son, my sulhye'

step-son, your (1) xnyxulhye'

twins lhtudu'alts'e

uncle 'utai

uncle, someone's 'uz'e

uncle (father's brother, father's sister's husband, mother's sister's husband) -tai

uncle (mother's brother) -z'e

uncle (mother's brother), my suz'e

uncle (mother's brother), our nez'e

uncle (mother's brother), their hubuz'e

uncle (mother's brother), your (1) xnyxuz'e

uncle (mother's brother), his/her buz'e

uncle other than mother's brother, our netai

uncle other than mother's brother, their hubutai

uncle other than mother's brother, your (1) ntai

uncle other than mother's brother, his/her butai

uncle other than mother's brother, my stai

widow tsandelh

widower tsandelh

wife 'at

wife, my s'at

wife, his/her bu'at

wives 'atke

wives, our ne'atke

wives, their hubu'atke

10. Clan and Potlatch

clan, one's father's buts'aha'elts'utne

clan song dedohshun

dance with one's partner in balhats, a 'indumanuk

hereditary chief keyah whuduchun

leader, village, the first among equals of the clan heads keyoh whuduchun

noble in clan system, female ts'ekeza'

noble in the clan system, of either sex 'uza'

noble in the clan sytsem, male duneza'

potlatch balhats

song, clan dedohneyun

song, clan dedohshun

song, face-saving kayanahudlishun

song, longing k'ehucudlishun

song, personal duneyun

song, public netsiyun

song, ritual duyun-ishun

song, shaming lhuhoonultsasshun

song, sung during the distribution of fruit at a balhats maiyun

song, sung to embarass someone 'udulhtsiyun

song, sweetheart 'indumanukshun

sweetheart song 'indumanukshun

11. Names

11.1. Given

11.2. Male

name of trickster culture hero 'Usdas

11.3. Family

Ketlo Ketloh

12. Ethnic Terms

12.1. Peoples, Tribes, Nations

Aboriginal person yunka dune

Babine person Babine whut'en

Babine person Nadot'en

Beaver Indian Tsat'en

Carrier person dakelh

Cree person Dushin

Fraser Lake people Nadlehbunne

Indian, non-Athabaskan 'utna

Indigenous person yunkaot'en

Indigenous person yunkawhut'en

Tahltan person Talht'en

Takla person Nadot'en

white person bosdun

white person meljah

white person nasdlez

13. People

adolescent, female t'et

adolescent, male chilh

ancestors dusneke

ancestors, my sdusneke

ancestors, our nedusneke

ancestors, their hubudusneke

ancestors, your (1) xnyxudusneke

ancestors, his/her budusneke

baby tsalhts'ul

baby uskiyaz

baby, little tsalhts'ulyaz

baby boy dune tsalhts'ul

baby girl ts'eke tsalhts'ul

boy 'uskidune

boy duneyaz

boy skuidune

boy, baby dune tsalhts'ul

boy, small skuiduneyaz

boys skehdune

Christian person Yak'usda bughunek k'une 'ut'en-un

cannibal dune uyi

captive lhuna

child ski

child skui

child, small 'uskiyaz

child, small skuiyaz

children 'uskehne

children skeh

children, my suzkeh

children, our nezkeh

children, their hubuzkeh

children, your (1) xnyxuzkeh

children, his/her buzkeh

different people lhelhyoone

elder hoonyan

elders netsowhuzulhne

female elder, very respected ts'eketi

friend t'eke

friend, my st'eke

friend, our net'eke

friend, their hubut'eke

friend, your (1) nt'eke

friend, his/her but'eke

ghost dunezoolh

ghost naoodnilh

girl skuits'eke

girl ts'ekeyaz

girl, baby ts'eke tsalhts'ul

girl, small skuits'ekeyaz

girls skehts'ekoo

girls ts'ekooyaz

hunchback 'uk'o'

hunchback k'o'

man dune

man, old duneti

man, older dunecho

men dunene

orphan tsunah

people dunene

person dune

relative t'eke

seniors netsawhudelhzulhne

seniors netsowhuzulhne

tall man duneyez

tall person dunecho

tall woman ts'ekeyez

tramp kegudsatalh

triplets lhulhcho

twins lhtudu'alts'e

woman ts'eke

woman, large ts'ekecho

woman, respected ts'ekecho

woman, young noh

women ts'ekoo

young man chilh

young men chilhuke

young woman t'et

14. Roles

14.1. Occupations, Ranks, etc.

boss moodih

boss, my smoodih

boss, our nemoodih

boss, their hubumoodih

bum kejudustl'ah

chef ts'uyi ghaghuna

chief, generally with reference to band chief dayi

doctor yoobeduyun

guard, security huwunli

hereditary chief keyah whuduchun

knowledgable person moodih

lawyer dune ba yalhduk

leader, village, the first among equals of the clan heads keyoh whuduchun

magician nedulhdehun

messenger khunek nu'a

messenger nawhulnuk

nurse ndudaneghunli

police officer neilhchuk

priest nawhulnuk

RCMP officer neilhchuk

secretary 'uk'une'uguz

servant 'ustlen

servant, my se'ustlen

servant, our ne'ustlen

servant, their hube'ustlen

servant, your (1) xnyxe'ustlen

servant, his/her be'ustlen

soldiers bahne

soldiers nubah

subject, my se'ustlen

subject, our ne'ustlen

subject, their hube'ustlen

subject, your (1) xnyxe'ustlen

subject, his/her be'ustlen

subject (of ruler) 'ustlen

teacher dune hodulh'eh

watchmen huwunline

15. Body

15.1. External Anatomy

Adam's Apple -yih k'ut nez'ai

ankle -kechunnah

ankle, my skechunnah

ankle, your (1) nkechunnah

ankle, his/her bukechunnah

ankles, our nekechunnah

ankles, their hubukechunnah

antler -de'

antlers 'ude

anus -boo'

anus -tsul

anuses, their hubutsul

arm -gan

arm, my sgan

arm, your (1) ngan

arm, his/her bugan

arm pit -chak'ests'ah

arms, our negan

arms, their hubugan

arms, your (2+) nahgan

back -t'ak

back, lower -t'ak yusts'i

back, my lower yusts'i st'ak

back, our net'ak

back, the small of my syunt'ak

back, the small of the -yunt'ak

back, your (1) nt'ak

back, your (1) lower yusts'i nt'ak
back, his/her but'ak
back, his/her lower yusts'i but'ak
back my st'ak
back of hand -lat'ak
backbone -yunts'un
backs, our lower yusts'i net'ak
backs, their hubut'ak
backs, their lower yusts'i hubuta'k
beak -da
belly -but
belly -chan
belly button -tsakelhgwus
bodies, our neyust'e
bodies, their hubuyust'e
body 'uyust'e
body -yust'e
body, my syust'e
body, your (1) nyust'e
body, his/her buyust'e
bones of fish 'ughak
braided hair -tsigha lhu'ool
breast 'utsoo
breast -ts'oo
breast, my sts'oo
breast, your (1) nts'oo
breast, his/her buts'oo
breasts, our nets'oo
breasts, their hubuts'oo
bum -tl'ah
bum, my stl'ah
bum, your (1) ntl'ah
bum, his/her butl'ah
bums, our netl'ah
bums, their hubutl'ah
buttock -tl'ah
calf, my swuda
calf, your (1) nwuda
calf, his/her buwuda
calf of leg -wuda
calves, our newuda
calves, their hubuwuda
cheek -nimbus
chest -dzik'ut
chest, my sdzik'ut
chest, our nedzik'ut
chest, their hubudzik'ut
chest, your (1) ndzik ut
chest, his/her budzik'ut
chin -zets'un
chin, my szets'un
chin, your (1) nzets'un
chin, his/her buzets'un
chins, our nezets'un
chins, their hubuzets'un
claw of rear paw -kengi
crotch -tl'it
crotch, his/her butl'it
crown of head 'utsitah
dog's paw lhike
down (of birds) chus
ear -dza
ear canal -dzek
earlobe -dzabal
elbow, my snints'uzti
elbow -nints'uzti
elbow -yints'iz
elbow, your (1) nnints'uzti

elbow, your (1) nyints'iz
elbow, his/her bunints'uzti
elbows, our nenints'uzti
elbows, their hubunints'uzti
eye -na
eye, a single nak'uz
eye, my sna
eye, your (1) unna
eye, his/her buna
eyebrows -nats'osdooz
eyebrows, my snats'osdooz
eyebrows, our nenats'osdooz
eyebrows, their hubunats'osdooz
eyebrows, your (1) unnats'osdooz
eyebrows, his/her bunats'osdooz
eyelashes -nabagha
eyelashes, my snabagha
eyelashes, our nenabagha
eyelashes, their hubunabagha
eyelashes, your (1) unnabagha
eyelashes, his/her bunabagha
eyelid -nabalos
eyes, fish lhookna
eyes, our nena
eyes, their hubuna
eyetooth 'unaghoo
eyetooth -naghoo
face -nen
face, my snen
face, your (1) unnen
face, his/her bunen
faces, our nenen
faces, their hubunen
feet -ke
feet, the tops of our nekent'ak
feet, the tops of their hubukent'ak
feet, his/her own duke
fin yenube
finger -lasge
finger, index -lasgek
finger, my slasge
finger, my index slasgek
finger, small -lasgeyaz
finger, your (1) nlasge
finger, his/her bulasge
fingernail -lagi
finger spaces -latl'i
finger spaces, my slatl'i
finger spaces, our nelatl'i
finger spaces, their hubulatl'i
finger spaces, your (1) nlatl'i
finger spaces, his/her bulatl'i
fingernail 'ulagi
fingernail -lagi
fingernail, his bulagi
fingernail, my slagi
fingernail, your (1) nlagi
fingernail, his/her own dulagi
fingernails, our nelagi
fingernails, their hubulagi
fingers -latsuk
fingers, our nelasge
fingers, their hubulasge
fish eyes lhookna
fish head lhooktsi
fish tail lhookchetl'a
fist tuchus
flesh -tsun

foot -ke

foot, the top of my skent'ak

foot, the top of your (1) nkent'ak

foot, the top of his/her bukent'ak

foot, top of -kent'ak

forehead -nint'ak

forehead, my snintak

forehead, your (1) nnintak

foreheads, our nenintak

fur 'ugha

fur -gha

fur, his/her own dugha

genitalia, female -k'al

groin -tl'et

groin -tl'it

groin, my stl'et

groin, your (1) ntl'et

groin, his/her butl'et

groins, our netl'et

groins, their hubutl'et

gums -ghootsun

gums, my sghootsun

gums, our neghootsun

gums, their hubughootsun

gums, your (1) nghootsun

gums, his/her bughootsun

hair 'ugha

hair, facial -dagha

hair, head 'utsigha

hair, my head stsigha

hair, our head netsigha

hair, their head hubutsigha

hair, your (1) head ntsigha

hair, his/her head butsigha

hair, his/her own dugha

hair braid -tsigha lhu'ool

hair of body -gha

hair of head -tsigha

hand, back of -lat'ak

hand, palm of -lak'et

hand, the back of my slat'ak

hand, the back of your (1) nlat'ak

hand, the back of his/her bulat'ak

hands, the backs of our nelat'ak

hands, the backs of their hubulat'ak

head -tsi

head, back of -tsit'ak

head, back of -tsitl'ah

head, back of the -tsint'ak

head, fish lhooktsi

head, my stsi

head, the back of your (1) ntsint'ak

head, the back of your (1) ntsitl'ah

head, the back of my stsint'ak

head, the back of my stsit'ak

head, the back of my stsitl'ah

head, the back of your (1) ntsit'ak

head, the back of his/her butsint'ak

head, the back of his/her butsit'ak

head, the back of his/her butsitl'ah

head, the top of my stsidah

head, the top of your (1) ntsidah

head, the top of his/her butsidah

head, top of -tsidah

head, your (1) ntsi

head, his/her butsi

heads, our netsi

heads, the backs of ours netsint'ak

heads, the backs of ours netsit'ak

heads, the backs of ours netsitl'ah

heads, the backs of their hubutsint'ak

heads, the backs of their hubutsit'ak

heads, the backs of their hubutsitl'ah

heads, the tops of our netsidah

heads, the tops of their hubutsidah

heads, their hubutsi

heel -kelatsul

heel -kentsul

heel -ketsul

heel, my skelatsul

heel, my skentsul

heel, your (1) nkelatsul

heel, your (1) nkentsul

heel, his/her bukelatsul

heel, his/her bukentsul

heels, my sketsul

heels, our nekelatsul

heels, our nekentsul

heels, our neketsul

heels, their hubukelatsul

heels, their hubukentsul

heels, their hubuketsul

heels, your (1) nketsul

heels, his/her buketsul

hide 'uzuz

hide, my szuz

hide, our nezuz

hide, their hubuzuz

hide, your (1) nzuz

hide, his/her buzuz

horn -de'

horns 'ude

index finger, my slasgek

knee -gwut

knee, back of -gwutt'uk

knee, my sgwut

knee, the back of my sgwutt'uk

knee, the back of your (1) ngwutt'uk

knee, the back of his/her bugwutt'uk

knee, your (1) ngwut

knee, his/her bugwut

kneecap -gwutduts'un

kneecap -gwuttsi

kneecap, my sgwuduts'un

kneecap, my sgwuttsi

kneecap, your (1) ngwutduts'un

kneecap, his/her bugwutduts'un

kneecaps, our negwutduts'un

kneecaps, their hubugwutduts'un

knees, our negwut

knees, the backs of our negwutt'uk

knees, the backs of their hubugwutt'uk

knees, their hubugwut

larynx -yih k'ut nez'ai

leg -kechun

leg, calf of -wuda

leg, my skechun

leg, upper -wuz

leg, your (1) nkechun

leg, his/her bukechun

legs ’ukechun

legs, back of -kechunt’ak

legs, our nekechun

legs, the back of my skechunt’ak

legs, the back of our nekechunt’ak

legs, the back of their hubukechunt’ak

legs, the back of your (1) nkechunt’ak

legs, the back of his/her bukechunt’ak

legs, their hubukechun

lips -da

lips, my sda

lips, our neda

lips, their hubuda

lips, your (1) nda

lips, his/her buda

lips, his/her own duda

little finger -lasts’ah

lungs, our nedes

moustache -dagha

moustache, my sdagha

moustache, your (1) ndagha

moustache, his/her budagha

moustaches, our nedagha

moustaches, their hubudagha

mouth -zek

mouth, my szek

mouth, roof of -zek dusts’ai

mouth, roof of -zek onduk

mouth, your (1) unzek

mouth, his/her buzek

mouths, our nezek

mouths, their hubuzek

nape of my neck, the sts’ilchunt’ak

nape of the neck -ts’ilchunt’ak

nape of your (1) neck, the nts’ilchunt’ak

nape of neck, the buts’ilchunt’ak

napes of our necks, the nets’ilchunt’ak

napes of their necks, the hubuts’ilchunt’ak

navel -tsakelhgwus

navel, my stsakelhgwus

navel, your (1) ntsakelhgwus

navel, his/her butsakelhgwus

navels, our netsakelhgwus

navels, their hubutsakelhgwus

neck -ts’ilchun

neck, my sts’ilchun

neck, nape of -ts’ilchunt’ak

neck, your (1) nts’ilchun

neck, his/her buts’ilchun

necks, our nets’ilchun

necks, their hubuts’ilchun

nipple -ts’oola

nipples, my sts’oola

nipples, our nets’oola

nipples, their hubuts’oola

nipples, your (1) nts’oola

nipples, his/her buts’oola

nostril -nik
palm, my slak'et
palm, your (1) nlak'et
palm, his/her bulak'et
palm of hand -lak'et
palms, our nelak'et
palms, their hubulak'et
paw, dog's lhike
penis 'utsukw
penis lait'i
penis -tsukw
penis, boy's kw'ulh
penis, his butsukw
penis, his (boy's) bukw'ulh
penis, my stsukw
penis, someone's 'utsukw
penis, your (1) ntsukw
penises, our netsukw
penises, their hubutsukw
pinkie -lasts'ah
pinkie, my slasts'ah
pinkie, your (1) nlasts'ah
pinkie, his/her bulasts'ah
pinkies, our nelasts'ah
pinkies, their hubulasts'ah
pubic hair -tl'egha
pubic hair, my stl'egha
pubic hair, our netl'egha
pubic hair, their hubutl'egha
pubic hair, your (1) ntl'egha
pubic hair, his/her butl'egha
rectum -boo'
rectum -tsul
rectum, my sboo'
rectum, my stsul
rectum, your (1) mboo'
rectum, your (1) ntsul
rectum, his/her buboo'
rectum, his/her butsul
rectums, our neboo'
rectums, our netsul
rectums, their hububoo
rectums, their hubutsul
roof of the mouth -zek dusts'ai
scrotum -ghezzus
scrotum, my sghezzus
scrotum, your (1) nghezzus
scrotum, his/her bughezzus
scrotums, our neghezzus
scrotums, their hubughezzus
shin -gwutyuschun
shins, my sgwutyuschun
shins, our negwutyuschun
shins, their hubugwutyuschun
shins, your (1) ngwutyuschun
shins, his/her bugwutyuschun
shoulder -wus
shoulders, my swus
shoulders, our newus
shoulders, their hubuwus
shoulders, your (1) nwus
shoulders, his/her buwus
skin 'uzuz
sole, my sketl'ah
sole, your (1) nketl'ah
sole, his/her buketl'ah
sole of foot -ketl'ah
soles, our neketl'ah

soles, their hubuketl'ah

spine -yunts'un

stomach -but

stomach, bottom of the -buttl'ah

stomach, my sbut

stomach, the bottom of my sbuttl'ah

stomach, the bottom of your (1) mbuttl'ah

stomach, the bottom of his/her bubuttl'ah

stomach, your (1) mbut

stomach, his/her bubut

stomachs, our nebut

stomachs, the bottoms of our nebuttl'ah

stomachs, the bottoms of their hububuttl'ah

stomachs, their hububut

tail, fish lhookchetl'a

teeth -ghoo

teeth, my sghoo

teeth, our neghoo

teeth, their hubughoo

teeth, your (1) nghoo

teeth, his/her bughoo

teeth, his/her own dughoo

thigh -wuz

thighs, my swuz

thighs, our newuz

thighs, their hubuwuz

thighs, your (1) nwuz

thighs, his/her buwuz

throat, my szesdak

throat, your (1) nzesdak

throat, his/her buzesdak

throat (interior) -zesdak

throats, our nezesdak

throats, their hubuzesdak

thumb -ninchuz

thumb, my sninchuz

thumb, your (1) unninchuz

thumb, his/her buninchuz

thumb, his/her own duninchuz

thumbs, our neninchuz

thumbs, their hubuninchuz

toe -kelamai

toe, big -kelamaicho

toe, big -kelascho

toe, fourth -kelamaiyazts'unts'e

toe, little -kelamaiyaz

toe, long -kelamaiyez

toe, middle -kelamaiyezts'unts'e

toe, my big skelamaicho

toe, my big skelascho

toe, my fourth skelamaiyazts'unts'e

toe, my little skelamaiyaz

toe, my long skelamaiyez

toe, my middle skelamaiyezts'unts'e

toe, your (1) big nkelamaicho

toe, your (1) big nkelascho

toe, your (1) fourth nkelamaiyazts'unts'e

toe, your (1) little nkelamaiyaz

toe, your (1) long nkelamaiyez

toe, your (1) middle nkelamaiyezts'unts'e

toe, his/her big bukelamaicho

toe, his/her big bukelascho

toe, his/her fourth bukelamaiyazts'unts'e

toe, his/her little bukelamaiyaz

toe, his/her long bukelamaiyez

toe, his/her middle bukelamaiyezts'unts'e

toe spaces, our neketl'i

toe spaces, their hubuketl'i

toe spaces, your (1) nketl'i

toe spaces, his/her buketl'i

toenail -kelagi

toenail -kengi

toenail, my skelagi

toenail, my skengi

toenail, our nekengi

toenail, your (1) nkelagi

toenail, your (1) nkengi

toenail, his/her bukelagi

toenail, his/her bukengi

toenails, our nekelagi

toenails, their hubukelagi

toenails, their hubukengi

toes, my skelamai

toes, our nekelamai

toes, our big nekelamaicho

toes, our big nekelascho

toes, our fourth nekelamaiyazts'unts'e

toes, our little nekelamaiyaz

toes, our long nekelamaiyez

toes, our middle nekelamaiyezts'unts'e

toes, spaces between -ketl'i

toes, their hubukelamai

toes, their big hubukelamaicho

toes, their big hubukelascho

toes, their fourth hubukelamaiyazts'unts'e

toes, their little hubukelamaiyaz

toes, their long hubukelamaiyez

toes, their middle hubukelamaiyezts'unts'e

toes, your (1) nkelamai

toes, his/her bukelamai

toes spaces, my sketl'i

tongue -tsoola

tongue, your (1) ntsoola

tongue, my stsoola

tongue, his/her butsoola

tongues, our netsoola

tongues, their hubutsoola

tooth -ghoo

tooth, wisdom -naghoo

wisdom tooth -naghoo

wrist -lachunnah

wrist -lat'ah

wrist, my slachunah

wrist, my slat'ah

wrist, your (1) nlachunah

wrist, your (1) nlat'ah

wrist, his/her bulachunah

wrist, his/her bulat'ah

wrists, our nelachunah

wrists, our nelat'ah

wrists, their hubulachunah

wrists, their hubulat'ah

15.2. Internal Anatomy

Adam's apple -zoolts'un

adenoids, my syesdak nez'ai

aorta 'udzich'ooz

arteries, our nedzidool

arteries, their hubudzidool

artery -dzidool

artery -zkaich'oozcho

artery, my sdzidzool

artery, my suzkaich'oozcho

artery, our nezkaich'oozcho

artery, their hubuzkaich'oozcho

artery, your (1) ndzidool

artery, your (1) xnyxuzkaich'oozcho

artery, his/her budzidool

artery, his/her buzkaich'oozcho

backbone -t'akts'un

backbone -yun

backbone, my st'akts'un

backbone, my syunts'un

backbone, the area surrounding -dayun

backbone, your (1) nt'akts'un

backbone, your (1) nyunt'sun

backbone, his/her but'akts'un

backbone, his/her buyunts'un

backbone region, my sdayun

backbones, our net'akts'un

backbones, their hubut'akts'un

backbones, their hubuyunts'un

backbones, our neyunts'un

bladder, gall -tl'uz

bladder, my sluzzus

bladder, your (1) nluzzus

bladder, his/her buluzzus

bladder (urinary) -lhuzzus

bladders, our neluzzus

bladders, their hubuluzzus

bowel -ts'ikcho

brain -tsinghai

brain, my stsinghai

brain, your (1) ntsinghai

brain, his/her butsinghai

brains, our netsinghai

brains, their hubutsinghai

breast bone -t'ayukts'un

cartilage 'ughuts

cartilage -ghuts

cheek bone -nimbusts'un

coccyx -chets'un

collar bone -gant'uk

collar bone -t'ukts'un

collarbone, our net'ukts'un

collarbone, their hubut'ukts'un

collarbone, your (1) nt'ukts'un

collarbone, his/her but'ukts'un

colon 'utsutle

cranium tsits'un

eardrum -dzek whudadilhchooz

eggs 'ughez

eggs -ghez

fat 'uk'a

fat k'a

femur -wuzts'un

femur, my swuzts'un

femur, your (1) nwuzts'un

femur, his/her buwuzts'un

femurs, our newuzts'un

femurs, their hubuwuzts'un

fish guts lhooktsik

flesh, my ussun

gall bladder -tl'uz

gall bladder, his/her butl'uz

gall bladder, my stl'uz

gall bladder, your (1) ntl'uz

gall bladders, our netl'uz

gall bladders, their hubutl'uz

guts -ts'ik

guts, fish lhooktsik

heart -dzi

heart, my sdzi

heart, your (1) ndzi

heart, his/her budzi

hearts, our nedzi

hearts, their hubudzi

hipbone, my sk'its'un

hipbone, your (1) nk'its'un

hipbone, his/her buk'its'un

hipbones, our nek'its'un

hipbones, their hubuk'its'un

intestine, large -ts'ikcho

intestine, my large sts'ikcho

intestine, your (1) large nts'ikcho

intestine, his/her large buts'ikcho

intestines -ts'ik

intestines, my sts'ik

intestines, our nets'ik

intestines, our large nets'ikcho

intestines, their hubuts'ik

intestines, their large hubuts'ikcho

intestines, your (1) nts'ik

intestines, his/her buts'ik

kidney -kw'uz

kidneys, my skw'uz

kidneys, our nekw'uz

kidneys, their hubukw'uz

kidneys, your (1) nkw'uz

kidneys, his/her bukw'uz

large intestine -ts'ikcho

larynges, our nezoolts'un

larynges, their hubuzoolts'un

larynx -zoolts'un

larynx, my szoolts'un

larynx, your (1) nzoolts'un

larynx, his/her buzoolts'un

liver 'uzut

liver -zut

liver, my szut

liver, your (1) nzut

liver, his/her buzut

livers, our nezut

livers, their hubuzut

lungs -des

lungs, my sdes

lungs, their hubudes

lungs, your (1) ndes

lungs, his/her budes

mind -ni

mind, my seni

mind, our neni

mind, your (1) xnyxeni

mind, his/her beni

muscle nut'uk

muscle, my snut'uk

muscle, your (1) unnut'uk

muscle, his/her bunut'uk

muscles, our nenut'uk

muscles, their hubunut'uk

pancreas -zumcho

pancreas, my szumcho

pancreas, your (1) nzumcho

pancreas, his/her buzumcho

pancreases, our nezumcho

pancreases, their hubuzumcho

pelvic bone -k'itsun'

pelvic region, my sk'i

pelvic region, your (1) nk'i

pelvic region, his/her buk'i

pelvic region -k'i

pelvic regions, our nek'i

pelvic regions, their hubuk'i

peritoneum 'uchanyoo

rib -chak

ribs, my schak

ribs, our nechak

ribs, their hubuchak

ribs, your (1) nchak

ribs, his/her buchak

sinus cavities, our neninchustah

sinus cavities, their hubuninchustah

sinus cavity -ninchustah

sinus cavity, my sninchustah

sinus cavity, your (1) nninchustah

sinus cavity, his/her buninchustah

skull -tsints'un

skull tsits'un

skull, my stsints'un

skull, your (1) ntsints'un

skull, his/her butsints'un

skulls, our netsints'un

skulls, their hubutsints'un

spine -t'akts'un

spine -yun

sternum -t'ayukts'un

stomach, my sbut

tailbone -chets'un

tailbone, my schets'un

tailbone, your (1) nchets'un

tailbone, his/her buchets'un

tailbones, their hubuchets'un

testicles 'ughez

tailbones, our nechets'un

testicles -ghez

testicles, my sghez

testicles, our neghez

testicles, their hubughez

testicles, your (1) nghez

testicles, his/her bughez

thigh bone -wuzts'un

tonsils -zesdazum

tonsils, my szesdazum

tonsils, our nezesdazum

tonsils, their hubuzesdazum

tonsils, your (1) nzesdazum

tonsils, his/her buzesdazum

trachea -zool

tripe 'utsutle

urinary tract -lhuzkah

urinary tract lhuzk'et

urinary tract, my sluzkah

urinary tract, their hubuluzkah

urinary tract, your (1) nluzkah

urinary tract, his/her buluzkah

urinary tracts, our neluzkah

uvula -zesdak'al

uvula -zesdalos

uvula -zesdamai

uvula, my szesdak'al

uvula, my szesdalos

uvula, my szesdamai

uvula, your (1) nzesdak'al

uvula, your (1) nzesdalos

uvula, your (1) nzesdamai

uvula, his/her buzesdak'al

uvula, his/her buzesdalos

uvula, his/her buzesdamai

uvulas, our nezesdak'al

uvulas, our nezesdalos

uvulas, our nezesdamai

uvulas, their hubuzesdak'al

uvulas, their hubuzesdalos

uvulas, their hubuzesdamai

vein -zkaich'ooz

veins, my suzkaich'ooz

veins, our nezkaich'ooz

veins, their hubuzkaich'ooz

veins, your (1) xnyxuzkaich'ooz

veins, his/her buzkaich'ooz

wind pipe -zool

wind pipe, my szool

wind pipe, your (1) nzool

wind pipe, his/her buzool

wind pipes, our nezool

wind pipes, their hubuzool

15.3. Bodily Fluids and Secretions

bile dultso

blood 'uzkai

blood -zkai

blood, my suzkai

blood, our nezkai

blood, their hubuzkai

blood, your (1) xnyxuzkai

blood, his/her buzkai

excrement tsan

feces tsan

feces, my stsan

feces, our netsan

feces, their hubutsan

feces, your (1) ntsan

feces, his/her butsan

milt lhooktl'uz

milt -tl'uz

pee lhuz

perspiration tsentoo

poop tsan

pus khuz

saliva -zekw

saliva, my suzekw

semen, fish -tl'uz

shit tsan

slime, fish -tl'us

sperm, fish lhooktl'uz

sweat tsentoo

urine lhuz

urine, my sluz

urine, our neluz

urine, their hubuluz

urine, your (1) nluz

urine, his/her buluz

16. Condition of Body

bald, he/she is tsinkwun

facial hair, I have dunusghai

facial hair, they have hudunughai

facial hair, we (3+) have ts'udunughai

facial hair, he/she has dunughai

fat, he/she has become again nanilk'a

fat, he/she is nelhuk'a

fat, he/she is not chanilk'ah

gaining weight, I am whenlhuk'a

hairy, it is dunughai

hairy, they are hudughai

hairy, we (2) are didughai

hairy, he/she is lhghai

hungry, I am sdaningi

hungry, you (1) are chahoonzum

hungry, you (1) are nye'elts'ul

hungry, he/she is oodaningi

obese, he/she has become again nanilk'a

obese, he/she has not become nachanlhuk'a

obese, he/she is nelhuk'a

obese, he/she is not chanilk'ah

offset, they (teeth) are dzohtsulyaz tink'us unli

peach-fuzz on his face, he/she has dunuyo

shaggy, it is dunuyo

skinny, I am sdaningi

skinny, he/she is budaningi

skinny, he/she is oodaningi

tan, I got a nisdutsil

walked in a loop, he/she has not nuchaiyal

weak, it is lait'ah

17. Bodily Functions

breathing, I am usjiz

burp, we (3+) 'uztukw'ak

burp, you (1) 'utinkw'ak

burp, you (2+) 'utahkw'ak

farts, he/she tutl'it

gagging, he/she is tukoch

go to the washroom, you (1) will 'az natanda

pee, I am going to natesdudluz

pee, I am going to natesdluz

pee, they will natedudluz

pee, we (3+) will naztedudluz

pee, you (1) will natandudluz

shat, I have nasusdutsan

shit, I am going to natesdutsun

shit, we (3+) will naztedutsun

shit, you (1) will natandutsun

shit, you (2+) will natehdutsun

18. Motion of Parts of the Body or Without Displacement

blinking, he/she is nak'ubul

blinks at her in dislike, she yunak'ult'ah

bucks, it (animal) tullhuk

closes both eyes and then opens them again, he/she oonats'ilh

put his foot inside, he/she dadel'ez

19. Medicine

19.1. Diseases and Conditions

cancer dadacho

cold, he/she has a kw'us 'uyinla

hangover, he/she has a nedotoo yundulhda

smallpox natsauneltsut

19.2. Symptoms

aches, it bulh'ults'ulh

aches, it bulh'unults'ulh

bled to death, he/she 'uskai yuzelhghi

bleeding, he/she is 'uskai suli

breath, I am out of syiz lhutulh

breath, I am short of syiz dilhdukw

chills, he/she has buts'utelhk'az

coughing, I am dulhukwus

coughing, he/she is dulkwus

cross-eyed, he/she is buna duguz

earache, I have an sdzek hah'elts'ul

earache, he/she has an budzek hah'elts'ul

gall stone tl'uztse

headache, I have a stsi nduda

pain, I cause (s.o.) ndulhda

pain, he/she has bulh'ults'ulh

pain, he/she has severe ootsas

pimple netsa uneltsut

sick, he/she is nduda

sore, it is nduda

sore, it is whunduda

sore feeling kwenduda

19.3 Medicines

aspirin netsiyoo

aspirin ntsich'a

Bromo selzer too undunulmulh

cod liver oil lhookghe

cough medicine kwusyoo

headache medicine netsiyoo

herbal medicine duchunyoo

medicine yoo

medicine, herbal duchunyoo

ointment lhootyoo

pill ts'untulwis

poison lubezo

19.4. Infrastructure

doctor yoobeduyun

health centre yoobayah

19.5. Verbs

clotted, it (blood) has dul suli

sick, they are hunduda

sick, we (3+) are ts'unduda

sick, you (1) are ndinda

19.6. Miscellaneous

ailment dada

illness dada

20. Reproduction and Sex

attracted (sexually) to me, he/she is suneztulh

born, he/she was whuzdli

brothel 'ul'en bayah

chicken egg ligokghez

egg, chicken ligokghez

egg, fish 'uk'oon

eggs 'ughez

eggs -ghez

fish egg 'uk'oon

kicked me in the face, he/she suneztulh

laying eggs, it (fish) is tesjaz

married, he is 'at ut'i

married, she is ki ut'i

married to him/her, he/she is yughusda

milt lhooktl'uz

milt -tl'uz

pregnant, I am ulhuchan

pregnant, she is ulchan

pregnant, you (2+) are ulhuchan

roe, fish 'uk'oon

semen, fish -tl'uz

sperm, fish lhooktl'uz

testicles 'ughez

testicles -ghez

triplets lhulhcho

twins lhtudu'alts'e

21. Water, Snow and Ice

21.1. General

cold water too nezk'uz

deadhead tehduchun

dew nadelhudzo

eddy 'ok'et

water, cold too nezk'uz

water, my sutoo

water, our netoo

water, their hubutoo

water, your (1) xnyxutoo

water, your (2+) nahtoo

water, his/her butoo

water, his/her yutoo

water, his/her own dutoo

waves on water tatsi

whirlpool 'ok'et

21.2. Bodies of Water

bank of body of water too busk'ut

bay tl'ah

centre of lake bunniz

confluence of rivers lheghidli

confluence of rivers lheidli

lake bun

ocean yatoo

pond tadez'ai

puddle tadez'ai

river 'ukoh

river koh

river took'oh

river bank toobus

river bank, underpart of overhanging toobust'ah

seep whuzainli

shore of river kohtaba

water hole took'et

21.3. Snow and Ice

avalanche 'unt'oh

frost so

frost up high dahzo

hailstone 'inlootsan

ice, a piece of lhum

ice over a surface tun

icicle kw'uzghoo

snow yus

snow up high dahyus

21.4. Motion, etc.

overflowed, it dadembun

22. Places

22.1. Rivers

Angly Creek Bundzikoh

Canyon Creek Noozkoh

Knapp Creek Belhk'akoh

Morris Creek Tats'utnai

Nautley River Nadlehkoh

Ormond Creek Noozkoh

Sutherland River Natl'alikoh

Sutherland River Natl'ilikoh

Ormond Lake River Choostl'okoh

22.2. Lakes

Alf Lake Chunbaz Bunk'ut

Angly Lake Bundzi Bunk'ut

Barlow Lake Duk'ai Hooni Bunk'ut

Burns Lake tselhk'azbunk'ut

Chowsunkut Lake Chawts'unk'ut

Dry William lake Wilyum sugi

Drywilliam Lake Wilyum sugi

Etcho Lake Ts'oon'ai Bunk'ut

Fraser Lake Nadlehbun

Fraser Lake Nadlehbunk'ut

Klez Lake Lhez Bunk'ut

Leg Lake Tsenabulh'a Bunk'ut

Mary Lake Basghelh Bunk'ut

Oona Lake Yoonoo Bunk'ut

Ormond Lake Choostl'o

Peta Lake Nanguz Bunk'ut

Stag Lake Yests'e Bunk'ut

Tatin Lake Tanyiz Bun

Top Lake Hoolhtan Bunk'ut

Trout Lake Duk'ai Hooni Bunk'ut

22.3. Mountains

Fraser Mountain K'atsizyus

Greer Mountain Deduk Usk'en

Mount Greer Deduk Usk'en

Mouse Mountain Lhkw'etsilhchola

Mouse Mountain Neyilasts'ah

Pitka Mountain Buzdaiyus

peak of Greer Mountain Skenicho

Saddle Mountain Buzdaiyus

unnamed hill Chundooyus

22.4. Towns

Nautley Nadleh

Stellakoh village, old name Meljayah

Vanderhoof Tehk'eilchuk

22.5. Other

Alexandria Stellayukh

Belgatse Belhk'achek

Cheslatta village Chestl'ada

Crowe Canyon Yausilh

Devil's Hairpin Liyubyints'izti

Endakoh River Endakoh

embankment along Nechakoh nooghun

Lejac Tseyaz

meadow tl'ok'et

meadow tl'ok'ut

Old Fort Fraser K'atsizchun

Peterson's Beach Forestry Campsite Sainaditi

Peterson's beach and point Nuswhutisdi

pasture tl'ok'ut

point of land on Fraser Lake below Mouse Mountain Lhkw'etsilhchola

Stellakoh River Stellakch

Stellakoh village Stellakoh

shady spot 'uscheh

territory keyah

trapline keyah

village keyah

Yensischuck 'Indzik'et

23. Land, Landforms, Rocks and Soils

23.1. General Terms

beach taba

brushy, it is whuduts'ul

bush duchuntah

forest duchuntah

shore taba

23.2. Landforms

bank of body of water too busk'ut

base of a mountain dzulhcheh

bay tl'ah

canyon duyuk

cave 'an

cave 'u'an

cave tse'an

cliff that descends into water tadunet'ai

den 'an

hillside whenun

island noo

mountain dzulh

mountain, big dzulhcho

mouth, river tachek

peaks, mountain tah

peninsula sdughak'ut

reef tehnoo

river mouth tachek

summit tl'adak

swamp whulhtsultah

swamp yak'ut

waterfall duyuk

23.3. Rocks and Soils

cinnabar tseskaik'ejil

duff chehhaindzool

gravel tseyaz

mud lhetl'us

obsidian tselgai

rock tse

sand sai

stone tse

24. Forest

bush tintah

duff chehhaindzool

forest tintah

25. Astronomy

aurora borealis yawhudoos

dark, it is tsaholhgus

eclipse sa yenin'ai

meteorite shower kwuncho nainkat

meteorite shower sumcho nainkat

meteorites sum nainkat

meteorites sum nakat

morning star sumcho

northern lights yawhudoos

sets, the sun sa k'ah naidutut'ih

setting, the sun is sa k'ah naidutat'en

setting, the sun is yuk 'utez'ai

shooting stars sum nainkat

shooting stars sum nakat

star sum

stars, shooting kwuncho nainkat

stars, shooting sum nainkat

stars, shooting sum nakat

stars, shooting sumcho nainkat

sun sa

sundogs sadzak'uilke

sunrise, red sa ilk'un han'ai

sunset, red lhk'un nasain'aih

sunset, red nasain'ai hoolk'un

26. Weather and Atmosphere

blizzard has begun, a yootsiz tezts'i

cloud kw'us

cloud yat

cloudy, it is kw'us hooni

cloudy, it is lheyatdendi

cold, it (weather) turned nawniduk'az

cold, it got hoonk'az

cold, it is whunezk'uz

cold, it is whuzk'uz

downpour, there is chan nainli

drizzling, it is nawhulhtiyaz

drizzling, it is nawhulhtsit

fog 'a

foggy, it is 'a naints'ut

foggy, it is naints'ut

frosty clear night, it is a yawtelhah

good, it (weather) is whudinzoo

hailing, it is 'inlootsan nanukat

hailstone 'inlootsan

hot, it is whunilghaz

hot and sunny, it is sa dendi

lightning strikes detnik oolhgis

overcast and stormy, it (the sky) becomes whudintsuk

overcast and stormy, it (the sky) is whudintsi'

pouring down, rain is chan nainli

rain chan

rainbow, there is a detnik na'nanguz

raining into water, it is tawhulhti

scattered clouds, there are ninawhunde

sky yat

sky, clear yazai

snow, it is blowing yus bulh nilhts'i

snowing, it is najas

snowing, it is nawhujas

sunny, it is ae hasadanat

sunny and hot, it is sa dendi

thunders, it detnik 'utni

waterspout nilhts'i tainya

27. Wind

blew in, a draft danilhk'az

blow, the wind has begun to tezts'i

blowing, it (wind) is nilhts'i

blowing into, it (wind) is danilhts'i

blowing into, it (wind) is dats'i

east wind ndazde nilhts'i

north wind yootsiz nilhts'i

snow, it is blowing yus bulh nilhts'i

south wind yoo'az nilhts'i

tornado nilhts'idus

wind nilhts'i

wind, east ndazde nilhts'i

windy, it is nilhts'i

28. Natural Events

avalanche 'unt'oh

buried in an avalanche or landslide, he/she is buk'enahoolhah

calamity khundughun

earthquake yun nughutnah

happened, it 'uhooja

happened, it has not nduchahoonil

happened, something bad dzah 'uhooja

happening, it is nduwhuneh

what happened? dahooja

what will happen? dawtenilh

29. Human Events

calamity khundughun

30. Clothing and Adornment

30.1. Verbs

dressed up, they are soonahduja

dressed up, we (3+) are soonazduja

dressed up, you (1) soonadinja

put on (st), I ghunadulhu'oo'

put on (st), he/she ghunadulhu'oo'

put your shoes on, you (1) kenaindutsi

30.2. Clothing

apron k'uztsan

belt se

belt, my suze

blue jeans doso tl'asus

bra ts'oozus

brassiere ts'oozus

breechcloth tsan

cap ts'ah

cap ts'oh

cape luglok

clothes naih

clothes, old naihjut

coat dzoot

cowboy hat ts'ahbul

dress tl'asus

dress, my stl'asus

dress, pants netl'asus

dress, their hubutl'asus

dress, your (1) ntl'asus

dress, his/her butl'asus

gloves butsuk hooni

handkerchief dzezoh

hat ts'ah

hat ts'oh

hat, cowboy ts'ahbul

hat, my cowboy sts'ahbul

hat, our cowboy nets'ahbul

hat, their cowboy hubuts'ahbul

hat, wide-brimmed ts'ahbul

hat, your (1) cowboy nts'ahbul

hat, his/her cowboy buts'ahbul

house, your (1) nyah

jacket, skin 'uzuz dzoot

jeans, blue doso tl'asus

kerchief dzezoh

leggings ketsih

mittens bat

mittens, their hububat

mitts, her bubat

mitts, my sbat

mitts, your (1) mbat

mitts, his/her bubat

moccasins kesgwut

moccasins, old, worn-out kesgwutch'ul

mukluks kesgwutcho

pajamas bulh sdutez

panties ts'ekoo tl'asus

pants dune tl'asus

pants tl'asus

pants, my stl'asus

pants, our netl'asus

pants, their hubutl'asus

pants, your (1) ntl'asus

pants, his/her butl'asus

rain coat toodzoot

scarf tsinezdelya

shawl lasal

shirt dzoozt'an

shirt, my sdzoozt'an

shirt, our nedzoozt'an

shirt, their hubudzoozt'an

shirt, your (1) ndzoozt'an

shirt, his/her budzoozt'an

shoe kegon

shoe laces ketl'oolh

shorts tl'abesu'ai

skin jacket 'uzuz dzoot

skirt ts'itukdukw

slip deyoh tl'asus

socks ketul

sombrero ts'ahbal

toque jook

toque, my sjook

toque, our nejook

toque, their hubujook

toque, your (1) njook

toque, his/her bujook

underwear deyoh naih

vest dzootdukw

wallet sooniyazuz

30.3. Materials and Parts

beads kw'usul

burlap doso

button nanezmaz

canvas doso

cloth dech'ulh

denim doso

pocket lubos

porcupine quills used for embroidery lhbai

30.4. Jewelry

earrings dzekw'ul

30.5. Miscellaneous

barefoot ke'et

brush, clothes naih benaldzooh

clothes brush naih benaldzooh

clothes line naih ba whenant'uk

eyeglasses nak'ezdulya

31. Skins and Cloth

31.1. Skins

cured hide 'uzuz

deer skin yests'ezuz

hide 'uzuz

hide -zuz

hide, moose dunizuz

hide, my szuz

hide, our nezuz

hide, their hubuzuz

hide, your (1) nzuz

hide, his/her buzuz

moose hide dunizuz

skin 'uzuz

skin -zuz

skin, deer yests'ezuz

31.2. Tanning Tools and Materials

frame, stretching busdi

scraper, bone skin be'ulje

stretching frame busdi

31.3. Tanning Verbs

cased, I ghanususguz

31.4. Cloth and Fabric

canvas mandah

31.5. Parts of Garment

button hole nanezmazk'et

32. Sewing, Weaving, Knitting

knitting (u.o.), I am 'ustl'oo

knitting (u.o.), he/she is 'utl'oo

needle yabatsoh

pin 'unibek

pin, straight 'unubek

sinew 'uts'eh

thread yabatseh

33. Containers

back pack 'unizus

bag 'uzus

bag, paper dustl'uszuz

basket tilh

basket, large bark chalhyal

basket, my sutilh

bassinet 'uski ts'ai

box chunkhelh

box, cardboard dustl'us chunkelh

bullet pouch k'azus

can testilh

can, jerry tes hoongwun

canister set be'usdudzak

cedar chest hoongwun

chest, cedar hoongwun

coffee pot lugafi 'oosa'

coffee pot lugafi be'bizih

coffee pot lugafi be'udlez

coffin dune besulhti

coffin nezek besulhti

cupboard be'uzdla

cupboard, food ts'uyi besula

cupboard, my sbe'uzdla

cupboard, our nebe'uzdla

cupboard, their hubube'uzdla

cupboard, your (1) xnyxube'uzdla

cupboard, his/her bube'uzdla

dresser be'uzdla

dresser naih besula

flour sack lhes 'uzus

flower pot t'an behanuyeh

food cupboard, my sts'uyibesula

food cupboard, our nets'uyibesula

food cupboard, their hubuts'uyibesula

food cupboard, your (1) nts'uyibesula

food cupboard, his/her buts'uyibesula

jerry can tes hoongwun

kettle 'oosa'

kettle betanizilh

pail 'oosa'

paper bag dustl'uszuz

pocket lubos

pot 'oosa'

pot, coffee lugafi be'udlez

pouch zus

pouch, bullet k'azus

purse 'ezdlai

purse soonuy uti zuz

purse soonuyazuz

rice sack rice 'uzus

sugar sack soogah 'uzus

tea pot ludi 'oosa'

thermos be'nezul

trunk duchun khelh

vase t'an bedezdla

wallet sooniyazuz

wash basin bela'ts'uldeh

waste basket bena'ulkuk

34. Rope, Cable, Chain, Cord

rope tl'oolh

rope, my stl'oolh

twine used for stringing nets, making dipnets, and so forth 'ubalt'i

35. Hunting

35.1. Gear

arrowhead noondai

bullet k'a

bullet pouch k'azus

pouch, bullet k'azus

rifle 'ulhti

rifle shell k'a

rifle lhti

35.2. Verbs

going after (hunting, trapping), he/she is ka'uneh

hunt (st), they will ka'hutet'ilh

hunt (st), we (2) will ka'tadut'ilh

hunt (st), we (3+) will ka'uztet'ilh

hunt (st), you (1) ka'tant'ilh

hunt around (in the bush), I am going to nutelhudzulh

hunted (st), they ka'het'en

hunted (st), we (2) ka'adut'en

hunted (st), we (3+) ka'ts'et'en

hunted (st), you (1) ka'ant'en

hunting, I am going nutesdzulh

hunting, I am going to go telhudzulh

hunting (st), they are ka'hut'en

hunting (st), we (2) are ka'idut'en

hunting (st), we (3+) are ka'ts'ut'en

hunting (st), you (1) are ka'int'en

36. Firearms

rifle, big 'ulhticho

rifle lhti

37. Trapping

37.1. Gear

anchor for rabbit snare nal'oo

beaver snare tsabilh

deadfall trap gooh

rabbit snare gambilh

snare bilh

trap, deadfall gooh

tree from which rabbit snare is hung nache'untut'ah

trigger for rabbit snare haltoo'

37.2. Miscellaneous

bait 'uni

trapline yuntah

38. Fishing

38.1. Gear

anchor, fish net tsetoos'ai

fish hook jus

fish net lhombilh

fish net anchor tsetoos'ai

fish net anchor pole lhoombilhchun

fish net weights lhoombilhtse

fish spear dagwut

fish trap 'uk'oondzai

fishing pole juschun

floats for fish net dolulh ga

hook sah

net, fish lhombilh

pole, fishing juschun

spear, fish dagwut

trap, fish 'uk'oondzai

trawling line nubalhtl'ool

twine used to string net balt'i

weight at end of fish net tse toos'ai

38.2. Verbs

fishing with a hook, he/she is 'oolhjas

hooked (fish), we (2) seduljus

hooked (fish), we (3+) ts'ulhjus

hooked (st:fish), I sulhjus

hooked it (fish), they huyulhjus

hooked it (fish), he/she yulhjus

set net, I te'nusla

set net, I te'usdle

set net, I will te'teslilh

set net, he/she te'ninla

38.3. Miscellaneous

bait 'uni

39. Gathering Plants

39.1. Picking Berries

picking (berries), I am oonusyin

picking (berries), we (2) oonidujin

picking (berries), he/she is oonuyin

40. Farming, Gardening, and Ranching

40.1. Gear

seeder behooyeznuleh

swather (machine for mowing hay) tl'o bedudut'as

40.2. Verbs

harvest, I hanusle

harvest, we (3+) hats'unule

harvest (u.o.), I ha'nusle

harvest (u.o.), they ha'hunule

harvest (u.o.), we (2) ha'nidulye

harvest (u.o.), we (3+) ha'ts'unule

harvests (u.o.), he/she ha'nule

milk, I am going to tesjas

milking it, he/she is yulhjas

plow, he/she will nawtelhch'ul

plowed, he/she nawhalhch'ul

40.3. Miscellaneous

fertilizer hanuyeh buts'uyi

haystack lubuyos

41. Horse

41.1. Related

tracks, horse yeztlik'ah

42. Logging

42.1. Verbs

felled, you (1) nadalhghez

43. Institutions, Organizations, Businesses, etc.

brothel 'ul'en bayah

cafe ts'uyi bayah

health centre yoobayah

hotel sdutez

post office dustl'us dakat

44. Commerce

44.1. Money

money chigamin

money sooniya

money soonuya

quarter gwada'

44.2. Verbs

broke, he/she is k'untuk

buy, I will ooteskulh

buying (st), I am oosket

buying (st), he/she is ooket

buying (u.o.), they are 'uhooket

buying (u.o.), we (2) are 'ooduket

buyng (u.o.), we (3+) are 'uts'ooket

costly, it is dezti

earned it, they whuch'ahinelhde

earned it, we (2) whuch'anidulde

earned it, we (3+) whuch'aznelhde

earned it, he/she whuch'ainelhde

earning income, he/she is 'udutl'us

expensive, it is dezti

owes him (st), he/she yuts'uwhulh'ai

penniless, he/she is k'untuk

44.3. Miscellaneous

credit jaboon

45. Tools, Machines and Instruments

awl 'utsoh

axe tsetsilh

brush, clothes naih benaldzooh

camera benek'edugus

camera benek'ehugis

can opener bedahudut'as

carburetor yesjiz

clock tsadzi

clothes brush naih benaldzooh

dipper betadzih

dipper too benuduka

dipper (for water) toobets'ukaih

grammar beyatuk

grindstone tsek'az

hammer be'ul'uz

handle ti

handle, its buti

knife 'utes

knife tes

knife, my se'utes

knife, our netes

nail lugloo

poker, stove koonangus

safety pin dughul

saw be'dudughut

scissors lheyih

spade lubel

stove poker koonangus

switch (for whipping) tut'oh

telephone beyatuk

telephone nilhdza beyaztuk

thermometer kw'uzbenul'en
things we (3+) use bene'ts'ut'en
use, things we (3+) be'ts'ut'en
watch tsadzi

46. Parts and Components

board dzihtel
handle ti
handle, its buti

47. Buildings and Rooms

bar tadutnai
barn tl'obayah
barn, cow musdusbayah
bathroom tsanbayah
bed tesk'et
bedroom tesk'et
bedroom ts'uztez ba whuz'ai
chicken coop ligokbayah
church lugliz
community hall, Nautley Nadlehyah
dance hall nudaih ba yah
dining room 'udai-a
dining room 'uts'uyi ba whuz'ai
food storage building tsuyi ba yah
health centre yoobayah
home yah
hotel sdutez
house koo
house yah
house, my skoo
house, my syah
house, our nekoo
house, their hubukoo
house, their hubuyah
house, your (1) nkoo
house, his/her bukoo
house, his/her buyah
kitchen ts'uyi ghutna-a
kitchen, my sts'uyi ghutna-a
kitchen, our nets'uyi ghutna-a
kitchen, their hubuts'uyi ghutna-a
kitchen, your (1) nts'uyi ghutna-a
kitchen, his/her buts'uyi ghutna-a
living room dune delts'i
living room, my sdune delts'i
living room, our nedune delts'i
living room, their hubudune delts'i
living room, your (1) ndune delts'i
living room, his/her budune delts'i
outhouse tsanbayah
outhouse, my stsanbayah
outhouse, our netsanbayah
outhouse, their hubutsanbayah
outhouse, your (1) ntsanbayah
outhouse, his/her butsanbayah
room, dining 'uts'uyi ba whuz'ai
smoke house slenyah
stable for cattle musdusbayah
stable for horses yeztlibayah

48. Structures

48.1. Caches

cache hawukaih

cache, food tsachun

food cache tsachun

48.2. Racks

rack for drying food nin'ai

48.3. Parts

attic bunt'ah

attic yakwut

attic, my sbunt'ah

attic, our nebunt'ah

attic, their hububunt'ah

attic, your (1) mbunt'ah

attic, his/her bubunt'ah

boards, floor yundzihtel

ceiling bunt'ah

chimney lusooni

chimney lhut be'hanajut

door dadentan

doorknob dadentanghunez'ai

doorknob, my sdadentanghunez'ai

doorknob, our nedadentanghunez'ai

doorknob, their hubudadentanghunez'ai

doorknob, your (1) ndadentanghunez'ai

doorknob, his/her budadentanghunez'ai

doorknobs dadentanghunezdla

doorknobs, my sdadentanghunezdla

doorknobs, our nedadentanghunezdla

doorknobs, their hubudadentanghunezdla

doorknobs, your (1) ndadentanghunezdla

doorknobs, his/her budadentanghunezdla

doorway dadentank'et

doorway dati

doorway datuk

doorway, my sdati

doorway, our nedati

doorway, their hubudati

doorway, your (1) ndati

doorway, his/her budati

eaves bunt'ah

entrance dati

entrance whudah

fence post nanestl'oo kechun

floor yun

floor, my syun

floor, our neyun

floor, their hubuyun

floor, your (1) nyun

floor, his/her buyun

floor boards yundzihtel

flooring yunlatel

flooring, my syunlatel

flooring, our neyunlatel

flooring, their hubuyunlatel

flooring, your (1) nyunlatel

flooring, his/her buyunlatel

ground yun

keyhole luglik'et

porch buntsilh

porch dadeltel

porch, my sdadeltel

porch, our nedadeltel

porch, their hubudadeltel

porch, your (1) ndadeltel

porch, his/her budadeltel

roof bun

roof, my sbun

roof, our nebun

roof, peaked ch'usghak

roof, their hububun

roof, your (1) mbun

roof, his/her bubun

wall sih

wall, my susih

wall, our nesih

wall, their busih

wall, your (1) nsih

wall, his/her busih

window dadent'az

window, my sdadent'az

window, our nedadent'az

window, their hubudadent'az

window, your (1) ndadent'az

window, his/her budadent'az

48.4. Miscellaneous

dam 'ulh

fence nanestl'oo

49. Electricity

electrical outlet or plug 'ana'dutsih

50. Temperature

boiling (u.o.), he is 'undunulhmulh

boiling it (water), they are hindunulhmulh

boiling it (water), he/she is yundunulhmulh

chilly, it is whunik'az

cold, it got hoonk'az

cold, it is nezk'uz

cold, it is nezk'uz

cold, it is nink'az

cold, it is whuzk'uz

cold water too nezk'uz

died of exposure, he/she dli yuzelhghi

exposure, he/she died of dli yuzelhghi

froze, it ustun

heat sazul

heat waves sul

heater beoonezul

heats (st), nulhghaz

heats (st), he/she whunulhghaz

hot, it has gotten nawnilghaz

hot, it is nilghaz

hot, it is nulwus

hot, it is whunilghaz

hot, it will start to get nawntelwus

hot, they are hunulwus

hot, we (3+) are ts'unulwus

hot, you (1) are nilwus

killed by frost, it was ustun

thermometer kw'uzbenul'en

water, cold too nezk'uz

51. Aircraft and Flight

airplane benuts'ut'ah

helicopter kw'ut nabadunelgih

52. Boats and Travel on Water

bailing out (st:boat), I am betaszih

canoe ts'i

canoe, spruce bark 'ulats'i

go by boat, I will not chatuzeskel

going by boat, I am not chaseskelh

land in a boat, I yaskih

land in a boat, they yahukih

land in a boat, we (2) yaidukih

land in a boat, we (3+) yats'ukeh

lands in a boat, he/she yakih

oar chos

paddle (canoe) chos

passed him in a boat, I beghoosuski

raft khinyus

set off by boat, I have not whechaskel

53. Roads and Trails

animal trail khunaiti

border-crossing niti

branches, it (road) lhk'eti

bridge kw'ut yan wheti

bridge sus

end of road niti

foot path duneti

fork in road lhk'eti

fork in road, three-way tsewhehudilhkw'ah

highway ti

rabbit trail gahti

road ti

trail ti

54. Transportation

54.1. Land

baby buggy uski be'nunulbaz

bicycle benuts'ulgaih

cane tuz

cane, my stuz

cane, his/her butuz

car benuts'ugoo

car otomobil

exhaust pipe -tl'it

exhaust pipe, its butl'it

eye -na

groin -tl'it

headlight -na

stroller 'uski benunulbas

tail lights of vehicle butsul kwun dilhk'ai

vehicle benugoo-i

vehicle benuts'ugoo

vehicle hiyeneyulhgoo

vehicle, my benusgoo-i

vehicle, your (1) beningoo-i

wagon luwagun

54.2. Snow and Ice

skidoo yusk'ut nugoo

sled nugoo

snow machine yusk'ut nugoo

toboggan 'usdloos

55. Fire, Explosions, and Light

55.1. Nouns

ash lhustsis

coal kwuntset

coal tse dezk'un

dynamite oondunultalh

ember kwuntset

fire kwun

firecracker oodunaltalh-i

firewood tsuz

flame kwun

hearth tseba

light kwun

matches kwunyoo

poker, fire daskwun

poker, stove koonangus

skewer daskwun

smoke lhut

stove poker koonangus

tail lights of vehicle butsul kwun dilhk'ai

55.2. Verbs

flammable, it is dillhuk

flashing, it is whedunukw

ignite, you (1) nadilhk'aih

ignite again, you (1) nakwundilhk'aih

smoulders, it diyooh

56. Appliances

clothes dryer naih benadugih

cook stove 'ook'et'as

dryer, clothes naih benadugih

frying pan be'ut'es

heater beoonezul

microwave oven 'a be'ut'es

oven be'ut'es

oven, microwave 'a be'ut'es

pan, frying be'ut'es

stove, cook 'ook'et'as

stove, my cook se'ook'et'as

stove, our cook ne'ook'et'as

stove, their cook hube'ook'et'as

stove, your (1) cook xnyxe'ook'et'as

stove, his/her cook be'ook'et'as

television nul'en-i

television, my snul'en-i

television, our nenul'en-i

television, their hubunul'en-i

television, his/her bunul'en-i

televsion, your (1) xnyxenul'en-i

57. Furnishings and Personal Belongings

57.1. General

table ludab

table ook'utudai

table, my se'ook'utudai

table, my sludab

table, our ne'ook'utudai

table, our neludab

table, their hube'ook'utudai

table, their hubuludab

table, your (1) nludab

table, your (1) xnyxe'ook'utudai

table, his/her be'ook'utudai

table, his/her buludab

57.2. Personal Items

sheet lilik'usulhchooz

sheet, my slilik'usulhchooz

sheet, our nelilik'usulhchooz

sheet, their hubulilik'usulhchooz

sheet, your (1) nlilik'usulhchooz

sheet, his/her bulilik'usulhchooz

57.3. Household Items

ash tray lhustsis bekuk

bath tub betoonuts'uya

bed lili

bed skelh'at

bed, my stesk'et

bed, my sulili

bed, our nelili

bed, our netsk'et

bed, their hubulili

bed, their hubutesk'et

bed, your (1) xnyxelili

bed, your (1) xnyxutesk'et

bed, his/her bulili

bed, his/her butesk'et

bed spread lili k'ut subal

bedroom, my stesk'et

bedroom, our netsk'et

bedroom, their hubutesk'et

bedroom, your (1) xnyxutesk'et

bedroom, his/her butesk'et

blanket nalhti

blanket, my snalhti

blanket, our nenalhti

blanket, their hubunalhti

blanket, your (1) nnalhti

blanket, his/her bunalhti

broom betina whultsas

carpet yun subal

carpet yun sulhchooz

chair kw'usduda

chair kw'uts'uda

chair, easy kw'uts'udacho

chair, my skw'uts'uda

chair, my easy skw'uts'udacho

chair, my reclining sduneti kw'uts'uda

chair, our nekw'uts'uda

chair, our easy nekw'uts'udacho

chair, our reclining neduneti kw'uts'uda

chair, reclining duneti kw'uts'uda

chair, rocking kw'usduda nubalh

chair, their hubukw'uts'uda

chair, their easy hubukw'uts'udacho

chair, their reclining hubuduneti kw'uts'uda

chair, your (1) nkw'uts'uda

chair, your (1) easy nkwuts'udacho

chair, your (1) reclining nduneti kw'uts'uda

chair, his/her bukw'uts'uda

chair, his/her easy bukw'uts'udacho

chair, his/her reclining buduneti kw'uts'uda

chest of drawers bena'dukuk

chesterfield kw'uts'udayez

chesterfield, my skw'uts'udayez

chesterfield, our nekw'uts'udayez

chesterfield, their hubukw'uts'udayez

chesterfield, your (1) nkw'uts'udayez

chesterfield, his/her bukw'uts'udayez

clothes chest bena'dukuk

clothes washer be'tunadugus

comforter nalhticho

cooker, Indian slow huyutsa k'ez hik'ut kwun dilhk'ai

couch k'usdudacho

couch kw'usdudacho

couch kw'usdudayez

cupboard be'uzdla

cupboard, dish tsets'ai besula

cupboard, my sbe'uzdla

cupboard, our nebe'uzdla

cupboard, their hubube'uzdla

cupboard, your (1) xnyxube'uzdla

cupboard, his/her bube'uzdla

curtain dadent'az ts'oh sula

curtain, window dadent'azts'asdla

curtains dadent'az ts'osdla

curtains, my sdadent'azts'asdla

curtains, our nedadent'azts'asdla

curtains, their hubudadent'azts'asdla

curtains, your (1) ndadent'azts'asdla

curtains, his/her budadent'azts'asdla

cushion tsi'alhyaz

detergent latl'us

dresser be'uzdla

dresser naih besula

easy chair kw'uts'udacho

floor covering yun sulchooz

floor covering yunsubul

floor covering, my syunsubul

floor covering, my syunsulchooz

floor covering, our neyunsubul

floor covering, our neyunsulchooz

floor covering, their hubuyunsubul

floor covering, their hubuyunsulchooz

floor covering, your (1) nyunsubul

floor covering, your (1) nyunsulchooz

floor covering, his/her buyunsubul

floor covering, his/her buyunsulchooz

flower pot t'an behanuyeh

freezer be'utun

glass, drinking betats'utnai

linoleum yun subal

mattress 'ustes

mattress madus

mattress, my se'ustes

mattress, our ne'ustes

mattress, their hube'ustes

mattress, your (1) xnyxe'ustes

mattress, his/her be'ustes

mattress kw'ut ts'uzti

medicine cabinet yoo besula

mirror tooznal'en

outlet, electrical 'ana'dutsih

pillow tsi'alh

pillow, my stsi'alh

pillow, our netsi'alh

pillow, their hubutsi'alh

pillow, your (1) ntsi'alh

pillow, his/her butsi'alh

quilt lhtanalkat

recliner kw'usduda t'az whebulh

reclining chair duneti kw'uts'uda

refrigerator be'nezk'uz

rocking chair kw'usduda nubalh

rug yunlhutel

soap latl'us

stereo bets'ujun

stove sto

table cloth ludab k'usulhchooz

toilet tsan bayah

towel lusooma

washer, clothing be'tunadugus

waste basket bena'ulkuk

58. Paper, Books, Writing Instruments

paper dustl'us

paper, our (3+) nedustl'us

paper, your (2+) nahdustl'us

59. Substances

abalone shell mulla

ashes lhez

carbon dioxide benuts'ugoo tl'it

card board dustl'us dutsun

charcoal t'es

charcoal, our net'es

cinnabar tseskaik'ejil

coal tse dezk'un

dirt lhez

dirt tsun

earth (soil), dust lhez

foam hawus

mercury nudubun

mercury too dulk'un

mother-of-pearl mulla
mud lhetl'us
obsidian tselgai
rock tse
sand sai
stone tse
water too

60. Cooking

60.1. Utensils and Tools

coffee pot lugafi 'oosa'
coffee pot lugafi be'bizih
coffee pot lugafi be'udlez
frying pan be'ut'es
kettle 'oosa'
kettle betanizilh
ladle tazul benuduka
oven be'ut'es
pail 'oosa'
pan, frying be'ut'es
poker, fire daskwun
pot 'oosa'
pot, coffee lugafi be'udlez
pot, tea ludibot
skewer daskwun
tea pot ludi 'oosa'
tea pot ludibot

60.2. Machines and Appliances

heater stocho
heater, my sestocho
heater, our nestocho
heater, their big hubustocho
heater, your (1) xnyxestocho
heater, his/her bustocho
microwave oven 'a be'ut'es
oven, microwave 'a be'ut'es
stove, your (1) big xnyxestocho
stove, air-tight chalhyalsto
stove, big stocho
stove, my big sestocho
stove, our big nestocho
stove, their big hubustocho
stove, his/her big bustocho

60.3. Verbs

boil, it came to a handunelmulh
boiling, it (liquid) is ndunulmulh
boiling (st), I am uslez
boiling (u.o.), he/she is 'undunulhmulh
boiling (u.o.), I am 'undunulhmulh
boiling it (water), they are hindunulhmulh
boiling it (water), he/she is yundunulhmulh
bringing (st) to a boil, he/she is handunelhmulh
broiling (st), I am ulhts'ul
broiling (st), we (2) are idults'ul
broiling (st), we (2) are ts'ulhts'ul
broiling it, she is yulhts'ul
broiling it, they are huylhts'ul
kneading it, he/she is yunulhlhus
plucked it, he/she hayanyuz

plucking (st), I am hasyis

plucking it, he/she is hayuyis

stewing (st), I am uslez

60.4. Miscellaneous

smoke house slenyah

61. Eating

61.1. Utensils

bowl, soup tazul bedzuz

chopsticks duchun be'ts'uyi

cup lubot

fork be'ooget

fork beooduget

fork lufooset

fork luhoos

glass, drinking toobets'utnai

knife 'utes

knife, my se'utes

knife, our netes

plate lasyet

plate tsets'ai

spoon tsunts'alh

61.2. Verbs

ate, I esyi

ate, they hanyi

ate, we (2) naduji

ate, we (3+) ts'anyi

ate, you (1) anyi

ate, he/she anyi

ate (u.o.), I na'esdai

ate (u.o.), we (2) na'adudai

ate (u.o.), we (3+) 'uts'anyi

ate (u.o.), we (3+) na'ts'edai

breast feed him/her, I be'ulht'uwk

breast-feeding him/her, she is ye'ulht'ukw

chewing, you (1) are duni'ulh

chewing, you (1) are ilh'ulh

chewing, he/she is duna'ulh

chews (u.o.), he/she 'a'alh

choked on something solid, he/she 'untelhah

choked on somthing liquid, he/she 'untelhuyiz

crunches in his mouth, he/she ulhgoos

drink liquor, they tahutnai

drink liquor, they will tatetnilh

drink liquor, we (3+) will taztenilh

drink liquor, you (1) will tatnilh

eat, I do not chasusyi

eat, I may not chaoosyi

eat, I will tesyi

eat, I will not chatuzesyi

eat, they huyi

eat, they do not chahusyi

eat, they will huteyi

eat, they will not chahutesyi

eat, we (2) idudai

eat, we (2) odudai

eat, we (2) will tadudai

eat, we (3+) ts'uyi

eat, we (3+) do not chats'usyi

eat, we (3+) will ts'uteyi

eat, we (3+) will not chaztesyi

eat, you (1) inyi

eat, you (1) will tanyi

eat, you (1) will not chatuzanyi

eat, you (2) do not chasinyi

eat, you (2+) ahyi

eat, you (2+) do not chazahyi

eat, you (2+) will tehyi

eat, you (2+) will not chatuzehyi

eat, he/she does not chasyi

eat, he/she will not chatesyi

eat (st), I usyi

eat (st), I may wusyi

eat (st), they may hooyi

eat (st), we (3+) may ts'ooyi

eat (st), we (3+) will naztedilh

eat (st), you (1) may onyi

eat (st), you (1) will natandilh

eat (st), he/she may wuyi

eat (st), he/she will teyi

eat (u.o.), he will 'uteyilh

eat (u.o.), I 'usyi

eat (u.o.), I na'usdai

eat (u.o.), I may 'oosyi

eat (u.o.), I may na'oosdai

eat (u.o.), I will 'utesyilh

eat (u.o.), I will na'tesdilh

eat (u.o.), we (2) na'idudai

eat (u.o.), we (2) will na'tadudilh

eat (u.o.), we (3+) 'uts'uyi

eat (u.o.), we (3+) na'ts'udai

eat (u.o.), we (3+) will na'uztedai

eat (u.o.), we (3+) will 'uzteyilh

eat (u.o.), you (1) 'inyi

eat (u.o.), you (1) na'indai

eat (u.o.), you (1) may na'ondai

eat (u.o.), you (1) will 'utanyilh

eating, he/she is uyi

eating (u.o.), they are 'uhuyi

eating (u.o.), we (2) are 'iduji

eating (u.o.), he/she is 'uyi

feed you, they are going to nghate'alh

sucked on me, it has suzt'ukw

sucking, he/she is ulht'ukw

62. Food

62.1. General

bannock lhes sut'e

bannock sto lhes sut'e

bannock, oven lhes sto bez sut'e

bear grease susghe

bread lhes

bread, fry lhes sut'e

bread, yeast liba

fat, black bear susk'a

fish lhook

fish soup lhook tazul

flour lhes nududzaih

food ts'uyi

fried meat 'utsun sut'e

grease khe

grease, black bear susghe

grease, my sghe

jam mai be'ulha

jam mai nesdliz

jam mai nudutleh

meat, fried 'utsun sut'e
meat soup 'utsun tazul oil khe
potato chips lubudakt'ooz
soup, fish lhook tazul
soup, meat 'utsun tazul

62.2. Meat

bacon gugoos k'atsun
bacon lubegin
beaver meat tsatsun
beef musdustsun
brisket 'ujoohtsun
colon 'utsutle
dried beaver meat tsaikaih
dried beaver meat tsatsun sugi
dried meat 'utsun sugi
ears, rabbit gahdzo
meat 'utsun
meat, bird dut'aitsun
meat, black bear sustsun
meat, deer yests'etsun
meat, dried 'utsungui
meat, dried 'utsun sugi
meat, dried beaver tsaikaih
meat, dried beaver tsatsun sugi
meat, ground 'utsun duneldus
meat, my se'utsun
meat, rabbit gahtsun
meat, raw t'eh
meat, his/her be'utsun
moose meat dunitsun
peritoneum 'uchanyoo
pork gugoostsun
rabbit meat, dried gahgi
tripe 'utsutle
venison yests'etsun

62.3. Seafood

dried salmon talook sugi
eggs, fish -k'oon
fish, dried lhook sugi
fish, reconstituted dried nalhtsul
fish eggs -k'oon
fish split once and dried k'ak
grease, oolichan sleghe
salmon, dried talook sugi

62.4. Plant

banana mooki
beets 'buzkai nawhudleh
beets lhits'e
berries, dried mai nezgi
berries, dried nezdzak
cabbage 'utancho
dried berries mai nezgi
dried berries nezdzak
fruit leather maitlus
leather, fruit maitlus
onion tl'otsun
oranges ooyutalhai
potato lubudak
turnips lusoosam

62.5. Beverages

alcohol nedotoo
booze nedotoo

coffee lugafi
drink, white man's kwuntoo
fire water kwuntoo
fruit juice maitoo
juice, fruit maitoo
liquor nedotoo
liquor, hard too ulhtus
milk lilet
milk, powdered lilet sugi
pop too lhuk'i
tea ludi
tea, my sludi
tea, their hubuludi
wine ts'ekootoo
pie bedutleh
pitch dzeh
porridge mus
raspberries bindak khastl'ah
roe, fish 'uk'oon
salt lusel
soup tazul
soup, hamburger 'utsun tazul
stew tazul
sugar lusook
sugar soogah
sugar, my sulusook
testicles 'ughez
tripe chalhzus

62.6. Miscellaneous

butter dooda
chicken egg ligokghez
cookie lhes lhuk'i
cracklings k'ohch'az
cream dalghen
egg, chicken ligokghez
egg, fish 'uk'oon
eggs 'ughez
fat 'uk'a
fish egg 'uk'oon
gum, chewing dzeh
hamburger soup 'utsun tazul
honey hoolht'oghe
Indian popcorn k'ohch'az
mustard skitsan
oatmeal mus
pepper sulhts'i

63. Recreational Drugs

63.1. Materials

alcohol nedotoo
booze nedotoo
drink, white man's kwuntoo
fire water kwuntoo
liquor nedotoo
marijuana beni hoolah 'ut'ot
tobacco dek'a
tobacco ts'ut'ot
tobacco, my sdek'a

63.2. Utensils

tobacco pipe dek'atse

63.3. Verbs

smoke, I 'ust'ot

smoking marijuana, he/she will be t'antet'ot

64. Warfare

64.1. Weapons

bullet k'a

club, war belhyal

rifle shell k'a

rifle lhti

spear ts'oh

64.2. Miscellaneous

captive lhuna

captures him/her, he/she yulhna

soldiers bahne

soldiers nubah

65. Traditional Technology

65.1. Baskets

rim of a basket bal'ah

66. Music

bell luglos

drum tughul

sang, I esjun

sing, I usjun

sing, I did not chasjun

sing, I do not chasusjun

sing, I will 'utesjun

sing, I will tesjun

sing, they are going to hutejun

sing, we (2) will 'utadujun

sing, we (3+) will 'uztejun

sing, we (3+) will not ts'utesjun

sing, you (1) will 'utanjun

sing, you (2+) will 'utehjun

sing, you (2+) will tehjun

sing, he/she will 'utejun

sing, he/she will tejun

song shun

song, my syun

song, our neyun

song, ritual duyun-ishun

song, he/she broke into shun hadan'ai

67. Recreation

67.1. Toys

ball nukuk

jump rope ye'ubul

67.2. Games

cards, playing lugar

gambling, I am 'aslih

gambling, they are 'ahulih

gambling, we (2) are 'aidudlih

gambling, we (3+) are 'ats'ulih

gambling, he/she is 'alih

lahal nehu'a

67.3. Sports

ball, large nukukcho

ball, small nukukyaz

baseball, he/she is playing nundunult'a

67.4. Dancing

dance, I nusdaih

dance hall nudaih ba yah

dance in the Indian way, I 'ulhudzus

dance in the Indian way, they 'uhuldzus

dances in the Indian way, he/she 'uldzus

dancing, they are nahuwhulyeh

dancing, we (3+) are uznudaih

dancing, he/she is nudaih

dancing, he/she is nuwhulyeh

dancing Indian-style, he/she is nuldzus

tap-dancing, he/she is duke be 'ulhghalh

68. Gambling

gamble, we (3+) will 'aztelih

gamble, you (1) will 'atanlih

gamble, you (2+) will 'atehlih

69. Religion

69.1. Traditional

God Yoodughi

ghost dunezoolh

ghost naoodnilh

medicine man duyun

omen me

shaman duyun

shaman, female duyun ts'eke

spiritual traveler duyun

sweat lodge tsezul ba yah

therapist duyun

69.2. Christian

angel lizas

angel yat'en

angels yat'en-ne

bless myself, I nunadulhunih

bless ourselves, we (2) nunadidulnih

bless ourselves, we (3+) nunazdulnih

bless themselves, they nunahudulnih

blesses himself, he/she nunadulnih

Christian person Yak'usda bughunek k'une 'ut'en-un

church lugliz

cross tulalhgus

devil liyab

devil netsudule

Easter Zooldubak

God Yak'usda
Holy Spirit Ndoni
heaven yak'uz
hell kwunchoyuk
messenger nawhulnuk
Pentecost Ndonidzen
Protestant luminis
priest nawhulnuk

69.3. Miscellaneous

pray, I tenadusdli
pray, we (3+) tenazdudli
pray, we (3+) may tenazdoodli
pray, you (1) tenadindli
pray, you (1) may tenadondlih
pray, you (2+) tenadahdli
prays, he/she tenadudli

70. Burial, Mourning, Cremation, Memorials

buried here, he is whuyudilhti
buried him/her, they hidilhti
graveyard ts'unk'ut

71. Emotions

afraid, I am nulhujut
afraid, they are hunuljut
afraid, we (2) are niduljut
afraid, we (3+) are uznuljut
afraid, he/she is nuljut
afraid of it, I am benlhujut
afraid of it, they are hiyenuljut
afraid of it, we (2) are beniduljut
afraid of it, we (3+) are beznuljut
afraid of it, he/she is yenuljut
amazed, I am sba hooncha
amused, I am dlonus'i
amused, I am sahulhgi
amused, I am sba huwonenik
amused, I am sbadlo
angry, he is huske
angry, I am hunlhuch'e
angry, he/she is hunilch'e
anxious, I am hubanelhu'it
anxious, I am whenlhujut
cling to (st), I ka hoosdutun
cranky, he/she is khusduke
embarassed, I am yooya susli
exult, they 'uhoolhkes
exults, he/she 'oolhkes
guilty, he/she feels tinch'a'unt'en
happy, I am hoonust'i
happy, they are huhoont'i
happy, we (3+) are ts'uhoont'i
happy, he/she is hoont'i
jealous, he/she is 'oolnih
love you (1), I nk'esi'
mad, he/she is hunilch'e
misses you (1), he/she nk'enentai
ordeal, he/she endured an dzah nusuzut
panicked, he/she khunantesja
quick to anger, he/she is 'a buzkai tujulh

suffered greatly, he dzah nusuzut
suffering, I am dzah nuszut
taken aback, I am sba hooncha
worried, I am ni usli

72. Mental States

bewildered, I am sba we'hooja
believe it, we (2) nahba 'alha'hoont'ah
believe it, I sba 'alha'hoont'ah
believe it, they huba 'alha'hoont'ah
believe it to be true, we (3+) neba 'alha'hoont'ah
believes it to be true, he/she ba 'alha'hoont'ah
bewildered, we (2) are nahba we'hooja
bewildered, we (3+) are neba whe'ooja
bewildered, he/she is ba we'hooja
busy, I am sba ut'en lhai
busy, they are huba't'en lhai
busy, we (2) are nahba't'en lhai
busy, we are neba't'en lhai
busy, he/she is ba't'en lhai
delusions of grandeur, he/she has duba nadesuti
fainted, I skawtezut
fainted, he/she bukawetezut
hungry, I am sye'elts'ul
hungry, you (1) are chahoonzum
hungry, you (1) are nye'elts'ul
poor, he is tel'en
poor, I am telhu'en
poor, they are tehul'en
poor, we (2) are teidul'en
thirsty, I am taoosde
thirsty, they are tahoosde
thirsty, you (1) are taoonde
wretched, I am telhu'en
wretched, they are tehul'en
wretched, we (2) are teidul'en
wretched, he/she is tel'en

73. Character Traits and Aptitudes

acquisitive, he/she is whudzaih
bossy, he/she is dune dunuyuz
braggart, he/she is a lhena'duduwhults'it
brainy, he/she is beni hooni
cranky, he/she is khusduke
dishonest, he/she is duzah
foolish, I am masdli
foolish, they are mahudli
foolish, he/she is madli
generous, he/she is bula ncha
generous, he/she is danzoo
glutton, he/she is a buchahooncha
good-natured, he/she is budzi nzoo
grasping, he/she is whudzaih
honest, he/she is ts'ihnus hoolh'i
humble, he/she is yuk 'ududildzun
meek, he/she is cha'dudilti
merciful, he/she is te'ninzun
merciless, he/she is budzi hoolah

opinated, he/she is dubat'eoonaoonudzun

proud, he/she is not cha'dudilti

quarrelsome, to be -zek taoodetnuk

quick to anger, he/she is 'a buzkai tujulh

smart, he/she is beni hooni

stubborn, to be -tsi dunuts'un

turned around, he/she was ni lhe'nintananesja

74. Abstract Concepts

old age jan

75. Perception, Sensation, Knowing, Believing

blind, I am chawzes'en

blind, he/she is chawes'en

body odour, he/she has bujootsun

caught sight of, he/she telh'en

checks on (st), he/she nanlhu'en

checks on (st), he/she nanlhu'ih

checks on (st), he/she nawhnul'en

checks on (st), he/she nawhnul'ih

checks on it, he/she nainul'en

checks on it, he/she nainul'ih

cold, he/she feels nasdli

look at, you (1) nilh'en

look at (st), we (2) nidul'en

look at it, they hinilh'en

looking at, I am nulh'en

looking at each other, they are lhahunul'en

looking at each other with admiration, they are lhahuntoolhe'en

looking at me, they are sahunilh'en

looking at me, he/she is sunilh'en

looking at them, they are hubanilh'en

looking at you (1), we (3+) are xnyxuznilh'en

looking at you (1), he/she is xnyxunilh'en

premonition, he/she had a nus whan'en

see, I cannot chawzes'en

see, he/she cannot chawes'en

see you (1) again, I may naxnyxoost'en

sees, he/she in'en

sees (wh-class object), he/she hon'en

taking a deliberate look at each other, they are hubuhunilh'en

wonder, I subah

76. Sound

applauds, he/she lhookat

barking, it is yutse

blowing, it (siren, whistle) is uzulh

burp, I natuskwa'

burp, we (3+) 'uztukw'ak

burp, you (1) 'utinkw'ak

burp, you (2+) 'utahkw'ak

burps, he/she natukwa'

claps, he lhookat

crackling, it is dulk'us

crying, they are ghunadutsa

crying, we are ghunasdutsa

crying, you are ghuninedutsa

drumming, it (Ruffed Grouse) is 'udutsut

drumming, it is 'utsut

echo dazul

farts, he/she tutl'it

going around talking loudly, he/she is nuduldlut

hollering, he/she is dulghi

hoop and holler so that the mountains resound, they tsehudilhts'ai

make noise so as to announce myself, I nadewelhuts'ah

make noise so as to announce ourselves, we (3+) nadewe'ts'elts'ah

make noise so as to announce themselves, they na'hududelhts'ah

noisy, it (wh-class) is taoodetnuk

noisy, it is tawdetnuk

orates, he/she duldlut

quiet, I am t'edusnih

quiet, it is t'ewdusnih

quiet, he/she became t'edinil

quietly nagoostliyaz

says, he/she ni

speak, I yalhduk

speak, I will yatelhduk

speak, they will yahutelhduk

speak, we (2) yaiduldu k

speak, we (2) will yataduldu k

speak, we (3+) yats'ulhduk

speak, we (3+) will yaztelhduk

speak, you (1) yailhduk

speak, you (1) will yatalhduk

speaking, they are yahulhduk

speaks, he/she yalhtuk

speaks very loudly, he duldlut

spoke, they yahalhduk

spoke, we (3+) yats'alhduk

stereo bets'ujun

talking loudly, he/she is going in a loop nuduldlut

thunders, it detnik 'utni

whistling yooyooz

77. Colours, Transparency

black, it is dulhgus

black, it is dunulhgus

black, it is whudulhgus

blue, it is duldzan

blue, it is dunuldzan

blue, it is whuduldzan

blue, it is not chanildzan

brown, it is yun dot'en

fade, it will be'dutedulh

faded, it be'didil

green, it is t'an dohot'en

green, it is t'an dot'en

green, it is t'an doonat'en

grey, it is dulgi

grey, it is whudulgi

off-white, it is dulba

orange, it is kwun dohot'en

orange, it is kwun doonat'en

pale, it is t'alint'ah

pink, it is 'inchooh isdak dohot'en

pink, it is 'inchooh isdak dot'en

pink, it is 'inchooh isdak doonat'en

purple, it is 'ilhtsultoo dohot'en

purple, it is 'ilhtsultoo dot'en

purple, it is 'ilhtsultoo doonat'en

purple, it is maitoo dot'en

red, it is dulk'un

red, it is dunulk'un

red, it is whudulk'un

white, it is nulyul

white, it (d-class) is dulyul

white, it (generic) is lhuyul

white, it is whulyul

yellow, it is dultl'uz

yellow, it is dunultl'uz

yellow, it is sa dot'en

78. Shapes and Configurations

pointed, it is dujooz

sharp, it is desgwut

sharp, it is dujooz

79. Tastes and Odours

bad-tasting, it is ts'ulki'

bitter, it is ts'ulki'

good, it tastes lhuk'i

smells good, it sooltsun

sour, it is dunink'ooz

sweet, it (generic) is lhuki

sweet, it is lhuk'i

sweet, it is nulk'i

tastes good, it nulk'i

weak, it is lait'ah

80. Descriptive Terms

80.1. Dimensional Terms

big, it is dincha

big, they are hincha

big as me, he/she is as sundulcho

big it is!, how lhe'ulcho

broad, it is hoontel

broad, it is ntel

broad as this, it is njahooltel

deep, it (body of water) is takhulh

deep, it (snow) is dincha

heavy, it is ndaz

how cold is it? dahoolk'uz

large, it is hooncha

large, it is ncha

large, it is nincha

long, it (generic) is nyez

long, it is ninyez

long, it is dinyiz

long, it is hoonyiz

short, it is dindukw

short, it is hoondukw

short, he/she is ndukw

small, it is ntsol

small it is!, how lhe'ultsolyaz

tall, I am usyez

tall, I am usyiz

tall, it (generic) is nyez

tall, it is hoonyiz

tall, they are hinyez

tall, we (2) are idujez

tall, we (3+) are ts'inyez

tall, you (1) are inyez

tall, you (2+) are uhyez

tall, he/she is nyez

tall, he/she is nyiz

tall, he/she is comparatively 'ilyez

wide, it is hoontel

wide, it is ntel

wide as this, it is njahooltel

80.2. Miscellaneous

bad, it is dintsi

bad, it is hoontsi'

bad, it is nintsi

bad, they are hintsi

barren, it is (nothing grows there) whudanzaz

beautiful, it is dinzoo

beautiful, he/she is ts'ekezoo

becalmed, it is dezghel

bloated, it is telhudzool

boring, it is whuts'oodutnik

busy, he/she is hubat'en lhai

calm, it is dezghel

cheap, it is chaditi

dangerous, it is beoonujut

difficult, it is huwawhulna

difficult for me, it is sbahuwawhulna

dirty, it (generic) is dutsun

dirty, it is dunutsun

dirty, it is whudutsun

easy, it is wachahoolnah

elastic, it is nuduguz

expensive, it is not chaditi

fresh from the store, it is talh.uki

fruitful, it is belhk'a

furry, it is dughai

good, I am uszoo

good, it is dinzoo

good, it is ninzoo

good, it is not chadizoo

good, it is not chahoozoo

good, it is not chaizoo

good, it is not chanizoo

good, they are not chahizoo

good, he/she is nzoo

greasy, it is khe unli

hairy, it is whudughai

hairy, he/she is dughai

handsome, he/she is dunezoo'

hard, it became duts'un suli

hard, it became whuduts'un suli

hard, it is dunuts'un

hard, it is duts'un

hard, it is huwawhulna

hard, it is whuduts'un

hardened, it duts'un suli

hardened, it whuduts'un suli

near, it is nilhdukw

noisy, it (wh-class) is taoodetnuk

powerful, it is beoonujut

pretty, it is nzoo

rough, it is dutsiz

shaggy, he/she is duyo

soft, it is dutleh

strong, it is ulhtus

ugly, she is ts'eketsi'

ugly, he/she is dunetsi'

unyielding, it is dunuts'un

81. Numbers

eight (abstract) lhk'udiwh

eight (locative) lhk'udidun

eight [generic] lhk'utdink'i

eight [human] lhk'utdine

eight [multiplicative] lhk'udit

eighty [generic] lhk'utdit lanezi

eleven [generic] lanezi 'on'at 'ilhuk'i

fifteen [generic] lanezi 'on'at skwunlai

fifty [generic] skwunlat lanezi

five [abstract] skwunlawh

five [generic] skwunlai

five [human] skwunlanun

five [locative] skwunladun

five [multiplicative] skwunlat

forty [generic] dit lanezi

four [abstract] diwh

four [generic] dink'i

four [human] dinun

four [locative] didun

four [multiplicative] dit

fourth dit

many, they are hulan

nine [abstract] 'ilhohoolawh

nine [generic] 'ilhah hoolah

nine [human] 'ilhah hoolanun

nine [locative] 'ilho hooladun

nine [multiplicative] 'ilha hoolat

ninety [generic] 'ilhah hoolah lanezi

numerous, they are hulan

once 'ilhah

one hundred [generic] lanezi lanezi

one [abstract] 'ilhowh

one [generic] 'ilhuk'i

one [human] 'ilhunun

one [locative] 'ilhuk'i

one [multiplicative] 'ilhah

seven [abstract] lhtak'alt'iwh

seven [generic] lhtak'alt'i

seven [human] lhtak'alt'inun

seven [locative] lhtak'alt'idun

seven [multiplicative] lhtak'alt'it

seventy [generic] lhtak'alt'it lanezi

six [abstract] lhk'utakiwh

six [generic] lhk'utak'i

six [human] lhk'utanun
six [locative] lhk'utadun
six [multiplicative] lhk'utat
sixty [generic] lhk'utat lanezi
ten [abstract] laneziwh
ten [generic] lanezi
ten [human] lanezinun
ten [locative] lanezidun
ten [multiplicative] lanezit
thirty [generic] tat lanezi
three [abstract] tawh
three [generic] tak'i
three [human] tanun
three [locative] tadun
three [multiplicative] tat
thrice tat
twenty [generic] nat lanezi
twice nat
two (locative) nadun
two [abstract] nawh
two [generic] nankah
two [generic] nankah
two [human] nanun
two [multiplicative] nat

82. Quantity

82.1. Quantifiers

half lhulcho
half full, it is lhulcho dezbun
halfway lhulcho
have many, they have huldlai
how many of them? dahuneltsuk
how many of us? dazneltsuk
how many of you? danelhutsuk
how much? daltsuk
many, there are whulai
many, they (generic) are lhai
many, they are nulai
many (places) lhat-un
many as, there are as njaneltsuk
many times lhat
many ways, in lhelhdun k'un'a
many [abstract] lhawh
many [human] lhane
many [human] lhanun
many [multiplicative] lhat
most 'uk'enus
most of them k'us hulan
numerous, they are dulai
numerous, they are whulai
times, many lhat

82.2. Miscellaneous

both of them nanehult'ah
half of it buk'uz
kinds, it is of (st) decah
little bit, a ntsolyaz
many as, there are as ndaneltsuk
nothing soo gak
one side of it buk'uz
some more 'uyoocha
two, they are nanehult'ah

83. Holidays

Easter Zooldubak

Pentecost Ndonidzen

84. Days of the Week

Friday 'Utsun Ts'usyih

Friday Wanderdi

Monday Dimos k'elh'az

Monday Landidzen

Saturday Dzenwhuyaz

Saturday Sumdi

Sunday Dimosdzen

Thursday Whulhditdzen

Tuesday Whulhnatdzen

Wednesday Whulhtatdzen

85. Months

April Talhtsi Ooza

August Talook Ooza

December Dzen Dilhdukw

December Sacho Din'ai

February Ts'uzyen Hoh Ts'utsun nun

February Yuske'nighus

January Sachonun

July Shen Whuniz

June Duk'ai Ooza

March Tsintel Ooza

May Dugoos Ooza

November Banghan Nuts'uke

October Bet Ooza

September Gestlah Ooza

86. Time

86.1. Day

day dzen

day after tomorrow bunde ombun

day before yesterday hulhda whutsah da

yesterday hulhda

86.2. Seasons

autumn dak'et

fall dak'et

midsummer dayun

spring 'olulh

summer shen

winter khit

86.3. Units

age yusk'ut

age, your (1) xnyxeyusk'ut

age, beyusk'ut

month sanun

week yanilhghel

year yusk'ut

years, your (1) xnyxeyusk'ut

86.4. Concepts and Adverbs

after, shortly k'adliyaz

always 'awhulyiz

frequently lhghun

frequently lhghunilhdukw

future, in the nus te

long ago 'uda

long time, for a sa

now k'an

now k'andit

occasionally whulutah

right now k'an 'awet

shortly after k'adliyaz

slowly sa

still 'awhuz

today k'anditdzen

today k'andzen

tomorrow bunde

86.5. Miscellaneous

afternoon dzetniz hukw'elh'az

clock tsadzi

daytime dzenes

evening hulhgha

evening, this k'an hulhgha

evening, tomorrow bunde lhgha

evening, yesterday lhda hulhgha

every morning bundada totsuk

last time yeda'

last year ghada

midnight tuzniz

morning bundada

morning, this bunda

morning, this k'an bundada

morning, tomorrow bunde bundada

morning, yesterday lhda bundada

next time 'on'at

next year 'on'at yusk'ut

next year nade

night lh'ek

noon dzetniz

passing, (time) is whudezulh

tomorrow morning bunde bundada

watch tsadzi

year, last ghada

year, next 'on'at yusk'ut

year, next nade

87. Education, Teaching, and Learning

explain, I whunulhtun

explain to them, I whuhubunulhtun

explains to me, he/she whusunulhtun

learn, we (3+) will hoozdutel'eh

learn, you (1) will hoodutal'eh

learn, you (2+) will hoodutelhu'eh

learning, he/she is hodul'eh

show how, I whunulhtun

show them how, I whuhubunulhtun

shows me how, he whusunulhtun

teacher dune hodulh'eh

teaching, he/she is hodulh'eh

teaching me, he/she is whusodulh'eh

teaching them, he/she is hobudulh'eh

teaching you (1), he/she is whuxnyxodulh'eh

teaching him/her, he/she is whuyodulh'eh

88. Government

chief, generally with reference to band chief dayi

hereditary chief keyah whuduchun

king lelwe

leader, village, the first among equals of the clan heads keyoh whuduchun

servant 'ustlen

servant, my se'ustlen

servant, our ne'ustlen

servant, their hube'ustlen

servant, your (1) xnyxe'ustlen

servant, his/her be'ustlen

subject, my se'ustlen

subject, our ne'ustlen

subject, their hube'ustlen

subject, your (1) xnyxe'ustlen

subject, his/her be'ustlen

subject (of ruler) 'ustlen

trapline yuntah

89. Justice, Punishment, Legal System, Equity

lawyer dune ba yalhduk

police officer neilhchuk

prisoner tsak'etdin'ai

RCMP officer neilhchuk

90. Manipulations of Objects

handling, he/she is nunulle

pick it up, you (1) dudi'a

stretch, we (3+) may ts'oobuz

stretch, you (1) may ombuz

stretch (st), I may noosbuz

stretch (st), I may wusbuz

stretched it, they huyuzbuz

stretching it, he/she is yubuz

turn (st), you (1) dilhghis

unscrew, you (1) hadilhghis

90. Miscellaneous Actions

abandoned (s.o.), I untesna

abandoned me, he/she unsutesna

baby-sit (s.o.), they ghuhinli

care of (st), they take ghuhinli

careful (of yourself), you (1) are khindli

careful with it, I am wa'oosdli

careful with it, they are higha'ooli

careful with it, we (2) are wa'oodudli

careful with it, we (3+) are wa'ts'ooli

careful with it, he/she is yugha'ooli

did something, you (1) 'inja

did something, he/she 'uja

doing, they are 'uhut'en

draft, they caused a dahalhts'i

dug (with a shovel), he/she hahonkai

effort, he/she is making an yudats'ein'ai

fenced us in, they nehaneztl'oo

happened to you (1), something 'inja

happened to him/her , something 'uja

hobo-ing around, he/she is nuduzas

messing around, he/she is nuduzas

work, we (3+) will 'uztet'ilh

work, you (1) will 'utant'ilh

work, he/she will 'utet'ilh

working, they are 'uhut'en

working, he/she is 'ut'en

working hard, we are usnedut'en

working hard, you (1) are undendut'en

91. Sameness and Difference

different people lhelhyoone

92. Aspectual, Evidential, and Modal

cannot ghait'ah

hard for him/her, it is too yughulait'ah

used to inle

was, he/she inle

93. Tending, Caring For, Protecting, Mistreating

guarding (s.o.), I am ghusli

guarding (s.o.), we (2) are ghidudli

guarding (s , we (3+) are ghuts'inli

guarding him/her, they are highunli

guarding him/her, he/she is yughunli

94. Bodily Care and Grooming

bath, I give (s.o.) a toonulhte

bath, I take a toonusya

bath, they (3+) take a toonuhudelh

bath, they give him/her a toonuhuyulhte

bath, we (2) take a toonit'az

bath, we (2) will give someone a toonidulte

bath, we (3+) carry toonuts'udilh

bath, we (3+) give (s.o.) toonuts'ulhte

bath, you (1) take a tooninya

bath, you (3+) take a toonahdilh

bath, he/she gives him a bath toonuyulhte

bath, he/she takes a toonuya

clippers, hair behananudughas

cold cream beznulha

comb tseldzook

curlers, hair benaznuldooz

dandruff tsitsis

hair oil netsigha bets'ulha

makeup, they put on nehndunulhtsi

makeup, we (3+) put on nehts'undunultsi

makeup, you (1) put on nehndunilhtsi

mirror tooznal'en

oil, hair netsigha bets'ulha

95. Carrying and Sending

bring you (1) (st), I will nts'untelhchoos

bring you (1) (st), I will nts'untelhdoh

bring you (1) (st), I will nts'untes'alh

bring you (1) (st), I will nts'untesdzih

bring you (1) (st), I will nts'utelhtelh

bring you (1) (st), I will nts'utes'alh

bring you (1) (st), I will nts'uteskalh

bring you (1) (st), I will nts'utestilh

bring you (1) (st), I will nts'utestleh

bring you (1) (st) on my back, I will nts'utesghilh

dipping it up, he/she is yukaih

drag to you (1) (st), I will nts'utesgus

dragging (st), I am usgus

dragging (st), we (2) are idugus

dragging (st), we (3+) are ts'egus

dragging it, they are huyegus

dragging it, he/she is yegus

drive (st) to you (1), I will nts'utelhgooh

lead (s.o.) to you (1), I will nts'untelhdlooh

load carried on back khelh

pack khelh

packed (st) back home, I whusanasasdughi

packed it back home, he/she whusanayedughi

packing, we (2) are edughilh

packing, he/she is eghilh

packing (u.o.), I am 'esghilh

packing (u.o.), they are 'uheghilh

packing (u.o.), he/she is 'eghilh

packing around (u.o.), I am ne'usghe

packing around (u.o.), he/she is ne'ughe

packing back, he/she is na'eghilh

packing back, he/she is

natesdughi

roll (st) to you (1), I will nts'untelhbus

96. Correcting, Repairing, and Damaging

breakable, it is du'ek

brittle, it is k'udentuk

brittle, it is k'unuduwul

chipped, it got hadanwul

flat, it (tire) is buniltal

healing, it is najih

patches it, he/she naiyulhkut

rots out, it hawhulhjuk

smashed, it is easily k'unuduwul

smashed into many pieces, it is yadanwul

97. Success and Failure, Finishing

overwhelmed, he/she is 'oot'e ghainil

98. Errors

fooled, they got nahunet'a

fooled, he/she got nenet'a

mistake, you (1) made a nananda

mistake, he/she is making a nanut'a

99. State and Change of State

am, I 'ust'ah

are, they 'uhint'ah

are, you (1) 'int'ah

are, you (1) 'int'ah

be, you (1) may 'ont'e

be, you (1) may naondle

become, I have susdli

former inle

is, he unli

is, it 'int'ah

is not, it chailah

late (deceased) inle

melt, it has started to nawhenilghen

melt, it will natelghen

melted, it has nalhughen

melting, it is nalgheh

melting (st), I am nalhgheh

melting (u.o.), he/she is na'ulhgheh

melting it, he/she is naiyulhgheh

used to inle

was, he/she inle

100. Growing and Aging

aged, they have hunesjan

aged, we (3+) have ts'unesjan

aged, he/she has nesjan

branches from it, it

buts'ahainalhyi

growing, it (plant) is hanuyeh

growing, he/she (child) is nuyeh

growing (st:plant), he/she is hanulhyeh

growing back, it (plant) is hananulyeh

grows out of it, it buts'ahainalhyi

old, it (wh-class) is whujut

old, he/she is nesjan

old age jan

over-ripe, it is 'on'un hanit'ai

101. Good, Bad, Improving, Deteriorating, Beauty

good, it is hoonzoo

good, they are hinzoo

good, we (2) are idudzoo

good, we (3+) are ts'inzoo

good, you (1) are inzoo

good, you (2+) uhzoo

102. Cleaning and Polishing, Dirtying, Cleanliness

cleaning cloth naih bewhuna'udeh

cloth, cleaning naih bewhuna'udeh

wash, you (1) toonaingus

washing, he/she is nawhugus

washing for himself, he/she is nawhudugus

103. Smearing, Painting, etc.

greasing, he/she is ulhlhah

paint (st), I dustl'us

paint (st), I whudustl'us

smeared, it has been talhutloh

smearing, we (2) are idullhah

smearing, he/she is talhlhoh

smearing it, he/she is yulhlhah

104. Wetting and Drying, Soaking, Fumigating

damp, it is sulhtsulyaz

dampen (st), I nanulhtsul

dampens it, he/she nainulhtsul

dry, they are nadesjak

dry, we (2) are nadusdujih

dry, you (1) are nadenjak

drying poles 'ujooh

rack for drying food nin'ai

wet, they are nalhutsul

wet, we (3+) are naznelhtsul

wet, you are naneltsul

105. Combining, Mixing and Attaching

braided, it has been lhu'ool

braiding it, she is naiyulh'ool

knot lhuschuz

put on your seatbelt, you (1) denawhudinges

tie yourself down, you (1) denawhudinges

106. Communication

advising him/her, he/she is yughatni

ask (s.o.), we (3+) ts'oodulhkut

ask him/her, they huyoodulhkut

asks me, he/she soodulhkut

asks him/her, he/she yoodulhkut

bragging about himself, he/she is dena'whudults'it

grammar beyatuk

history 'udada

hollering, he/she is dulghi

language khunek

language khunik

legend 'udada

message khunek

messenger khunek nu'a

messenger nawhulnuk

orates, he/she duldlut

priest nawhulnuk

says, he/she ni

speak, I yalhduk

speak, I will yatelhduk

speak, they will yahutelhduk

speak, we (2) yaiduldduk

speak, we (2) will yataduldduk

speak, we (3+) yats'ulhduk

speak, we (3+) will yaztelhduk

speak, you (1) yailhduk

speak, you (1) will yatalhduk

speaking, they are yahulhduk

speaks, he/she yalhtuk

speaks (language), he/she k'uyalhtuk

speaks broken Carrier, he/she k'uneduyus

speaks very loudly, he/she duldlut

spoke, they yahalhduk

spoke, we (3+) yats'alhduk

story 'udada

talk dirty, to -zek hoontsi'

talking to you (1), I am nts'uyasduk

telephone beyatuk

telephone nilhdza beyaztuk

television nul'en-i

television, my snul'en-i

television, our nenul'en-i

television, their hubunul'en-i

television, his/her bunul'en-i

televsion, your (1) xnyxenul'en-i

told me, he/she sudani

told them, he/she hubu dani

told us, he/she ne dani

told us (2), he/she nawhudani

told you (1), he/she nyudani

told you (2+), he/she nahdani

told him/her , he/she yudani

watch what you say neghunik whuts'inli

word khunek

word khunik

write, you (1) 'uk'une'inguz

writing, they are 'uk'une'huguz

writing, we (3+) are 'uk'unets'uguz

107. Language and Linguistics

grammar beyatuk

telephone beyatuk

108. Concealment, Avoidance, Deception

cheats, he/she ha'dunet'ah

cheats me, he/she saha'dunet'ah

109. Conflict, Competition, Argument, Disputation

fight, we are going to beztedulh

fight, you (1) are going to betandulh

fight, you (2+) are going to betehdulh

go to war, we (3+) may ts'utooba

110. Contact

collided with (st), it sulhkulh

collided with one another, he/she lhuhoolkal

collides with it, it yulhkulh

111. Causation and Transformation

drying (st), we (2) are idulgi

112. Cutting, Crushing, Grinding, Scraping, Tearing and Pinching

chopped a hole in it, he/she hukwunintsel

chopped it to pieces, he/she yaidantsel

chopping (u.o.), he/she is 'utselh

crumple (st), I dunusyus

grade (st:road), he/she will nawtezo

grading (st:road), he/she is nawhuzo

grinding (st), he/she is dunulhdus

knife 'utes

knife, my se'utes

mowing, I am ust'as

planing, he/she is dughas

planing (u.o.), we (2) are 'udidughas

plow (st:road), he/she will nawtezo

plowing (st:road), he/she is nawhuzo

sawing (st), we (2) are didughut

sawing (st), he/she is dughut

sawing (u.o.), we (3+) are 'uzdughut

113. Desire and Attempting

begging, he/she is datsa

hunger dai

hungry, I am chawhuszum

hungry, I am schahoonzum

hungry, I am sdaningi

hungry, they are chahuhoonzum

hungry, we (2) are chahoodudzum

hungry, we (3+) are chats'uhoonzum

hungry, he/she is oodaningi

skinny, I am sdaningi

skinny, he/she is oodaningi

want (st), I hukwa'nuszun

want (st), I ka'nuszun

want (st), they ka'huninzun

want (st), we (2) ka'nidudzun

want (st), we (3+) ka'uzninzun

want (st), you (1) ka'ninzun

want (st), you (2+) ka'nahzun

want it, I ooka'nuszun

want it, they hika'ninzun

want it, we (2) ooka'nidudzun

want it, we (3+) ooka'ts'uninzun

wants (st), he/she ka'ninzun

114. Dwelling, Camping, Spending the Night

ant hill 'andihken

bear den, black sus'an

beaver lodge tsaken

beehive hoolht'ot'o

den, black bear sus'an

home yah

home village kooz

home village, your (1) nkooz

hotel sdutez

house koo

house yah

house, my skoo

house, our nekoo

house, their hubukoo

house, your (1) nkoo

house, his/her bukoo

lodge, beaver tsaken

lodge, muskrat tsek'etken

meadow tl'otelk'ut

muskrat lodge tsek'etken

reside, you (1) hoont'i

115. Emission

ejaculate, I am going to tesjaz

milk, I am going to tesjas

milking it, he/she is yulhjas

116. Expressions and Gestures

applauds, he/she lhookat

blinks at him/her in dislike, he/she yunak'ult'ah

claps, he/she lhookat

crying, they are ghunadutsa

crying, we are ghunasdutsa

crying, you are ghuninedutsa

pouting, he/she is nulch'e'

smiling, I am dlonuszun

smiling, they are dlohuninzun

smiling, we (3+) are dlots'uninzun

117. Exerting Force and Striking

attracted (sexually) to me, he/she is suneztulh

attracted (sexually) to him/her, he/she is yuneztulh

bothering him/her, he/she is yughaghuna

choked him/her, he/she yuze'usnik

choking him/her, he/she is yuze'unih

crush (st), I usyus

hammering (u.o.), he/she is 'unulh'uz

hammering it, he/she is yunulh'uz

kick (him) up the ass, you (1) tsul'aintal

kicked (d-class object), he/she deztulh

kicked a hole in it, I kw'unustal

kicked a hole in it, they hik'untal

kicked a hole in it, we (2) kw'unidutal

kicked a hole in it, he/she yuk'untal

kicked him in the face, he/she yuneztulh

kicked into a hole, he/she 'a'antal

kicked it (the door) in, he/she daidental

kicked me in the face, he/she suneztulh

kicked the door in, I danustal

kicked the door in, they dahidental

kicked the door in, we (2) danidutal

kicked the door in, we (3+) dazdental

kicked him/her , they huyuztulh

kicking (st), I am oostalh

kicking (st), we (2) are oodutalh

kicking (st), we (3+) are ts'ootalh

kicking it, they are huyootalh

kicking it, he/she is yootalh

slapped it thoroughly, he/she yayankat

spanked it thoroughly, he/she yayankat

118. Existence, Presence, Absence

absent, he/she is hoolah

absent, he/she is hooloh

disappeared, he/she hoolil

there is not hoolah

119. Holding and Keeping

cling to, I 'awhenadustl'as

clings to, he/she 'awhenadintl'as

held, I am being soontun

hold us, they newoontun

holding me, they are suhoontun

holds us, he/she nehoontun

120. Images and the Creation Thereof

draw, I 'uk'une'ulhtsi

paint, I 'uk'une'ulhtsi

121. Killing, Injuring, Healing, Living and Dying

alive, I am khusna

alive, they are hukhuna

alive, we (2) are khidutna

alive, we (3+) are ts'ukhuna

alive, he/she is khuna

bled to death, he/she 'uskai yuzelhghi con nezek besulhti

dead, the yaidlane

deceased uzdani

died, he/she dazsai

died of exposure, he/she dli yuzelhghi

drowned, they (3+) drowned too hubanghan

drowned, he/she too yuzelhghi

drowning, he/she is too yuzulhghe

exposure, he/she died of dli yuzelhghi

froze, it ustun

grave dune 'adilhti

kill each other, they (3+) lhuhudughan

killed (st), I selhghi

killed (st), we (2) sedulghi

killed by frost, it was ustun

killed each other, they (3+) lhuhedughan

killed it, they huyuzelhghi

killed it, he/she yuzelhghi

killed them (3+), I hubesghan

killed them (3+), they hubuhanghan

killed them (3+), we (2) hubadughan

killed them (3+), we (3+) hubuts'anghan

killed them (3+), he/she hubanghan

late, the uzdani

122. Lingering and Rushing, Hesitating, Waiting

procrastinating, he/she is nus whe'alh

waiting for me, he/she is sbalh'i

waiting for you (1), I am mbalh'i

123. Measurement

thermometer kw'uzbenul'en

124. Motion

advancing, we (3+) are nus ts'edulh

all [human] ts'iyane

arrived, he/she has whusainya

blowing into, it (wind) is danilhts'i

blowing into, it (wind) is dats'i

boat, I am not going in a loop in nuchasuske

boat, we (2) set off by wheiduki

boat, he/she did not set off by whechaikel

boat, he/she has gone around by nusuki

boat, he/she is not going by chaskelh

boat, he/she set off by wheinki

boat, he/she will go by tekelh

boat, he/she will not go by chateskel

bolted, it tsitelhya

bounced away, it whendunelduz

bouncing up and down, it is duk duntuldus

came back, I lhtananesja

came back, they (3+) lhtanahunesdel

came back, we (2) lhtananet'az

came back, we (3+) lhtanaznesdel

came back, you (1) lhtananenja

came back, he/she lhtananesja

came from, you (1) hainya

came from, I hasya

capsized, I nadesghuz

capsized, it nadezghuz

dragging his feet, he/she is kedeldzolh

dripping, it is nadulyul

drive, I will tesgooh

drive, I will not chatuzesgooh

drive, they will hutegooh

drive, they will not chahutesgooh

drive, we (2) will tadugooh

drive, we (2) will not chatuzadugoo

drive, you (1) will tangooh

drive, you (1) will not chatuszangooh

drive, he/she will tegooh

driven in a loop, I have nususgoo

driven in a loop, they have nuhuzgoo

driven in a loop, they have not nuchahigooh

driven in a loop, he/she has not nuchaigooh

driving, I am usgooh

driving, they are hegooh

driving, they are not chahesgooh

driving, they have set off whehangoo

driving, we (2) are idugooh

driving, we (2) are not chasidugooh

driving, we (2) set off wheidugoo

driving, you (1) are ingooh

driving, you (1) are not chasangooh

driving, he/she is ugooh

driving, he/she is not chasgooh

driving away, they are natedugooh

driving away, we (3+) are naztedugooh

driving away, you (1) are natandugoo

driving in a loop, they are nuhugoo

driving in a loop, we (3+) nuts'ugoo

driving in a loop, you (1) are ningoo

drove in a loop, we (2) nusidugoo

drove in a loop, you (1) nusingoo

entered, I danusya

entered, they (3+) dahunindil

entered, we (2) danidut'az

entered, we (3+) daznindil

entered, he/she daninya

entered your (2+) house, he/she nahghudaninya

entered your (2+) house, they (3+) nahghundanindil

entered your (2+) house, we (3+) nahghudazdindil

everyone ts'iyane

exited, I tinusya

exited, they (3+) tihunindil

exited, we (2) tinit'az

exited, we (3+) tiznindil

exited, he/she tininya

falling from the trees, it (snow) nadeh

fell, you (1) nadilhghis

fell down, it nalts'ut

fell down, he/she nadelduz

floating around, we (2) are nidudlat

floating around, he/she is nulat

flowing out of, it is yahainli

fly home, they will nahutet'ah

fly in a loop, they will nahutet'ah

fly in a loop, they will natet'ah

fly in a loop, he/she will nutet'ah

fly in a loop, he/she will nutet'ah

flying in a loop, they are nuhut'ah

flying in a loop, he/she is nut'ah

flying in a loop, he/she is nut'ah

go around by boat, they are not going to nuchahateskel

go around by boat, they are not going to nucha:teske

go around by boat, they are not going to nucha:teskel

go around by boat, he/she is going to nutekelh

go around by boat, he/she is not going to nuchateske

go around by boat, he/she is not going to nuchateskel

go around in a boat, I am not

going to nuchatuzeske

go around in a boat, we (2) may noduke

go around in a boat, we (2) may not nuchasoduke

go by boat, I will teskelh

go by boat, I will not chatuzeskel

go by boat, they will hutekelh

go by boat, they will not chahuteskelh

go by boat, we (2) will tadukelh

go by boat, we (3+) will uztekelh

go by boat, we (3+) will not chazteskel

go by plane, I will sulh 'utet'ah

go by plane, I will not sulh cha'test'ah

go by plane, they will hubhulh 'utet'ah

go by plane, we (2) will nawhulh 'utet'ah

go by plane, he/she will bulh 'utet'ah

going around by boat, I am nuske

going around by sleigh, they are ne'hulooz

going around talking loudly, he/she is nuduldlut

going back by boat, I am nasdukelh

going back by boat, they are nahedukelh

going back by boat, we (2) are naidukelh

going back by boat, we (3+) are nats'edukelh

going back by boat, you (1) are naindukelh

going back by boat, you (2+) are nahdukelh

going back by boat, he/she is nadukelh

going by boat, I am uskelh

going by boat, I am not going chaseskelh

going by boat, they are hekelh

going by boat, they are not chaheskelh

going by boat, we (2) are idukelh

going by boat, we (3+) are ts'ekelh

going by boat, we (3+) are not chats'eskelh

going by boat, you (1) are inkelh

going by boat, you (2+) are uhkelh

going by boat, he/she is ukelh

going by plane, I am sulh 'et'ah

going by plane, I am not sulh cha'est'ah

going by plane, they are hubulh 'et'ah

going by plane, we (2) are nawhulhutet'ah

going by plane, he/she is bulh 'et'ah

gone by boat in a loop, I have nususki

gone in a loop by boat, they have nuhuzki

gone in a loop by boat, they have not nuchahikel

gone in a loop by boat, we (3+) have nuts'uzki

gone in a loop by plane, I have sulh ne'uzt'a

gone in a loop by plane, they have hubulh ne'uzt'a

gone in a loop by plane, we (2) have nawhulh ne'uzt'a

gone in a loop by plane, he/she has bulh ne'uzt'a

gushes, it tujulh

hobo-ing around, he/she is nuduzas

jumping, it (fish) is halghok

jumps out of, he/she hallhat

jumps out of, he/she hallhuk

land in a boat, I yaskih

land in a boat, they yahukih

land in a boat, we (2) yaidukih

land in a boat, we (3+) yats'ukeh

lands in a boat, he/she yakih

left by boat, they whehanki

left by boat, we (3+) whets'anki

messing around, he/she is nuduzas

migrating, they are nukehududeh

migrating, he/she is nukedudeh

passed him in a boat, I beghoosuski

passed them driving, I hubeghoosusgoo

ran in a loop, I nusulhugai

ran in a loop, they (2) nuhulhugai

ran in a loop, we (2) nusidulgai

ran in a loop, we (3+) nuts'ulhughaz

ran in a loop, you (1) nusilgai

roll down, it will nts'ununtelbus

rolled over, it nadezghuz

rolled over (in vehicle), I nadesghuz

run, I will telhugih

run, they (2) will hutelgih

run, they (3+) will hutelwus

run, we (2) will tadulgih

run, we (3+) will uztelwus

run, we (3+) will uztelwus

run, you (1) will talgih

run, you (1) will talgih

running, I am ulhugih

running, we (2) are idulgih

running, we (3+) are ts'elwus

running, you (1) are ilgih

running, he/she is ulgih

running after it, they (3+) are hikatelhughaz

running after it, we (2) are kwatedulgai

running after it, we (3+) are kwaztelhughaz

running around, he/she is nulgaih

runs, he/she nulgaih

scurry around, we (3+) nukats'ulgas

set off by boat, I wheski

set off by boat, I have not whechaskel

set off by car, I whesgoo

set off by plane, I have not sulh whechait'ah

set off driving, they did not whechahigooh

set off on foot, they (3+) have whehandil

set off on foot, we (3+) whets'andil

skating around, he/she is nuzoot

skating around for pleasure, we (2) are nunidudzoot

slippery it is lhkut

smooth, it is lhkut

spurts, it tujulh

started on foot, we (3+) have uztezdil

talking loudly, he/she is going in a loop nuduldlut

toppled, it nananghiz

transporting me by boat, he/she is selhkelh

tripped and fell, he/she dughuntezghal

walk, I will tesyalh

walk, they (3+) will hutedulh

walk, they (3+) will not chahutesdulh

walk, we (2) will tat'us

walk, we (2) will not chatuzat'us

walk, we (3+) will uztedulh

walk, we (3+) will not chaztesdulh

walk, you (1) will tanyalh

walk, you (3+) will tehdulh

walk, he/she is going to teyalh

walk, he/she will not chatesyalh

walk in a loop, I will nutesya

walk in a loop, I will not nuchatuzesyal

walk in a loop, we (2) will nutat'us

walk in a loop, we (2) will not nuchatuzat'us

walk in a loop, we (3+) will nuztedulh

walk in a loop, we (3+) will not nuchaztesdil

walk in a loop, you (1) may nonya

walk in a loop, you (1) will nutanya

walk in a loop, you (1) will not nuchatuzanya

walk in a loop, he/she will nuteya

walk in a loop, he/she will not nuchatesya

walked in a loop, I nususya

walked in a loop, I have not nuchasyal

walked in a loop, they (3+) have nuhuzdil

walked in a loop, they (3+) have not nuchahidulh

walked in a loop, we (2) nusit'az

walked in a loop, we (2) have not nuchasit'az

walked in a loop, we (3+) have nuts'uzdil

walked in a loop, we (3+) have not nuchats'idulh

walked in a loop, you (1) nusinya

walked in a loop, you (1) have not nuchainyal

walked in a loop, you (2) have nusah'az

walked in a loop, he/she nusuya

walking, I am usyalh

walking, they (3+) are hedulh

walking, we (2) are it'us

walking, we (2) are not chasit'us

walking, we (2) set off wheit'az

walking, we (3+) are ts'edulh

walking, he/she did not set off whechaiyal

walking, he/she is tezya

walking, he/she is uyalh

walking, he/she is not chasyalh

walking, he/she set off wheinya

walking forward, we (3+) are nus ts'edulh

walking in a loop, I am nusya

walking in a loop, I am not nuchasusyah

walking in a loop, we (2) are nit'as

walking in a loop, we (2) are not nuchasit'as

walking in a loop, we (3+) are nuts'udilh

walking in a loop, we (3+) are not nuchats'usdil

walking in a loop, you (1) are ninya

walking in a loop, you (1) are not nuchasinyah

walking in a loop, he/she is nuya

walking in a loop, he/she is not nuchasyah

walking in a slow, stately manner, he/she is del'us

walking quickly, he/she is gal be 'ut'en

walking quickly, he/she is gal be nuye

went in a loop by boat, we (2) nusiduki

whips around, he/she nunutsus

125. Naming

call (st), we (3+) ts'utni

call it, they huyulhni

call you (1), we (3+) xnyxuts'ulhni

name 'oozi

name -oozi

name, your (1) xnyxoozi

name, his/her boozi

126. Helping, Obstructing, Hindering

helped me, you (2+) have snachalhuya

helped us, you (1) have nenachailya

helping la

helping me sla

helping them hubula

in aid of la

in aid of me sla

thank you (one person to more than one) slahja

thank you (one person to one person) snachailya

thank you (1 person to one person) sla inja

thank you (1), we (3+) nenachailya

thank you (2+), I snachalhuya

127. Location and Position

across yaztsih

buried, he/she is 'adilhti

corner koonghak

earth, on yun k'ut

floor yun

floor, my syun

floor, our neyun

floor, their hubuyun

floor, your (1) nyun

floor, his/her buyun

ground yun

home, at koonk'et

located, he/she is usda

midstream 'ukoh whuniz

on -k'ut

on earth yun k'ut

sitting, he/she is usda

there 'et

128. Posture and Assumption of Position

located, he is usda

sitting, I am susda

sitting, they (2) are huzke

sitting, they (3+) are hudelhts'i

sitting, we (2) are siduke

sitting, we (3+) are ts'udelhts'i

sitting, we (3+) are uzdelhts'i

sitting, you (1) are sinda

sitting, you (2) are sahke

sitting, you (3+) are delhuts'i

sitting, he/she is usda

stand amongst them, I hubutususyin

129. Possession and Change of Possession

gave it to me, he/she sghainin'ai

gave it to me, he/she sghainintan

gave me, he/she sghani'ai

gave me, he/she sghaninkai

gave me, he/she sghanintan

give (st) to me, you (1) slaintih

give me, he/she is going to sghante'alh

give me (st), he/she will sghadutetilh

have many, they have huldlai

hers buch'i'

his buch'i'

its buch'i'

mine sch'e'

mine sch'i'

mine, it is se'ilhdzun

property -ch'e'

theirs hubuch'i'

130. Attitudes

dislike him/her, I buts'udusnik

disliked, I am sts'udunik

full of it, I think he is bech'anuszun

honour (s.o.), I delhti

honour (s.o.), we (2) dedulti

honour (st), we (3+) uzdelhti

honour them, we (2) hubudedulti

honour them, we (3+) hubuzdelhti

honour him/her, they hidelhti

honours them, he/she hubudelhti

like (st), I hoonust'i

likes (st), he/she hoont'i

mad, he/she got hunlhuch'e

obedient, they are 'uk'une'hut'en

obey him/her, I buk'une'ust'en

obey him/her, we (2) buk'une'idut'en

obey him/her, we (3+) buk'une'ts'ut'en

obeys him/her, he/she yuk'une'ut'en

pity him/her, I wate'nuszun

respect them, you (2+) hubudulhti

131. Mental Activities

dreamt about something, he/she whunadlal

dreamt about him/her, I bunasdlal

look good for himself, he/she makes it dubanawhuldzit

think, I nuninuszut

think, I will nuninteszut

think, they nuninuhinzut

think, they will nunihuntezut

think, we (2) nuninidudzut

think, we (2) will nunintadudzut

think, you (1) nunininzut

think, you (1) will nunintanzut

think about it, I will hukw'unintelhdzut

think about it, they will hukw'unihuntelhdzut

think about it, we (2) will hukw'unintaduldzut

think about it, we (3+) will hukw'units'untelhdzut

think about it, you (1) hukwunilhdzut

think about it, you (1) will hukw'unintalhdzut

think that, we (3+) ts'uninzun

thinking about it, I am hukw'uninulhdzut

thinking about it, they are hukw'unihunilhdzut

thinking about it, we (2) are hukw'uniniduldzut

thinking about it, we (3+) are hukw'uniznilhdzut

thinking about it, you (1) are hukw'uninilhdzut

thought about it, I hukw'uninelhdzut

thought about it, we (3+) hukw'uniznalhdzut

thought about it, you (1) hukw'uninalhdzut

thought that, we (3+) uznanzin

thought that, you (1) nanzin

132. Putting

buried, we (3+) 'ats'alhti

buried, we (3+) whuyats'alhti

buried him/her, they hidilhti

buried him/her, he/she 'ayalhti

place (st), you (1) nininle

pouring it back into (st), he/she is denayulhdzeh

pouring it into (st), he/she is deyulhdzeh

pouring it into it, I am beyulhdzeh

pouring it into it, he/she is yeyulhdzeh

put them on each other, he/she lhk'einla

screw in, you (1) 'ailhghis

133. Taking, Removing, Capturing

scoop, he/she tries to ookaih

take from you (1), I am going to nghuteschulh

take it from you (1), they will nghahitelhchulh

134. Touching, Stroking, Rubbing

caressing her, he/she is yughadzedzi

massaging it, he/she is yulhguk

rubbing it, he/she is yulhguk

135. Resting and Sleeping

awake, he/she is whunih

bed tesk'et

bedroom tesk'et

dreaming, he/she is nute

fatigued, he/she is shih yuzelhghi

overslept, he/she whunandunesti

rest, you (1) nailyis

rest, you (1) naolyis

rest, you (1) will natalyis

resting, they are nahulyis

resting, we (3+) are nats'ulyis

sleep, I will go to nantestelh

sleep, they will hutetez

sleep, we (2) odutez

sleep, we (3+) will nants'untetus

sleep, we (3+) will uztetez

sleep, you (1) sinti

sleep, you (1) are going to nantantelh

sleep, you (1) may onte

sleeping, I am susti

sleeping, they are huztez

sleeping, we (2) are sidutez

sleeping, we (3+) are ts'uztez

sleepy, we are bulh nenideltsut

sleepy, you (1) are bulh nideltsut

sleepy, you (2+) are bulh nahnideltsut

slept, we (2) idutez

slept, we (3+) ts'antez

slept, you (1) inte

136. Searching, Enquiring, Pursuing, Finding, and Losing

ferrets things out, he/she ha'ulih

ferrets things out, he/she ha'ulih

going after it, I am kwatesya

going after it, they (3+) are hikatezdil

going after it, we (2) are kwatet'az

going after it, we (3+) are kwaztezdel

going after it in a car, I am kwatesgoo

going after it in a car, we (2) are kwatedugoo

going after it in a car, we (3+) are kwaztezgoo

looking for, you (1) are kuninta

looking for (st), I am hukwunusta

looking for (st), you (1) are hukwuninta

looking for me, they are skanuta

looking for me, he/she is skunuta

looking for them, he/she is hubukanuta

looking for us, he/she is nekahunuta

looking for us (2), he/she is nahkunuta

looking for you (1), he/she is nkunuta

looking for you (2+), he/she is nahkunuta

looking for him/her , you (1) are bukuninta

looking for him/her, he/she is yukunuta

running after it, I am kwatelhugai

running after it, they (3+) are hikatelhughaz

running after it, we (2) are kwatedulgai

running after it, we (3+) are kwaztelhughaz

137. Separating, Removing, Disassembling, and Breaking

breakable, it is du'ek

broken, it (rope) has sduch'us

broken in two, it is k'udentuk

broken into many pieces, it has yadatuk

chopped it to pieces, he/she yaidantsel

smashed into many pieces, it is yadanwul

split (st) into many pieces lengthwise, he/she lhtadelnat

split it in two lengthwise, he/she lhts'eidalnat

splitting (st), he/she is dulhnat

138. Social Relations

bachelor, he/she is bu'at hoolah

married, he is 'at ut'i

married, she is ki ut'i

married to him/her, he/she is yughusda

single, he/she is bu'at hoolah

139. Spatial Relations

above 'onduk

ahead nus

among one another lhtah

among them butah

among them (areas) whutah

around it bunat

base of, at the cheh

before tsah

before it [areal] whutsah

before me stsah

before them hubutsah

before him/her butsah

beside zih

beside it (areal) whuzih

beside me suzih

beside us (2) nahzih

beside us (3+) nezih k'ut

beside you (1) nzih

between mountains dzulhyeguz

between them hubeguz

between them [areal] wheguz

bottom, a 'utl'ah

bottom, its buttl'ah

centre whuniz

edge ba

far, it is nilhdza

far, it is not chanilhdza

far away place yaba

forward nus

front of me, in sbut

here njan

horizon yaba

next to it bughah

older than them hubutsah

on it buk'ut

on me sk'ut

on the snow yusk'ut

on top of each other lhk'ut

other side, on the 'onyaz ts'i

outside 'az

over it butus

over there whuz

overseas yaba

preceding tsah

under yah

upper dusts'i

within the house kooz

140. Logical Relations

depending on for subsistence ke

141. Temporal Relations

after (in time) k'elh'az

after it hukw'elh'az

after him/her buk'elh'az

before tsah

preceding tsah

142. Pronouns

by himself dich'ah
he 'en
he didut
her 'en
her didut
him 'en
him didut
himself dich'ah
I si
me si
ours nech'i'
ourselves whenich'ah
she 'en
she didut
them 'enne
themselves hudich'ah
themselves, by hudich'ah
they 'enne
us wheni
we wheni
you (2+) nawheni

143. Adverbs

after all ts'ante'ez
again doocha
almost 'ankw'us
apparently si
around 'oh
by myself sich'ah
different places lhelhdun
fast 'a
myself sich'ah
quickly 'a

144. Manner

by my hand sla be
manner, in that whuzun'a
manually sla be
properly soocho
quickly 'acho
quickly soo 'a
really soo
thus whuzun'a
way, this ndezuna
well soo

145. Speaker Attitude

definitely soo ts'ihun
maybe tusih
probably tusih

146. Subject Attitude

tentatively khuntusinilhholhdulh

147. Postpositions

aid of, in -la
among tah
because of gha

because of me sgha
because of them hubugha
by means of be
by reason of that 'ighun
for ba
for me sba
for us neba
greater than you (1) xnyxanus
less than -k'elh'ih
less than me sk'elh'ih
on behalf of ba
than -anus
together with each other lhulh
with, together -lh
with (by means of) be
with me sulh
with them hubulh
with us nelh
with him/her bulh
without it bu'et

148. Directionals

around hoh
blowing from the south nazde
downward nts'un
east, from the ndazde
east, to the nda
from ts'i
from away 'azdeh
from outside 'azdeh
left 'intl'us
left side 'int'lus tsih
lower yusts'i
north, from the nizde
northward ni
right nailhni
right side nalhni
south, from the ntsizde
southward nun
to ts'i
to ts'un
to me sts'un
west, from the nusde
west, to the nuk

149. Question

from where? nts'ez
how are they? dahint'ah
how are you (1)? daint'ah
how big is it? dalcho
how cold is it? dahoolk'uz
how far? dahooldzoh
how is he? dant'ah
how is it? dahot'e
how is it? dahoont'ah
how long a time? dahooldzoh
how long is it? dalyiz
how many (years of age) nts'oh
how many of them? dahuneltsuk
how many of us? dazneltsuk
how many of you? danelhutsuk
how much? daltsuk
how pretty is it? daldzoo
how? nts'ezun'a
question marker ih
what happened? dahooja

what will happen? dawtenilh

what? dant'i

what? di

whence? nts'ez

where nts'e

where?, from nts'ez

who? mbe

why? dika

150. Complementizers

only if 'et ze

151. Particles

also cha

emphatic particle si

focus la

not 'aw

only ze

question marker lak

too cha

152. Conjunctions

and 'ink'ez

either k'us

or k'us

plus 'on'at

153. Interjections

alright a'ah

come here! 'ane

come here! 'anih

dear goolh

don't touch it! ts'it

give me de'

hello hadih

hello! hadi'

here you are nah

no 'awundooh

okay a'ah

oops! hadi'

ouch! 'ayah

ouch! 'ayoh ho

sweetie goolh

that's all, the end 'awetze

well adih

what! hadi'

yes a

154. Affixes

154.1. Prefixes and Suffixes

areal object of postposition whu-

big -cho

first person dual possessor nah-

first person plural posssessor ne-

first person singular possessor s-

little -yaz

possessor, first person dual nah-

possessor, first person plural ne-

possessor, first person singular s-

possessor, second person duo-plural nah-

possessor, second person singular n-

possessor, third person duo-plural, for class 1 nouns hubu-

possessor, third person reflexive, for class 1 nouns du-

possessor, third person singular bu-

second person duo-

plural possessor nah-

second person singular possessor n-

small -tsol

small -yaz

third person singular possessor bu-

Root List by English Gloss

age	yan$_1$
angry	ch'eh$_1$
bad	tsi'$_1$
be	t'ah$_1$
become	li$_3$
big	cha$_1$
black	gus$_1$
blink	bul$_1$
blue	dzan$_1$
boat	ke$_2$
boil	mulh$_1$
braid	'ool$_1$
break	tuk$_2$
breathe	yis$_1$
buy	ket$_1$
calm	ghel$_2$
chew	'alh$_1$
circle	bas$_1$
cl-body-c	ti$_1$
cl-coc-c	kai$_1$
cl-euo-c	dzai$_1$
cl-fluffy-c	dɔ$_1$
cl-liquid-c	dzeh$_1$
cl-liquid-i-rough	yul$_3$
cl-lro-c	tan$_1$
cl-mdo-c	le$_1$
cl-mdo-i	kat$_1$
cl-sdo-c	'ai$_1$
cl-sdo-u	tl'es$_1$
cl-2df-c	chus[illegible]
cold	k'uz$_1$
crunch	goos$_1$
dance	daih$_1$
decay	jut$_1$
difficult	na$_2$
distant	dza$_1$
do	'en$_2$
drag	gus$_3$
dream	dlal$_1$
drink	nai$_1$
drive	goo$_1$
drum	ghalh$_1$
drum	tsut$_1$
eat	yi$_1$
fart	tl'it$_1$
feel cold	dli$_1$

fib	ts'it$_1$
fish jump	ghok$_1$
flick	k'us$_1$
flow	li$_1$
fly	t'oh$_2$
fragile	'ek$_1$
freeze	tun$_2$
frighten	joot$_1$
full	bun$_1$
gamble	lih$_1$
gnaw	ghaz$_2$
good	zoo$_1$
grey	gi$_1$
hairy	ghai$_1$
hard	ts'un$_1$
have	t'i$_2$
heavy	daz$_1$
hold	tun$_1$
hook	jas$_2$
sick	da$_2$
sing	yun$_1$
sit-1	da$_1$
sit-2	ke$_1$
sit-3	ts'i$_1$
slap	kat$_3$
sleep-12	ti$_4$
sleigh	looz$_1$
slide	zoot$_1$
small	tsol$_1$
smear	tlah$_1$
smoke	t'ot$_1$
smoulder	yooh$_1$
soft	tle$_2$
speak	tuk$_1$
spool	ghal$_1$
stand	yin$_1$
star moves	'a$_3$
stretch	buz$_1$
strong	tus$_2$
suck on	t'ukw$_1$
take	choot$_1$
teach	tun$_3$
tear	ch'ul$_1$
think	zun$_1$
think-12	zit$_4$
valuable	ti$_2$
wait	'i$_1$
walk-1	ya$_1$
walk-2	'as$_1$
walk-3	dil$_1$
war	ba$_2$
white	yul$_1$
wide	tel$_1$
wind	ts'i$_2$
yellow	tl'uz$_1$

Root List, Dakelh

'a$_3$ For one or two heavenly bodies such as the sun, moon, and stars, to move. *Tag: star moves-12.*

'ai$_1$ To handle non-plural default objects in a controlled manner. *Tag: cl-sdo-c.*

'alh$_1$ To chew. *Tag: chew.*

'as$_1$ For two to walk on one pair of limbs. *Tag: walk-2.*

'eh$_1$ To learn, become accustomed to.
Tag: learn.

'ek$_1$ To be fragile, flimsy, easy. *Tag: fragile.*

'en$_1$ To see. *Tag: see.*

'en$_2$ To do. *Tag: do.*

'es$_1$ To make a non-violent motion with the foot. *Tag: move foot.*

'i$_1$ To wait. *Tag: wait.*

'ool$_1$ To braid. *Tag: braid.*

ba$_1$ To be off-white. This includes shades such as cream, tan, and blond. *Tag: off-white.*

ba$_2$ To engage in combat, fight in a war. *Tag: war.*

bas$_1$ To be circular or to move in a circle. *Tag: circle.*

bul$_1$ To blink. *Tag: blink.*

bun$_1$ To be or become full. *Tag: full.*

buz$_1$ To stretch on a frame or mechanical device, as a moosehide for tanning. *Tag: stretch.*

cha$_1$ To be big. *Tag: big.*

chan$_1$ To be pregnant. *Tag: pregnant.*

choot$_1$ To take. *Tag: take.*

chus$_1$ To handle a saliently two-dimensional flexible object, such as cloth, fur, or paper, in a controlled manner. *Tag: cl-2df-c.*

ch'eh$_1$ To be angry. *Tag: angry.*

ch'ul$_1$ To act violently on a soft material, e.g. to tear or split it, with a result that is often not complete separation. *Tag: tear.*

da$_1$ For one to sit. *Tag: sit-1.*

da$_2$ To be sick, sore. *Tag: sick.*

dah$_1$ To be of a kind, for there to be so many kinds. *Tag: kind.*

daih$_1$ To dance. *Tag: dance.*

daz$_1$ To be heavy. *Tag: heavy.*

dil$_1$ For three or more to walk on one pair of limbs. *Tag: walk-3.*

do$_1$ To handle fluffy stuff in a controlled manner. *Tag: cl-fluffy-c.*

dukw$_1$ To be short. *Tag: short.*

dlal$_1$ To dream about. *Tag: dream.*

dli$_1$ For someone to feel that he or she is cold. *Tag: feel cold.*

dli$_2$ To plead, pray. *Tag: plead.*

dza$_1$ To be distant. *Tag: distant.*

dzai$_1$ To handle effectively uncountable objects in a controlled manner. *Tag: cl-euo-c.*

dzan$_1$ To be blue. *Tag: blue.*

dzeh$_1$ To handle liquid in a controlled manner. *Tag: cl-liquid-c.*

dzus$_1$ To dance Indian-style. *Tag: indian-dance.*

gaih$_1$ For one or two people to run. *Tag: run-12.*

gi$_1$ To be grey. *Tag: grey.*

goo$_1$ To drive a land vehicle. The original meaning is to travel in a sleigh. It has been extended to include travel in a horse-drawn carriage and in motor vehicles. It is not used in reference to boats, or to aircraft except when they are taxi-ing on the ground. *Tag: drive.*

goos$_1$ To crunch, especially to eat by crunching. *Tag: crunch.*

guk$_1$ To rub with a gentle friction or press upon with a hard object. *Tag: rub.*

gus$_1$ To be black. *Tag: black.*

gus$_3$ To drag. *Tag: drag.*

ghai$_1$ To be hairy. *Tag: hairy.*

ghal$_1$ Denotes undulating, prompt, and semi-circular movements in a downward direction. *Tag: spool.*

ghalh$_1$ To strike in a rhythmic fashion, as in drumming or threshing grain. By extension from drums to other musical instruments, to play a musical instrument. *Tag: drum.*

ghan$_1$ To kill three or more. *Tag: kill-3.*

ghas$_1$ To scrape with a knife-like blade *Tag: scrape.*

ghaz$_1$ To be hot. *Tag: hot.*

ghaz$_2$ To gnaw. *Tag: gnaw.*

ghe$_1$ To pack on the back. *Tag: pack.*

ghe$_2$ For something frozen to melt or thaw, that is, to return to its natural liquid state. This verb is not used for the melting of substances that are normally solid, such as iron, nor is it used for the dissolution of substances in a solvent. *Tag: melt.*

ghel$_2$ To be calm. *Tag: calm.*

ghi$_1$ To kill one or two people or animals. *Tag: kill-12.*

ghis$_1$ To turn over or revolve on oneself. The motion need not carry on for a full cycle. *Tag: revolve.*

ghok$_1$ For fish to jump. *Tag: fish jump.*

ghut$_1$ To cut with a saw. *Tag: saw.*

jas$_1$ To secrete. *Tag: secrete.*

jas$_2$ To catch with a hook, normally fish. *Tag: hook.*

joot$_1$ To become frightened or cause to become frightened. *Tag: frighten.*

jooz$_1$ To be pointed, sharp. *Tag: pointed.*

jut$_1$ To decay. *Tag: decay.*

kai$_1$ To handle the contents of an open container in a controlled manner. This is also used for handling beds. *Tag: cl-coc-c.*

kat$_1$ For multiple default objects to move without themselves generating the motion, as when falling. *Tag: cl-mdo-i.*

kat$_3$ To clap, slap, move a flat, heavy object rapidly. *Tag: slap.*

ke$_1$ For two to sit. *Tag: sit-2.*

ke$_2$ To go by boat. *Tag: boat.*

ket$_1$ To buy, sell. *Tag: buy.*

k'aih$_1$ To ignite. *Tag: ignite.*

k'un$_1$ To be red. *Tag: red.*

k'us$_1$ To flick, usually with the finger. *Tag: flick.*

k'uz$_1$ To be cold to the touch. *Tag:*

cold.

lai$_1$ To be numerous. *Tag: many.*

le$_1$ To handle plural default objects in a controlled fashion. *Tag: cl-mdo-c.*

li$_1$ To flow. *Tag: flow.*

li$_3$ To become. *Tag: become.*

lih$_1$ To gamble. *Tag: gamble.*

looz$_1$ To go by sleigh. *Tag: sleigh.*

luz$_1$ To excrete fluid. Usually, to urinate, but also applicable to sweating and to skunks spraying. *Tag: piss.*

mulh$_1$ For liquid to boil, bubble. *Tag: boil.*

na$_1$ To be alive. *Tag: live.*

na$_2$ To be difficult. *Tag: difficult.*

nai$_1$ To drink. *Tag: drink.*

nih$_2$ To move the hand in a slow, non-violent manner. *Tag: move hand.*

ta$_1$ To seek, look for. *Tag: seek.*

tan$_1$ To handle long, rigid objects in a controlled manner. *Tag: cl-lro-c.*

tel$_1$ To be wide, broad. To extend. *Tag: wide.*

ti$_1$ To handle a body, animal or human, living or dead. *Tag: cl-body-c.*

ti$_2$ To be valuable, expensive, important, holy. *Tag: valuable.*

ti$_3$ To rain. *Tag: rain.*

ti$_4$ For one or two to sleep. *Tag: sleep-12.*

tuk$_1$ To speak. *Tag: speak.*

tuk$_2$ To break. *Tag: break.*

tulh$_1$ To kick. *Tag: kick.*

tun$_1$ To hold. *Tag: hold.*

tun$_2$ To freeze. *Tag: freeze.*

tun$_3$ To teach, show how, explain. *Tag: teach.*

tus$_2$ To be strong. *Tag: strong.*

t'ah$_1$ To be. *Tag: be.*

t'i$_1$ To like. *Tag: like.*

t'i$_2$ To have, to possess. With a secondary object, to treat as. *Tag: have.*

t'oh$_2$ To fly. *Tag: fly.*

t'ot$_1$ To smoke (e.g. cigarettes). *Tag: smoke.*

t'ukw$_1$ To suck on. To attach to a surface or nipple by creating a vacuum, possibly, but not necessarily, thereby drawing in liquid. *Tag: suck on.*

tlah$_1$ To smear. *Tag: smear.*

tle$_2$ To be soft, supple, limp. *Tag: soft.*

tl'es$_1$ To handle one or two default objects in an uncontrolled manner. *Tag: cl-sdo-u.*

tl'it$_1$ To break wind, to fart. *Tag: fart.*

tl'oo$_1$ To weave, tie, knit, knot. *Tag: knot.*

tl'uz$_1$ To be yellow. *Tag: yellow.*

tsan$_1$ To defecate, to shit. *Tag: shit.*

tsi'$_1$ To be bad. *Tag: bad.*

tsol$_1$ To be small in size, young. *Tag: small.*

tsut$_1$ To drum, in the sense in which a male Ruffed Grouse drums by flapping its cupped wings upward and forward. *Tag: drum.*

ts'i$_1$ For three or more to sit. *Tag: sit-3.*

ts'i$_2$ For the wind to blow. *Tag: wind.*

ts'it$_1$ To fib. *Tag: fib.*

ts'un$_1$ To be hard. *Tag: hard.*

ya$_1$ For one to walk on one pair of limbs. *Tag: walk-1.*

yan$_1$ To be or grow old. *Tag: age.*

yez$_1$ To be long, tall. *Tag: long.*

yi$_1$ To eat. *Tag: eat.*

yin$_1$ To stand. *Tag: stand.*

yis$_1$ To breathe. *Tag: breathe.*

yooh$_1$ To smoulder. *Tag: smoulder.*

yul$_1$ To be white. *Tag: white.*

yul$_3$ For liquid to flow turbulently. This applies not only to liquids in the strict sense but to other things that may be conceived of as liquids, such as sand, beads, and even a large quantity of loose potatoes. *Tag: cl-liquid-i-rough.*

yun$_1$ To sing. *Tag: sing.*

zit$_4$ For one or two to engage in mental activity, to think. *Tag: think-12.*

zoo$_1$ To be good, beautiful, clean, nice. *Tag: good.*

zoot$_1$ To slide or skate. *Tag: slide.*

zun$_1$ To think. *Tag: think.*

Stem List

\b 'a\b0	*IA*	cont	'ai,ÇÅ	cl-sdo-c
\b 'a\b0	*IA*	mom	'ai,ÇÅ	cl-sdo-c
\b 'ai\b0	*PA*	mom	'a,ÇÉ	star moves-12
\b 'ai\b0	*PA*	mom	'ai,ÇÅ	cl-sdo-c
\b 'alh\b0	*IA*	cust	'alh,ÇÅ	chew
\b 'alh\b0	*FA*	mom	'ai,ÇÅ	cl-sdo-c
\b 'as\b0	*IA*	cont	'as,ÇÅ	walk-2
\b 'as\b0	*IN*	cont	'as,ÇÅ	walk-2
\b 'az\b0	*PA*	cont	'as,ÇÅ	walk-2
\b 'az\b0	*PA*	mom	'as,ÇÅ	walk-2
\b 'az\b0	*PN*	cont	'as,ÇÅ	walk-2
\b 'eh\b0	*IA*	cont	'eh,ÇÅ	learn
\b 'ek\b0	*IA*	stat	'ek,ÇÅ	fragile
\b 'en\b0	*IA*	unsp	'en,ÇÅ	see
\b 'en\b0	*IA*	cont	'en,ÇÇ	do
\b 'en\b0	*PA*	unsp	'en,ÇÅ	see
\b 'en\b0	*PA*	cont	'en,ÇÇ	do
\b 'en\b0	*OA*	unsp	'en,ÇÅ	see
\b 'en\b0	*IN*	unsp	'en,ÇÅ	see
\b 'ez\b0	*PA*	mom	'es,ÇÅ	move foot
\b 'i\b0	*IA*	cont	'i,ÇÅ	wait
\b 'ih\b0	*IA*	cust	'en,ÇÅ	see
\b 'ilh\b0	*FA*	unsp	'en,ÇÇ	do
\b 'oo'\b0	*PA*	mom	tl'oo,ÇÅ	knot
\b 'ool\b0	*IA*	cont	'ool,ÇÅ	braid
\b 'ool\b0	*PA*	mom	'ool,ÇÅ	braid

\b 'ulh\b0	*IA*	**cont**	**'alh,ÇÅ**	**chew**
\b 'us\b0	*IA*	**prog**	**'as,ÇÅ**	**walk-2**
\b 'us\b0	*FA*	**cont**	**'as,ÇÅ**	**walk-2**
\b 'us\b0	*FA*	**mom**	**'as,ÇÅ**	**walk-2**
\b 'us\b0	*IN*	**prog**	**'as,ÇÅ**	**walk-2**
\b 'us\b0	*FN*	**cont**	**'as,ÇÅ**	**walk-2**
\b 'us\b0	*FN*	**mom**	**'as,ÇÅ**	**walk-2**
\b 'uz\b0	*IA*	**rep**	**tl'es,ÇÅ**	**cl-sdo-u**
\b ai\b0	*IA*	**unsp**	**yi,ÇÅ**	**eat**
\b ai\b0	*PA*	**unsp**	**yi,ÇÅ**	**eat**
\b ai\b0	*FA*	**unsp**	**yi,ÇÅ**	**eat**
\b ai\b0	*OA*	**unsp**	**yi,ÇÅ**	**eat**
\b ba\b0	*IA*	**stat**	**ba,ÇÅ**	**off-white**
\b ba\b0	*OA*	**mom**	**ba,ÇÇ**	**war**
\b bul\b0	*IA*	**unsp**	**bul,ÇÅ**	**blink**
\b bun\b0	*IA*	**stat**	**bun,ÇÅ**	**full**
\b bun\b0	*PA*	**mom**	**bun,ÇÅ**	**full**
\b bus\b0	*FA*	**mom**	**bas,ÇÅ**	**circle**
\b buz\b0	*IA*	**unsp**	**buz,ÇÅ**	**stretch**
\b buz\b0	*OA*	**unsp**	**buz,ÇÅ**	**stretch**
\b cha\b0	*IA*	**stat**	**cha,ÇÅ**	**big**
\b chan\b0	*IA*	**stat**	**chan,ÇÅ**	**pregnant**
\b cho\b0	*IA*	**compara-tive**	**cha,ÇÅ**	**big**
\b choos\b0	*FA*	**mom**	**chus,ÇÅ**	**cl-2df-c**
\b chulh\b0	*FA*	**unsp**	**choot,ÇÅ**	**take**
\b ch'e\b0	*IA*	**stat**	**ch'eh,ÇÅ**	**angry**
\b ch'e\b0	*PA*	**mom**	**ch'eh,ÇÅ**	**angry**

\b ch'ul\b0	*PA*	**unsp**	**ch'ul,ÇÅ**	**tear**
\b ch'ul\b0	*FA*	**unsp**	**ch'ul,ÇÅ**	**tear**
\b da\b0	*IA*	**unsp**	**da,ÇÅ**	**sit-1**
\b da\b0	*IA*	**stat**	**da,ÇÇ**	**sick**
\b dah\b0	*IA*	**stat**	**dah,ÇÅ**	**kind**
\b daih\b0	*IA*	**cont**	**daih,ÇÅ**	**dance**
\b daz\b0	*IA*	**stat**	**daz,ÇÅ**	**heavy**
\b del\b0	*PA*	**mom**	**dil,ÇÅ**	**walk-3**
\b dil\b0	*PA*	**cont**	**dil,ÇÅ**	**walk-3**
\b dil\b0	*PA*	**mom**	**dil,ÇÅ**	**walk-3**
\b dil\b0	*IN*	**cont**	**dil,ÇÅ**	**walk-3**
\b dil\b0	*FN*	**cont**	**dil,ÇÅ**	**walk-3**
\b dilh\b0	*IA*	**cont**	**dil,ÇÅ**	**walk-3**
\b doh\b0	*FA*	**mom**	**do,ÇÅ**	**cl-fluffy-c**
\b duk\b0	*IA*	**unsp**	**duk,ÇÅ**	**speak**
\b duk\b0	*PA*	**unsp**	**duk,ÇÅ**	**speak**
\b duk\b0	*FA*	**unsp**	**duk,ÇÅ**	**speak**
\b dukw\b0	*IA*	**stat**	**dukw,ÇÅ**	**short**
\b dulh\b0	*IA*	**prog**	**dil,ÇÅ**	**walk-3**
\b dulh\b0	*FA*	**cont**	**dil,ÇÅ**	**walk-3**
\b dulh\b0	*FA*	**mom**	**dil,ÇÅ**	**walk-3**
\b dulh\b0	*PN*	**cont**	**dil,ÇÅ**	**walk-3**
\b dulh\b0	*FN*	**mom**	**dil,ÇÅ**	**walk-3**
\b dlal\b0	*PA*	**unsp**	**dlal,ÇÅ**	**dream**
\b dli\b0	*IA*	**unsp**	**dli,ÇÅ**	**feel cold**
\b dli\b0	*IA*	**unsp**	**dli,ÇÇ**	**plead**
\b dlih\b0	*OA*	**unsp**	**dli,ÇÇ**	**pray**

\b dza\b0	*IA*	stat	dza,ÇÅ	distant
\b dza\b0	*IN*	stat	dza,ÇÅ	distant
\b dzan\b0	*IA*	stat	dzan,ÇÅ	blue
\b dzan\b0	*IN*	stat	dzan,ÇÅ	blue
\b dzeh\b0	*IA*	unsp	dzeh,ÇÅ	cl-liquid-c
\b dzih\b0	*FA*	mom	dzai,ÇÅ	cl-euo-c
\b dzus\b0	*IA*	cont	dzus,ÇÅ	dance
\b gai\b0	*PA*	cont	gaih,ÇÅ	run-12
\b gai\b0	*PA*	mom	gaih,ÇÅ	run-12
\b gaih\b0	*run*	cont	gaih,ÇÅ	run-12
\b gi\b0	*IA*	stat	gi,ÇÅ	grey
\b gih\b0	*IA*	prog	gaih,ÇÅ	run-12
\b gih\b0	*FA*	mom	gaih,ÇÅ	run-12
\b goo\b0	*PA*	cont	goo,ÇÅ	drive
\b goo\b0	*PA*	mom	goo,ÇÅ	drive
\b gooh\b0	*IA*	prog	goo,ÇÅ	drive
\b gooh\b0	*FA*	mom	goo,ÇÅ	drive
\b gooh\b0	*IN*	prog	goo,ÇÅ	drive
\b gooh\b0	*PN*	cont	goo,ÇÅ	drive
\b gooh\b0	*PN*	mom	goo,ÇÅ	drive
\b gooh\b0	*FN*	mom	goo,ÇÅ	drive
\b goos\b0	*IA*	cust	goos,ÇÅ	crunch
\b guk\b0	*IA*	cont	guk,ÇÅ	rub
\b gus\b0	*IA*	stat	gus,ÇÅ	black
\b ghai\b0	*IA*	stat	ghai,ÇÅ	hairy
\b ghal\b0	*PA*	mom	ghal,ÇÅ	spool
\b ghalh\b0	*IA*	unsp	ghalh,ÇÅ	drum

\b ghan\b0	*IA*	mom	ghan,ÇÅ	kill-3
\b ghan\b0	*PA*	mom	ghan,ÇÅ	kill-3
\b ghas\b0	*IA*	cont	ghaz,ÇÇ	gnaw
\b ghaz\b0	*IA*	unsp	ghaz,ÇÅ	hot
\b ghaz\b0	*IA*	stat	ghaz,ÇÅ	hot
\b ghaz\b0	*PA*	trans	ghaz,ÇÅ	hot
\b ghe\b0	*IA*	cont	ghe,ÇÅ	pack
\b ghe\b0	*IA*	unsp	ghi,ÇÅ	kill-12
\b gheh\b0	*IA*	cont	ghe,ÇÇ	melt
\b ghel\b0	*PA*	mom	ghel,ÇÇ	calm
\b ghen\b0	*PA*	mom	ghe,ÇÇ	melt
\b ghen\b0	*FA*	mom	ghe,ÇÇ	melt
\b ghi\b0	*PA*	mom	ghe,ÇÅ	pack
\b ghi\b0	*PA*	mom	ghi,ÇÅ	kill-12
\b ghilh\b0	*IA*	prog	ghe,ÇÅ	pack
\b ghilh\b0	*FA*	mom	ghe,ÇÅ	pack
\b ghis\b0	*IA*	mom	ghis,ÇÅ	revolve
\b ghiz\b0	*PA*	mom	ghis,ÇÅ	revolve
\b ghok\b0	*IA*	unsp	ghok,ÇÅ	fish jump
\b ghut\b0	*IA*	cont	ghut,ÇÅ	saw
\b ghuz\b0	*PA*	mom	ghis,ÇÅ	revolve
\b jas\b0	*IA*	cont	jas,ÇÇ	hook
\b jas\b0	*FA*	unsp	jas,ÇÅ	secrete
\b jaz\b0	*IA*	unsp	jas,ÇÅ	secrete
\b jooz\b0	*IA*	stat	jooz,ÇÅ	pointed
\b juk\b0	*IA*	cust	jut,ÇÅ	decay
\b jut\b0	*IA*	stat	joot,ÇÅ	frighten

\b kai\b0	*PA*	mom	kai,ÇÅ	cl-coc-c
\b kaih\b0	*IA*	cont	kai,ÇÅ	cl-coc-c
\b kat\b0	*IA*	dist	kat,ÇÅ	cl-mdo-i
\b kat\b0	*IA*	unsp	kat,ÇÉ	slap
\b kat\b0	*PA*	rep	kat,ÇÉ	slap
\b ke\b0	*IA*	unsp	ke,ÇÅ	sit-2
\b ke\b0	*IA*	cont	ke,ÇÇ	boat
\b ke\b0	*OA*	cont	ke,ÇÇ	boat
\b ke\b0	*IN*	cont	ke,ÇÇ	boat
\b ke\b0	*ON*	cont	ke,ÇÇ	boat
\b keh\b0	*IA*	mom	ke,ÇÇ	boat
\b kel\b0	*PN*	cont	ke,ÇÇ	boat
\b kel\b0	*PN*	mom	ke,ÇÇ	boat
\b kel\b0	*FN*	cont	ke,ÇÇ	boat
\b kel\b0	*FN*	mom	ke,ÇÇ	boat
\b kelh\b0	*IA*	prog	ke,ÇÇ	boat
\b kelh\b0	*FA*	cont	ke,ÇÇ	boat
\b kelh\b0	*FA*	mom	ke,ÇÇ	boat
\b kelh\b0	*IN*	prog	ke,ÇÇ	boat
\b ket\b0	*IA*	mom	ket,ÇÅ	buy
\b ki\b0	*PA*	cont	ke,ÇÇ	boat
\b ki\b0	*PA*	mom	ke,ÇÇ	boat
\b kih\b0	*IA*	cust	k,ÇÇ	boat
\b kulh\b0	*FA*	unsp	ket,ÇÅ	buy
\b k'aih\b0	*IA*	unsp	k'aih,ÇÅ	ignite
\b k'un\b0	*IA*	stat	k'un,ÇÅ	red
\b k'us\b0	*IA*	stat	k'us,ÇÅ	flick

\b k'uz\b0	*IA*	stat	k'uz,ÇÅ	cold
\b lai\b0	*IA*	stative	lai,ÇÅ	numerous
\b le\b0	*IA*	mom	le,ÇÅ	cl-mdo-c
\b li\b0	*IA*	cont	li,ÇÅ	flow
\b li\b0	*IA*	stat	li,ÇÉ	become
\b li\b0	*PA*	mom	li,ÇÉ	become
\b lih\b0	*IA*	cont	lih,ÇÅ	gamble
\b lih\b0	*FA*	cont	lih,ÇÅ	gamble
\b looz\b0	*IA*	cont	looz,ÇÅ	sleigh
\b luz\b0	*FA*	unsp	luz,ÇÅ	excrete
\b mulh\b0	*IA*	cont	mulh,ÇÅ	boil
\b na\b0	*IA*	cont	na,ÇÅ	live
\b na\b0	*IA*	stat	na,ÇÇ	difficult
\b nai\b0	*IA*	unsp	nai,ÇÅ	drink
\b nih\b0	*IA*	unsp	nih,ÇÇ	move hand
\b nilh\b0	*FA*	unsp	nai,ÇÅ	drink
\b ta\b0	*IA*	cont	ta,ÇÅ	seek
\b tal\b0	*IA*	mom	tulh,ÇÅ	kick
\b tal\b0	*PA*	mom	tulh,ÇÅ	kick
\b talh\b0	*IA*	rep	tulh,ÇÅ	kick
\b tan\b0	*PA*	mom	tan,ÇÅ	cl-lro-c
\b te\b0	*OA*	cont	ti,ÇÑ	sleep-12
\b tel\b0	*IA*	stat	tel,ÇÅ	wide
\b telh\b0	*FA*	mom	ti,ÇÅ	cl-body-c
\b telh\b0	*FA*	mom	ti,ÇÑ	sleep-12
\b ti\b0	*IA*	stat	ti,ÇÇ	valuable
\b ti\b0	*IA*	cont	ti,ÇÉ	rain

\b ti\b0	*IA*	**stat**	**ti,ÇÑ**	**sleep-12**
\b ti\b0	*PA*	**mom**	**ti,ÇÅ**	**cl-body-c**
\b ti\b0	*PA*	**mom**	**ti,ÇÅ**	**cl-body**
\b ti\b0	*PA*	**stat**	**ti,ÇÑ**	**sleep-12**
\b ti\b0	*IN*	**stat**	**ti,ÇÇ**	**valuable**
\b tih\b0	*IA*	**mom**	**tan,ÇÅ**	**cl-lro-c**
\b tilh\b0	*FA*	**mom**	**tan,ÇÅ**	**cl-lro-c**
\b tuk\b0	*PA*	**mom**	**tuk,ÇÇ**	**break**
\b tun\b0	*IA*	**cont**	**tun,ÇÅ**	**hold**
\b tun\b0	*IA*	**unsp**	**tun,ÇÉ**	**teach**
\b tun\b0	*PA*	**mom**	**tun,ÇÇ**	**freeze**
\b tus\b0	*IA*	**stat**	**tus,ÇÇ**	**strong**
\b t'a\b0	*PA*	**cont**	**t'oh,ÇÇ**	**fly**
\b t'ah\b0	*IA*	**stat**	**t'oh,ÇÅ**	**be**
\b t'ah\b0	*IA*	**prog**	**t'oh,ÇÇ**	**fly**
\b t'ah\b0	*FA*	**cont**	**t'oh,ÇÇ**	**fly**
\b t'ah\b0	*FA*	**mom**	**t'oh,ÇÇ**	**fly**
\b t'ah\b0	*IN*	**prog**	**t'oh,ÇÇ**	**fly**
\b t'ah\b0	*PN*	**mom**	**t'oh,ÇÇ**	**fly**
\b t'ah\b0	*FN*	**mom**	**t'oh,ÇÇ**	**fly**
\b t'e\b0	*OA*	**stat**	**t'oh,ÇÅ**	**be**
\b t'i\b0	*IA*	**unsp**	**t'i,ÇÅ**	**like**
\b t'i\b0	*IA*	**stat**	**t'i,ÇÇ**	**have**
\b t'oh\b0	*IA*	**cont**	**t'oh,ÇÇ**	**fly**
\b t'ot\b0	*IA*	**cont**	**t'ot,ÇÅ**	**smoke**
\b t'ot\b0	*FA*	**cont**	**t'ot,ÇÅ**	**smoke**
\b t'ukw\b0	*IA*	**unsp**	**t'ukw,ÇÅ**	**suck on**

\b t'ukw\b0	*PA*	unsp	t'ukw,ÇÅ	suck on
\b tleh\b0	*IA*	stat	tle,ÇÇ	soft
\b tl'it\b0	*IA*	unsp	tl'it,ÇÅ	fart
\b tl'oo\b0	*PA*	mom	tl'oo,ÇÅ	knot
\b tl'uz\b0	*IA*	stat	tl'uz,ÇÅ	yellow
\b tsan\b0	*PA*	unsp	tsan,ÇÅ	shit
\b tsi'\b0	*IA*	stat	tsi',ÇÅ	bad
\b tsol\b0	*IA*	stat	tsol,ÇÅ	small
\b tsut\b0	*IA*	cont	tsut,ÇÅ	drum
\b ts'it\b0	*IA*	cust	ts'it,ÇÅ	fib
\b ts'i\b0	*IA*	unsp	ts'i,ÇÅ	sit-3
\b ts'i\b0	*IA*	cont	ts'i,ÇÇ	wind
\b ts'i\b0	*PA*	mom	ts'i,ÇÇ	wind
\b ts'un\b0	*IA*	stat	ts'un,ÇÅ	hard
\b us\b0	*PA*	mom	jas,ÇÇ	hook
\b ya\b0	*IA*	cont	ya,ÇÅ	walk-1
\b ya\b0	*PA*	cont	ya,ÇÅ	walk-1
\b ya\b0	*PA*	mom	ya,ÇÅ	walk-1
\b ya\b0	*FA*	cont	ya,ÇÅ	walk-1
\b ya\b0	*OA*	cont	ya,ÇÅ	walk-1
\b ya\b0	*FN*	cont	ya,ÇÅ	walk-1
\b yah\b0	*IN*	cont	ya,ÇÅ	walk-1
\b yal\b0	*PN*	cont	ya,ÇÅ	walk-1
\b yal\b0	*PN*	mom	ya,ÇÅ	walk-1
\b yalh\b0	*IA*	prog	ya,ÇÅ	walk-1
\b yalh\b0	*FA*	mom	ya,ÇÅ	walk-1
\b yalh\b0	*IN*	prog	ya,ÇÅ	walk-1

\b yalh\b0	*FN*	mom	ya,ÇÅ	walk-1
\b yan\b0	*PA*	unsp	yan,ÇÅ	age
\b yez\b0	*IA*	stative	yez,ÇÅ	long
\b yi\b0	*IA*	unsp	yi,ÇÅ	eat
\b yi\b0	*PA*	mom	yi,ÇÅ	eat
\b yi\b0	*FA*	unsp	yi,ÇÅ	eat
\b yi\b0	*OA*	unsp	yi,ÇÅ	eat
\b yi\b0	*IN*	unsp	yi,ÇÅ	eat
\b yi\b0	*FN*	unsp	yi,ÇÅ	eat
\b yi\b0	*ON*	unsp	yi,ÇÅ	eat
\b yilh\b0	*FA*	unsp	yi,ÇÅ	eat
\b yin\b0	*IA*	unsp	yin,ÇÅ	stand
\b yiz\b0	*IA*	unsp	yis,ÇÅ	breathe
\b yooh\b0	*IA*	cust	yooh,ÇÅ	smoulder
\b yul\b0	*IA*	stat	yul,ÇÅ	white
\b yulh\b0	*IA*	mom	yul,ÇÉ	cl-liquid-i-rough
\b yun\b0	*IA*	cont	yun,ÇÅ	sing
\b yun\b0	*PA*	cont	yun,ÇÅ	sing
\b yun\b0	*FA*	cont	yun,ÇÅ	sing
\b yun\b0	*IN*	cont	yun,ÇÅ	sing
\b yun\b0	*PN*	cont	yun,ÇÅ	sing
\b zoo\b0	*IA*	stative	zoo,ÇÅ	good
\b zoo\b0	*IN*	stative	zoo,ÇÅ	good
\b zoot\b0	*IA*	cont	zoot,ÇÅ	slide
\b zun\b0	*IA*	cont	zun,ÇÅ	think
\b zut\b0	*FA*	unsp	zit,ÇÑ	think
\b zut\b0	*IA*	unsp	zit,ÇÑ	think

Stems Sorted By Root

Root	Tag	TMA	Aspect	Stem	Note
\b 'a$_3$\b0	**star moves-12**	*PA*	**mom**	**'ai**	
\b 'ai$_1$\b0	**cl-sdo-c**	*IA*	**cont**	**'a**	
		IA	**mom**	**'a**	
		PA	**mom**	**'ai**	
		FA	**mom**	**'alh**	
\b 'alh$_1$\b0	**chew**	*IA*	**cust**	**'alh**	
		IA	**cont**	**'ulh**	
\b 'as$_1$\b0	**walk-2**	*IA*	**cont**	**'as**	
		IA	**prog**	**'us**	
		PA	**cont**	**'az**	
		PA	**mom**	**'az**	
		FA	**cont**	**'us**	
		FA	**mom**	**'us**	
		IN	**cont**	**'as**	
		IN	**prog**	**'us**	
		PN	**cont**	**'az**	
		FN	**cont**	**'us**	
		FN	**mom**	**'us**	
\b 'eh$_1$\b0	**learn**	*IA*	**cont**	**'eh**	
\b 'ek$_1$\b0	**fragile**	*IA*	**stat**	**'ek**	
\b 'en$_1$\b0	**see**	*IA*	**cust**	**'ih**	
		IA		**'en**	
		PA		**'en**	
		OA		**'en**	
		IN		**'en**	

\b 'en$_2$\b0	do	*IA*	cont	'en	
		PA	cont	'en	
		FA		'ilh	
\b 'es$_1$\b0	move foot	*PA*	mom	'ez	
\b 'i$_1$\b0	wait	*IA*	cont	'i	
\b 'ool$_1$\b0	braid	*IA*	cont	'ool	
		PA	mom	'ool	
\b ba$_1$\b0	off-white	*IA*	stat	ba	
\b ba$_2$\b0	war	*OA*	mom	ba	
\b bas$_1$\b0	circle	*FA*	mom	bus	
\b bul$_1$\b0	blink	*IA*		bul	
\b bun$_1$\b0	full	*IA*	stat	bun	
		PA	mom	bun	
\b buz$_1$\b0	stretch	*IA*		buz	
		OA		buz	
\b cha$_1$\b0	big	*IA*	comparative	cho	
		IA	stat	cha	
\b chan$_1$\b0	pregnant	*IA*	stat	chan	
\b choot$_1$\b0	take	*FA*		chulh	
\b chus$_1$\b0	cl-2df-c	*FA*	mom	choos	
\b ch'eh$_1$\b0	angry	*IA*	stat	ch'e	
		PA	mom	ch'e	
\b ch'ul$_1$\b0	tear	*PA*		ch'ul	
		FA		ch'ul	
\b da$_1$\b0	sit-1	*IA*		da	
\b da$_2$\b0	sick	*IA*	stat	da	
\b dah$_1$\b0	kind	*IA*	stat	dah	

\b dzeh$_1$\b0	cl-liquid-c	*IA*		dzeh	
\b dzus$_1$\b0	dance	*IA*	cont	dzus	
\b dzai$_1$\b0	cl-euo-c	*FA*	mom	dzih	
\b gaih$_1$\b0	run-12	*IA*	prog	gih	
		PA	cont	gai	
		PA	mom	gai	
		FA	mom	gih	
		run	cont	gaih	
\b gi$_1$\b0	grey	*IA*	stat	gi	
\b goo$_1$\b0	drive	*IA*	prog	gooh	
		PA	cont	goo	
		PA	mom	goo	
		FA	mom	gooh	
		IN	prog	gooh	
		PN	cont	gooh	
		PN	mom	gooh	
		FN	mom	gooh	
\b goos$_1$\b0	crunch	*IA*	cust	goos	
\b guk$_1$\b0	rub	*IA*	cont	guk	
\b gus$_1$\b0	black	*IA*	stat	gus	
\b ghai$_1$\b0	hairy	*IA*	stat	ghai	
\b ghal$_1$\b0	spool	*PA*	mom	ghal	
\b ghalh$_1$\b0	drum	*IA*		ghalh	
\b ghan$_1$\b0	kill-3	*IA*	mom	ghan	
		PA	mom	ghan	
\b ghaz$_1$\b0	hot	*IA*		ghaz	
		IA	stat	ghaz	

		PA	trans	ghaz	
\b ghaz$_2$\b0	gnaw	*IA*	cont	ghas	
\b ghe$_1$\b0	pack	*IA*	cont	ghe	
		IA	prog	ghilh	
		PA	mom	ghi	
		FA	mom	ghilh	
\b ghe$_2$\b0	melt	*IA*	cont	gheh	
		PA	mom	ghen	
		FA	mom	ghen	
\b ghel$_2$\b0	calm	*PA*	mom	ghel	
\b ghi$_1$\b0	kill-12	*IA*		ghe	
		PA	mom	ghi	
\b ghis$_1$\b0	revolve	*IA*	mom	ghis	
		PA	mom	ghiz	
		PA	mom	ghuz	
\b ghok$_1$\b0	fish jump	*IA*		ghok	
\b ghut$_1$\b0	saw	*IA*	cont	ghut	
\b jas$_2$\b0	hook	*IA*	cont	jas	
		PA	mom	us	
\b jas$_1$\b0	secrete	*IA*		jaz	
		FA		jas	
\b joot$_1$\b0	frighten	*IA*	stat	jut	
\b jooz$_1$\b0	pointed	*IA*	stat	jooz	
\b jut$_1$\b0	decay	*IA*	cust	juk	
\b k$_2$\b0	boat	*IA*	cust	kih	
\b kai$_1$\b0	cl-coc-c	*IA*	cont	kaih	
		PA	mom	kai	

\b kat$_1$\b0	cl-mdo-i	IA	dist	kat	
\b kat$_3$\b0	slap	IA		kat	
		PA	rep	kat	
\b ke$_1$\b0	sit-2	IA		ke	
\b ke$_2$\b0	boat	IA	cont	ke	
		IA	mom	keh	
		IA	prog	kelh	
		PA	cont	ki	
		PA	mom	ki	
		FA	cont	kelh	
		FA	mom	kelh	
		OA	cont	ke	
		IN	cont	ke	
		IN	prog	kelh	
		PN	cont	kel	
		PN	mom	kel	
		FN	cont	kel	
		FN	mom	kel	
		ON	cont	ke	
\b ket$_1$\b0	buy	IA	mom	ket	
		FA		kulh	
\b k'aih$_1$\b0	ignite	IA		k'aih	
\b k'un$_1$\b0	red	IA	stat	k'un	
\b k'us$_1$\b0	flick	IA	stat	k'us	
\b k'uz$_1$\b0	cold	IA	stat	k'uz	
\b lai$_1$\b0	numerous	IA	stat	lai	
\b le$_1$\b0	cl-mdo-c	IA	mom	le	

\b li$_1$\b0	**flow**	*IA*	**cont**	**li**	
\b li$_3$\b0	**become**	*IA*	**stat**	**li**	
		PA	**mom**	**li**	
\b lih$_1$\b0	**gamble**	*IA*	**cont**	**lih**	
		FA	**cont**	**lih**	
\b looz$_1$\b0	**sleigh**	*IA*	**cont**	**looz**	
\b luz$_1$\b0	**excrete**	*FA*		**luz**	
\b mulh$_1$\b0	**boil**	*IA*	**cont**	**mulh**	
\b na$_1$\b0	**live**	*IA*	**cont**	**na**	
\b na$_2$\b0	**difficult**	*IA*	**stat**	**na**	
\b nai$_1$\b0	**drink**	*IA*		**nai**	
		FA		**nilh**	
\b nih$_2$\b0	**move hand**	*IA*		**nih**	
\b ta$_1$\b0	**seek**	*IA*	**cont**	**ta**	
\b tan$_1$\b0	**cl-lro-c**	*IA*	**mom**	**tih**	
		PA	**mom**	**tan**	
		FA	**mom**	**tilh**	
\b tel$_1$\b0	**wide**	*IA*	**stat**	**tel**	
\b ti$_1$\b0	**cl-body-c**	*PA*	**mom**	**ti**	
\b ti$_1$\b0	**cl-body**	*PA*	**mom**	**ti**	
\b ti$_1$\b0	**cl-body-c**	*FA*	**mom**	**telh**	
\b ti$_2$\b0	**valuable**	*IA*	**stat**	**ti**	
		IN	**stat**	**ti**	
\b ti$_3$\b0	**rain**	*IA*	**cont**	**ti**	
\b ti$_4$\b0	**sleep-12**	*IA*	**stat**	**ti**	
		PA	**stat**	**ti**	
		FA	**mom**	**telh**	

		OA	**cont**	**te**	
\b tuk$_2$\b0	**break**	*PA*	**mom**	**tuk**	
\b tulh$_1$\b0	**kick**	*IA*	**mom**	**tal**	
		IA	**rep**	**talh**	
		PA	**mom**	**tal**	
\b tun$_1$\b0	**hold**	*IA*	**cont**	**tun**	
\b tun$_2$\b0	**freeze**	*PA*	**mom**	**tun**	
\b tun$_3$\b0	**teach**	*IA*		**tun**	
\b tus$_2$\b0	**strong**	*IA*	**stat**	**tus**	
\b t'i$_1$\b0	**like**	*IA*		**t'i**	
\b t'i$_2$\b0	**have**	*IA*	**stat**	**t'i**	
\b t'oh$_1$\b0	**be**	*IA*	**stat**	**t'ah**	
		OA	**stat**	**t'e**	
\b t'oh$_2$\b0	**fly**	*IA*	**cont**	**t'oh**	
		IA	**prog**	**t'ah**	
		PA	**cont**	**t'a**	
		FA	**cont**	**t'ah**	
		FA	**mom**	**t'ah**	
		IN	**prog**	**t'ah**	
		PN	**mom**	**t'ah**	
		FN	**mom**	**t'ah**	
\b t'ot$_1$\b0	**smoke**	*IA*	**cont**	**t'ot**	
		FA	**cont**	**t'ot**	
\b t'ukw$_1$\b0	**suck on**	*IA*		**t'ukw**	
		PA		**t'ukw**	
\b tle$_2$\b0	**soft**	*IA*	**stat**	**tleh**	
\b tl'es$_1$\b0	**cl-sdo-u**	*IA*	**rep**	**'uz**	

\b tl'it$_1$\b0	fart	*IA*		tl'it	
\b tl'oo$_1$\b0	knot	*PA*	mom	'oo'	
		PA	mom	tl'oo	
\b tl'uz$_1$\b0	yellow	*IA*	stat	tl'uz	
\b tsan$_1$\b0	shit	*PA*		tsan	
\b tsi'$_1$\b0	bad	*IA*	stat	tsi'	
\b tsol$_1$\b0	small	*IA*	stat	tsol	
\b tsut$_1$\b0	drum	*IA*	cont	tsut	
\b ts'it$_1$\b0	fib	*IA*	cust	ts'it	
\b ts'i$_1$\b0	sit-3	*IA*		ts'i	
\b ts'i$_2$\b0	wind	*IA*	cont	ts'i	
		PA	mom	ts'i	
\b ts'un$_1$\b0	hard	*IA*	stat	ts'un	
\b ya$_1$\b0	walk-1	*IA*	cont	ya	
		IA	prog	yalh	
		PA	cont	ya	
		PA	mom	ya	
		FA	cont	ya	
		FA	mom	yalh	
		OA	cont	ya	
		IN	cont	yah	
		IN	prog	yalh	
		PN	cont	yal	
		PN	mom	yal	
		FN	cont	ya	
		FN	mom	yalh	
\b yan$_1$\b0	age	*PA*		yan	

\b yez$_1$\b0	long	*IA*	stat	yez	
\b yi$_1$\b0	eat	*IA*		ai	1
		IA		yi	
		PA	mom	yi	
		PA		ai	2
		FA		ai	3
		FA		yi	
		FA		yilh	
		OA		ai	4
		OA		yi	
		IN		yi	
		FN		yi	
		ON		yi	
\b yin$_1$\b0	stand	*IA*		yin	
\b yis$_1$\b0	breathe	*IA*		yiz	
\b yooh$_1$\b0	smoulder	*IA*	cust	yooh	
\b yul$_1$\b0	white	*IA*	stat	yul	
\b yul$_3$\b0	cl-liquid-i-rough	*IA*	mom	yulh	
\b yun$_1$\b0	sing	*IA*	cont	yun	
		PA	cont	yun	
		FA	cont	yun	
		IN	cont	yun	
		PN	cont	yun	
\b zoo$_1$\b0	good	*IA*	stat	zoo	
		IN	stat	zoo	

\b yez$_1$\b0	long	*IA*	stat	yez	
\b zoot$_1$\b0	slide	*IA*	cont	zoot	
\b zit$_4$\b0	think	*IA*		ut	
		FA		zut	
\b zun$_1$\b0	think	*IA*	cont	zun	

Notes on Stems

1	ai	This is a suppletive allomorph used only in D-Effect contexts.
2	ai	This is a suppletive allomorph used only in D-Effect contexts.
3	ai	This is a suppletive allomorph used only in D-Effect contexts.
4	ai	This is a suppletive allomorph used only in D-Effect contexts.

TOPICAL INDEX

Affix List

bu-	third person singular possessor
-cho	big
du-	third person reflexive possessor for class 1 nouns
hubu-	third person duo-plural possessor for class 1 nouns
n-	second person singular possessor
nah-	first person dual possessor
nah-	second person duo-plural possessor
ne-	first person plural possessor
s-	first person singular possessor
-tsol	small
whu-	areal object of postposition
-yaz	small, little

Index of Affixes by Gloss

whu-	areal object of postposition
-cho	big
nah-	first person dual possessor
ne-	first person plural possessor
s-	first person singular possessor
-yaz	little
nah-	possessor, first person dual
ne-	possessor, first person plural
s-	possessor, first person singular
nah-	possessor, second person duo-plural
n-	possessor, second person singular
hubu-	possessor, third person duo-plural, for class 1 nouns
du-	possessor, third person reexive, for class 1 nouns
bu-	possessor, third person singular
nah-	second person duo-plural possessor
n-	second person singular posessor
-tsol	small
-yaz	small
bu-	third person singular possessor

Index of Place Names

'Indzik'et	Yensischuck
'Usdas Tezdluz	A rock formation along the shore of Fraser Lake west of IR 1 in front of the present day Bill Evans' ranch It shows the impression of 'Usdas's body where he fell from the sky, as recounted in the legend of 'Usdas and the swans.
Azilenchonoo	The larger of the two small rock islands in the mouth of the Nadleh River where it flows out of Fraser Lake at Nadleh.
Basghelh	Mary Mountain
Basghelh Bunk'ut	Mary Lake
Belhk'achek	Belgatse
Belhk'akoh	Knapp Creek
Bundzi Bunk'ut	Angly Lake
Bundzikoh	Angly Creek
Buzdaiyus	Pitka Mountain
Chawts'unk'ut	Chowsunkut Lake
Chestl'ada	Cheslatta village
Choostl'o	Ormond Lake
Chunbaz Bunk'ut	Alf Lake
Chundooyus	Unnamed hill
Chustl'adak	The mountain near old Cheslatta village
Daideyus	A mountain between Nadleh and Nak'azdli
Datsan T'acho	The open, grassy hills on the north side of Fraser Lake, visible from Lejac.
Datsanchonun	The open, grassy hills on the north side of Fraser Lake, visible from Lejac
Deduk Usk'en	Mount Greer
Dek'e Yusk'ut	A small hill on the northeast side of Nadleh Village, where the water tower is now located.
Dek'ezyusk'ut	The mountain East of Nadleh
Dohcho	A mountain near the Endako mine
Dohyaz	A mountain near the Endako mine
Duk'ai Hooni Bunk'ut	Barlow Lake
Endakoh	Endakoh River
Ges Hutaghe	The trail from the old Nechako River bridge crossing to the canyon along the ridge of the hill.

Hoolhtan Bunk'ut	Top Lake
K'atsizchun	Old Fort Fraser
K'atsizyus	Fraser Mountain
K'ibet	A low area in a birch stand near Kenny Nooski's house.
Kw'eoodiltsuk	A mountain near Nadleh
Liyubyints'izti	Devil's Hairpin
Lhaghahutullhuk	An area traditionally used for broad-jumping contests on the old Nautley Road near Kenny Nooski's house.
Lhez Bunk'ut	Klez Lake
Lhkw'etsilhchola	Mouse Mountain
Lhkw'etsilhchola	Point of land on Fraser Lake below Mouse Mountain
Meljayah	Stellakoh village
Musdus Lhenul'elh	A ford on the Nechako River upstream of the Nautley River used before there were bridges or
Nadleh	ferries.
Nadlehbun	Nautley
Nadlehbunk'ut	Fraser Lake
Nadlehkoh	Fraser Lake
Nahnelt'o	Nautley River
	The location of the original fish weir on the
Nahudujun	Nautley River
Nanguz Bunk'ut	A mountain near Nadleh
Nataoozjaz	Peta Lake
	A lake on the south side of IR 1 Reported to be a place to avoid due to quicksand and poisonous water. This lake is a feature in the legend of a creature carried there from another small lake
Natl'alikoh	on the north side of Sugarloaf Mountain.
Natl'ilikoh	Sutherland River
Netats'anghi	Sutherland River
Neyilasts'ah	Sugarloaf Mountain
Noo'ai	Mouse Mountain
Noonghak	An island in Fraser Lake.
	A ford on the Nechako River upstream of the Nautley River used before there were bridges
Noonghun	or ferries.
	The embankment along the Nechako River on the east side of the new subdivision by
Noozkoh	Alana's house.
Nuswhutisdi	Ormond Creek
	Peterson's beach and point

Nyan Wheti	The trail from Nadleh village to Tsaɔoche "Sowchea" on Stuart Lake. This trail became the pack trail for the Northwest Company, then the Hudson's Bay Company, between 1806 and 1906, until it was replaced by the Dog Creek Trail, which skirted Saddle Mountain rather than crossing it. Used for travel, trade, communications, hunting, trapping, rabbit snaring, bark and plant gathering.
Nooghun	An embankment along the Nechakoh River.
Sainaditi	Peterson's Beach Forestry Campsite
Sdabiz	A place on the north shore of Fraser Lake.
Sgookoh	The creek that enters Fraser Lake about two miles West of Nadleh along the north shore.
Skenicho	Peak of Greer Mountain
Stellakoh	Stellakoh River
Stellakoh	Stellakoh village
Stellayukh	Alexandria
Shasti	A trail (draw) west of Serina Greene's house just north of the north end of the Nautley River bridge. This is where grizzly bears came down to feed on spawning salmon.
Shasti	The deep gully West of Nadleh
Talhedayui	A lake on the south side of IR 1. Said to be a place to avoid due to quicksand and poisonous water.
Tanyiz Bun	Tatin Lake
Tat'at	The eddy at the junction of the Nautley and Nechako rivers It is not as distinct due to the reduction of the flows since the construction of the Kenney Dam.
Tats'utnai	Morris Creek
Tehdune Yananestez	Smooth flat slabs of rock along the shores of IR 2 just east of Vinnedge Cove on Fraser Lake The site of a legend in which a large man sleeping on the rocks was surprised and dove into the water, never to reappear The stones are said to resemble men wearing cloaks.
Tehk'eilchuk	Vanderhoof

Tl'otelk'et	A point on the east end of Fraser Lake on the south side of the Nautley River where it flows out of Fraser Lake.
Tselhutulh	A distinct vertical crack in the lakeshore rocks on the North shore of Fraser Lake near Tehdune Yananestez.
Tsenabulh'a Bunk'ut	Leg Lake
Tseyaz	Lejac
Tsezoolh Dezdla	Some wave-formed rocks along the shore in Gala Bay.
Tsela Dujoz	A large rock on the North bank of the Nautley River below Matthew Ketlo's house. Said to resemble a rock that used to be at Stellako, later removed when the road was built, that Joz Dayi used to stand on to deliver speechesIn the 1940s when Leah Patrick's son Raymond was born, He was taken directly after his birth while still wet and rubbed on this rock to give him the strength of the rock This site is associated with the legend of 'Usdas and the Virgin.
Tselhk'azbunk'ut	Burns Lake
Tselhk'azkoh	The river that flows into Burns Lake.
Tselhk'azyus	The mountain outside Burns Lake.
Tselhts'aichoyus	An important moose and deer hunting area.
Ts'esdlol	A little mountain on the south side of the Nechako River aross from the mouth of Tats'utnai (Morris Creek).
Ts'oon'ai Bunk'ut	Etcho Lake
Wilyum sugi	Drywilliam Lake
Whududlun Whucho	The steep hill on the old road on the south side of the Nechako River leading to 'Indzik'et.
Yausilh	Crowe Canyon
Yests'e Bunk'ut	Stag Lake
Yoonoo Bunk'ut	Oona Lake
Yun Dulhdum	A location on Dog Creek Road just before Tachakoh "Three Mile Creek".

Index of English Place Names

Alexandria	Stellayukh
Alf Lake	Chunbaz Bunk'ut
Angly Creek	Bundzikoh
Angly Lake	Bundzi Bunk'ut
Barlow Lake	Duk'ai Hooni Bunk'ut
Belgatse	Belhk'achek
Burns Lake	tselhk'azbunk'ut
Canyon Creek	Noozkoh
Cheslatta village	Chestl'ada
Chowsunkut Lake	Chawts'unk'ut
Crowe Canyon	Yausilh
Devil's Hairpin	Liyubyints'izti
Dry William Lake	Wilyum sugi
Drywilliam Lake	Wilyum sugi
Endakoh River	Endakoh
Etcho Lake	Ts'oon'ai Bunk'ut
embankment along Nechako	nooghun
Fraser Lake	Nadlehbun
Fraser Lake	Nadlehbunk'ut
Fraser Mountain	K'atsizyus
Greer Mountain	Deduk Usk'en
island in Fraser Lake	Noo'ai
Klez Lake	Lhez Bunk'ut

Knapp Creek	Belhk'akoh
Leg Lake	Tsenabulh'a Bunk'ut
Lejac	Tseyaz
Mary Lake	Basghelh Bunk'ut
Morris Creek	Tats'utnai
Mount Greer	Deduk Usk'en
Mouse Mountain	Lhkw'etsilhchola
Mouse Mountain	Neyilasts'ah
Nautley	Nadleh
Nautley River	Nadlehkoh
Old Fort Fraser	K'atsizchun
Oona Lake	Yoonoo Bunk'ut
Ormond Creek	Noozkoh
Ormond Lake	Choostl'o
Peta Lake	Nanguz Bunk'ut
Peterson's Beach Forestry Campsite	Sainaditi
Peterson's beach and point	Nuswhutisdi
Pitka Mountain	Buzdaiyus
peak of Greer Mountain	Skenicho
point of land on Fraser Lake below Mouse Mountain	Lhkw'etsilhchola
Saddle Mountain	Buzdaiyus
Stag Lake	Yests'e Bunk'ut
Stellakoh River	Stellakoh
Stellakoh village	Stellakoh
Stellakoh village, old name	Meljayah
Sutherland River	Natl'alikoh
Sutherland River	Natl'ilikoh
Tatin Lake	Tanyiz Bun
Top Lake	Hoolhtan Bunk'ut
Trout Lake	Duk'ai Hooni Bunk'ut
unnamed hill	Chundooyus
Vanderhoof	Tehk'eilchuk
Yensischuck	'Indzik'et

Index of Loan Words

balhats	Chinook jargon
bosdun	English via Chinook jargon
chigamin	Nuuchanulth via Chinook jargon
dayi	Chinook jargon
gwada'	English
hat'al	Wet'suwet'en
joz	French
jaboon	English
lasal	French
lasyet	French
lelwe	French
liba	French
ligok	French
lilet	French
liyab	French
lizas	French
lubel	French
lubezo	French
lubos	French
lubot	French
lubudak	French
lubuyos	French
ludi	French
lufooset	French
lugafi	French
lugar	French
lugliz	French
luglos	French
lugloo	French
luhoos	French
lusel	French
lusook	French

lusooni	French	**soogah**	English
lusoosam	French	**sooniya**	Cree
lhaga'as	Gitksan	**soonuya**	Cree
mandah	Spanish	**Sumdi**	French
mus	English	**Wanderdi**	French
musdus	Cree	**Za**	French
		Zooldubak	French

Numbers

	Generic	Human	Multiplicative	Locative	Abstract
1	'ilhuk'i	'ilhunun	'ilhah	'ilhuk'i	'ilhuk'i
2	nankah	nanun	nat	nadun	nawh
3	tak'i	tanun	tat	tadun	tawh
4	dink'i	dinun	dit	didun	diwh
5	skwunlai	skwunlanun	skwunlat	skwunladun	skwunlawh
6	lhk'utak'i	lhk'utanun	lhk'utat	lhk'utadun	lhk'utakiwh
7	lhtak'alt'i	lhtak'alt'inun	lhtak'alt'it	lhtak'altidun	lhtak'altiwh
8	lhk'utdink'i	lhk'utdine	lhk'udit	lhk'udidun	lhk'udiwh
9	'ilhah hoolah	'ilhah hoolanun	'ilhah hoolat	'ilohooladun	'ilhohoolawh
10	lanezi	lanezinun	lanezit	lanezidun	laneziwh

Larger numbers are formed by adding decades with the conjunction *'on'at* "plus".

For example, 234 is *nat lanezit lanezi 'on'at tat lanezi 'on'at dink'i.*

Months

January	Sachonun
February	Yuske'nighus
March	Tsintel Ooza
April	Talhtsi Ooza
May	Dugoos Ooza
June	Duk'ai Ooza
July	Shen Whuniz
August	Talook Ooza
September	Gestlah Ooza
October	Bet Ooza
November	Banghan Nuts'uke
December	Sacho Din'ai

Days of the Week

Sunday	Dimosdzen
Monday	Dimosk'elh'az
Monday	Landidzen
Tuesday	Whulhnatdzen
Wednesday	Whulhtatdzen
Thursday	Whulhditdzen
Friday	'Utsun Ts'usyih
Friday	Wanderdi
Saturday	Dzenwhuyaz
Saturday	Sumdi

GRAMMAR SKETCH

Sound System

The consonants are shown below:

Labial	Alveolar	Lateral	Palatal	Velar	Labiovelar	Laryngeal
(p)	t			k	kw	'
b	d			g	gw	
	t'			k'	kw'	
	ts	tl	ch			
	dz	dl	j			
	ts'	tl'	ch'			
(f)	s		sh	kh		
	z			gh		
m	n		(ny)	ng		
w	r	l	y			
wh		lh				h

With the exception of /ny/, the sounds shown in parentheses are not native to the language but appear in some loans. ny is native to the language but appears only in a few second person singular morphemes. Glottal stop is distinctive both in the onset and in the coda.

Codas are limited to a single consonant, which, if a stop, must be unaspirated. The affricates do not appear in the coda, with the exception of /ts/, which is found in *-ghuts* "cartillage" and in the loanword *balhats* "potlatch". Onsets consist of at most two consonants, with the the first restricted to /lh/ and /s/. The nasals are syllabic in pre-conconsontal word-initial position, as in *ngan* "your arm" and *mbat* "your mitts".

Carrier has just six vowels: *a, e, i, o, u,* and *oo.* The vowel u is not found in word-final open syllables with the exception of a few postpositions, such as *ts'u* "to", which should probably be analyzed as incorporated into the following verb.

There are no underlying length contrasts, but surface phonemic length contrasts arise due to the operation of phonological rules. An example is *nuchateske* "he is not going to go around by boat" vs. *nucha:teske* "they are not going to go around by boat", where the long vowel of the latter results from the assimilation of the third person duo-plural subject marker /h/. Vowel length is not marked in the Carrier Linguistic System writing system but is marked in this dictionary by means of a colon.

Syntax

Carrier is a *head-final* language, in which the defining element of a phrase, the head, typically comes at the end of the phrase. The usual order of words is Subject Object Verb. The subject normally precedes the verb.

(1) **Ts'eke 'uyi**.

woman she-is-eating-something

A woman is eating something.

However, the verb "to say" is an exception to this generalization; it almost invariably precedes its subject, as in (2):

(2) **Hadih, ni 'uba.**

hello 3s-say-PA Father

Hello, said Father.

The object normally precedes the verb.

(3) **Musdus<u>ts</u>un usyi.**

beef I-am-eating

I am eating beef.

The subject normally precedes the object.

(4) **Ts'eke musdustsun uyi.**

woman beef he-is-eating

A woman is eating beef.

Instead of prepositions, Carrier has *postpositions,* which follow the noun phrase that they govern. In (5), the postposition ba "for", follows the noun *Bida* "Peter".

(5) **K'andzen njan Bida ba tl'obayah nets'uwhute'alh**

today here Peter for hayshed we-will-build

Today we are going to build a hayshed here for Peter.

Complementizers (subordinating conjunctions) follow their clause. In (6), the complementizer *whuch'a* "lest" follows the subordinate clause *nghoo doolhjut* "your teeth decay".

(6) **Nghoo doolhjut whuch'a toonaingus.**

your-teeth decay-OA lest you (1) clean

Brush your teeth lest they decay.

Subordinate clauses usually precede the main verb as in (6). However, subordinate clauses sometimes follow the main verb, separated from it by a pause, as an afterthought. In (7) the subordinate clause *untooldoh wheni* "so that it will rise" follows the main verb *ninuskai* "I put".

(7) **Lhes sto ninuskai untooldoh wheni**

dough stove I-put it-rise-OA so that

I put the dough by the stove so that it will rise.

Nouns

Most nouns have no distinct plural form: *tes* can mean either "knife" or "knives" depending on the context. With few exceptions the only nouns that have distinct plural forms are those that refer to people, and even most of these do not have plural forms. Most nouns that have plural forms are kinship terms. There is no productive means of forming plurals.

Most nouns that have a distinct non-singular form only have two forms, a singular and a *duo-plural*. A few nouns distinguish singular, *dual,* and *plural* forms, where the dual is used for two and the plural for three or more. These are all nouns derived from number-restricted verbs. An example is *nunaduda* "tourist", which has the dual *nunadut'as* "two tourists" and the plural *nunadudelh* "three or more tourists".

Possession of nouns is indicated by prefixation. There are four main classes of noun, exemplified by the paradigms below. Which nouns belong to which class is largely, though not entirely, predictable. Most nouns whose stem begins with a consonant other than glottal stop (') belong to class 1. Most nouns beginning with glottal stop or *h* belong to class 2. Most nouns beginning with a vowel belong to class 3. However, some nouns beginning with glottal stop belong to class 1, and a few nouns beginning with a consonant other than glottal stop belong to class 2. Class 4, which differs only slightly from class 3, contains only a small number of nouns. In most cases the noun class is not indicated. In these cases it is safe to assume that the noun belongs to the class predicted by the above rules. The class of a noun may be indicated in the dictionary entry by its *pclass.*

Within class 1 the second person singular prefix takes three forms. The usual form of the prefix is *n-*, which becomes *m-* before nouns beginning with *b.* Before nouns beginning with *m* the prefix is *nyu-*. Before nouns beginning with *n* the prefix is *nyu-* or *n*. Syllable /n/ alternates freely with *un*.

The abbreviations used below are as follows:

1s

First person singular, e.g. "my mittens"

1d

First person dual, e.g. "the mittens of the two of us".

1p

First person plural, e.g. "our mittens", where there are three or more of us.

2s

Second person singular, e.g. "your (one person) mittens.

2p

Second person plural, e.g. "your (two or more people) mittens.

3s

Third person singular, e.g. "his/her/its mittens".

3p

Third person plural, e.g. "their mittens".

Reflexive

Reflexive, e.g. "his/her own mittens". This is used when the clause has a third person singular subject that is the same as the possessor of the noun.

DJR

Disjoint Reference. This means "his/her/its" but is used in place of the plain third person singular or the reflexive when the clause has a third person singular subject that is not the same as the possessor of the noun.

Reciprocal

Reciprocal, meaning "each other's". This is very rare on nouns but not uncommon as the object of postpositions.

Plural Disjoint Reference

This is the plural obviative, meaning "his/her/its", used when the clause has a third person plural subject disjoint from the third person singular possessor of the noun.

Areal

The areal means "it's" and is used when the third person possessor is saliently areal or spatial.

Indefinite

This means "someone's" or "a(n)" and is used primarily with inalienably possessed nouns (mostly body parts and kinship terms), e.g. *'ugan*, "someone's arm, an arm".

The following paradigms are intended to illustrate the various sets of prefixes. The particular forms shown may not be commonly used. In particular, the indefinite prefix is rarely used with nouns that are not inalienably possessed. Thus, the indefinite form *'udayi* "someone's chief" is unlikely to be encountered and sounds rather odd. However, it illustrates the prefix that is used with inalienably possessed nouns of the same class, such as "arm". *'ugan* sounds much more natural.

Class 1 - *dayi* "chief"

1s	sdayi	**1p**	nedayi
2s	ndayi	**2p**	nahdayi
3s	budayi	**3p**	hubudayi
Reflexive	dudayi	**DJR**	yudayi
Reciprocal	lhdayi	**PDJR**	hidayi
Areal	whudayi	**Indefinite**	'udayi

Class 2 - *'oosa'* "pail"

1s	se'oosa'	**1p**	ne'oosa'
2s	nye'oosa'	**2p**	nawhe'oosa'
3s	be'oosa'	**3p**	hube'oosa'
Reflexive	de'oosa'	**DJR**	ye'oosa'
Reciprocal	lhe'oosa'	**PDJR**	huye'oosa'
Areal	whe'oosa'	**Indefinite**	'e'oosa'

Class 3 - *oozi* "name"

1s	soozi	**1p**	neoozi
2s	nyoozi	**2p**	nahoozi
3s	boozi	**3p**	huboozi
Reflexive	doozi	**DJR**	yoozi
Reciprocal	lhoozi	**PDJR**	huyoozi
Areal	hoozi	**Indefinite**	'oozi

Class 4 - *ulhtus* "sister"

1s	sulhtus	**1p**	nelhtus
2s	nyulhtus	**2p**	nawhulhtus
3s	bulhtus	**3p**	hubulhtus
Reflexive	dulhtus	**DJR**	yulhtus
Reciprocal	lhulhtus	**PDJR**	huyulhtus
Areal	whulhtus	**Indefinite**	'ulhtus

Some nouns have a different stem when possessed from when they unpossessed. For example, we corresponding to *lhi* "dog" we have *slik* "my dog", *bulik* "his or her dog", etc. Finally, the possessed stem of the noun may be listed in the *poss* field.

Some nouns may not be used as separate words without a possessive prefix. Such nouns are said to be *inalienably possessed.* For example, the stem meaning "arm" is *-gan*, which we can see by analyzing such forms as *sgan* "my arm", *ngan* "your arm", *bugan* "his arm" and so forth. *gan,* however, is not a word that can be used by itself. It is necessary to indicate whose arm it is, e.g. *sgan* "my arm". To refer to an arm without indicating whose it is, it is necessary to say *'ugan* "someone's arm".

The great majority of inalienably possessed nouns are body parts such as "arm" and kinship terms such as "sister". There are, however, a few other inalienably possessed nouns, such as *-k'ah* "tracks, footprints". When a noun is inalienably possessed, it is entered with a preceding hyphen. Thus, "arm" is entered as *-gan.* Such hyphens are ignored in alphabetization with the exception that forms with leading hyphens are listed after otherwise identical forms without leading hyphens.

Verbs

The Carrier verb is based upon a *stem*, which is invariably the last syllable of the verb. This is the part that contributes the basic meaning of the verb. For example, the stem of *nuske* "I am going around by boat" is *ke*. The same stem is seen in related forms, such as *nuts'uke* "we are going around by boat". We may associate the meaning "go by boat" with the stem *ke*.

The stem may change according to the tense and mode of the verb, whether it is affirmative or negative, and the aspect. For example, for some speakers, "I am going to go around by boat" is *nuteskelh*. Here the stem is *kelh*. The corresponding negative, meaning "I am not going to go around by boat", is *nuchatuzeske*. Here the stem is *ke*. To say "we customarily go around by boat" we say *nuts'ukih*. Here the stem is *kih*. The extent to which verb stems change with tense, mode, negation, and aspect varies from speaker to speaker.

The various stems may be associated with an abstract *root*. For example, we may say that the root meaning "to go around by boat" is *ke*. At least historically, the various stems were derived from a root by adding suffixes. For example, the future affirmative stem *kelh* is derived from the root ke by adding the suffix *lh*. The system for deriving stems from roots is complicated, and it is not entirely clear whether it is still functioning today as it once did. At the very least, the root is a useful name for a set of related stems.

Different roots may have identical stems. The stem *ket* is one of several stems associated with the root “to buy”, such as the Perfective Affirmative. It is also the Optative Affirmative stem of the root “to slip”. The relationship among roots, stems, and individual verb forms is shown by this diagram.

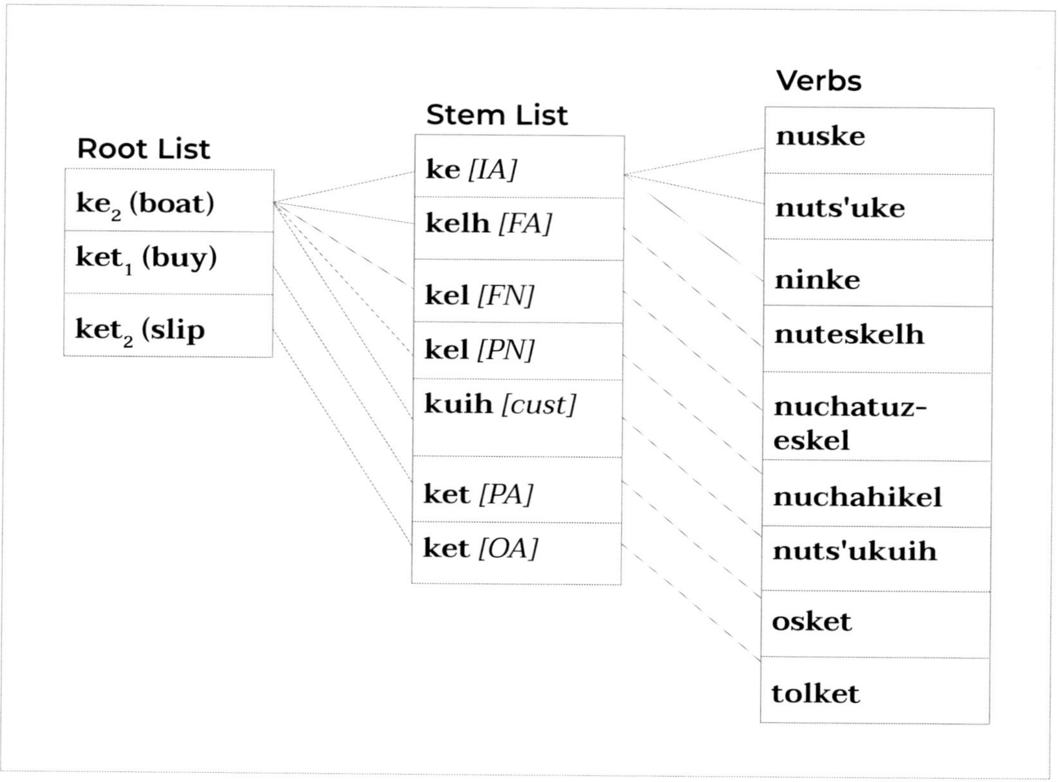

This dictionary contains a *Root List* and a *Stem List.* The Root List lists the roots known for the language and explains their meanings. Since there may be more than one root with the same form, roots have subscript numbers. For example, the root meaning “to go around by boat” is listed as ke_2 since there is another root, listed as ke_1, meaning “for two people to sit or be located”, as in the form *huzke* “they are sitting”.

An unusual feature of Carrier is the existence of *number-restricted verbs.* These are verbs whose subject, if intransitive, or object, if transitive, is restricted to a certain number. For example, paradigm of “to sit” combines three different verbs: one that means “for one to sit”, a second that means “for two to sit”, and a third that means “for three or more to sit”.

To Sit [Imperfective Affirmative]

	singular	dual	plural
1	s̲usda	s̲iduke	ts'udelhts'i
2	s̲inda	s̲ahke	delhuts'i
3	us̲da	huz̲ke	hudelhts'i

In some cases there is one verb for the singular and dual and another for the plural:

To Walk on All Pairs of Limbs [Imperfective Affirmative]

	singular	dual	plural
1	nulhuget	nidulget	nuts'ulh'as
2	nilget	nulhuget	nulh'as
3	nulget	nuhulget	nuhulh'as

With transitive verbs, the number restriction applies to the object rather than the subect. For example, there are two verbs "to kill", one for a singular or dual object, the other for a plural object. Thus we have: *yuz̲elhghi* "he killed him", *hubuz̲elhghi* "he killed them (2)", *hubanghan* "he killed them (3+)".

One of the main divisions of Carrier verbs is based on what valence prefix, if any, they have. The valence prefix is the prefix that immediately precedes the stem. A verb may have no valence prefix, in which case it is referred to as having *0-valence,* or it may have a *lh, l*, or *d.* The valence prefixes frequently combine with the preceding prefix. This is most frequently an inner subject prefix, so there appear to be several different sets of subject markers. In fact, there is just one, with the same subject marker realized differently according as it combines with the different valence prefixes. This is illustrated by the following examples of imperfective affirmative paradigms of verbs with each of the four valences.

0-Valance — To Cook in Hot Liquid or Steam

	singular	dual	plural
1	uslez	idudlez	ts'ulez
2	inlez	ahlez	ahlez
3	ulez	hulez	hulez

lh-Valance — To Roast

	singular	dual	plural
1	ulht'es	idult'es	ts'ulht'es
2	ilht'es	ulht'es	ulht'es
3	ulht'es	hulht'es	hulht'es

l-Valance — To Rest

	singular	dual	plural
1	nalhuyis	naidulyis	nats'ulyis
2	nailyis	nalhuyis	nalhuyis
3	nalyis	nahulyis	nahulyis

d-Valance — To Drink

	singular	dual	plural
1	usnai	idutnai	ts'utnai
2	intnai	utnai	utnai
3	utnai	hutnai	hutnai

These paradigms are the result of the interaction of a single underlying set of subject-marking prefixes with the various valence prefixes. The underlying forms of the subject markers are:

Underlying Forms of Subject Markers

	singular	dual	plural
1	s	idud	ts'
2	in	h	h
3	0	h	h

Before a vowel the first person plural subject prefix /ts'/ is always realized as /ts'/, e.g. *ts'ootakulh* "we will buy", **zootakulh*. However, immediately before a consonant it is optionally realized as /z/ and put into the coda of the syllable, e.g. *ts'utekelh* "we will go by boat" or *uztekelh*. The /z/ allomorph does not occur if the 1p prefix immediately precedes the verb stem: *'uts'u'yi* "we are eating" but *'*uzyi.*

To Dance [Imperfective Affirmative]

	singular	dual	plural
1	nusdaih	nidudaih	ts'unudaih
2	nindaih	nuhdaih	nuhdaih
3	nudaih	hunudaih	hunudaih

The subject markers do not all go in the same position. The first person plural *ts'*- and the third person duo-plural *h*- precede the others. This can be seen in a verb like the following in which another prefix comes in between.

Notice how the prefixes *ts'* and *h* precede the *n* while the others follow it. *ts'*- and *h*- are referred to as *outer subject markers* (s o.); *s*-, *in*-, *idud*-, and *h*- are referred to as *inner subject markers* (S_i).

The subject markers combine with the l-valence prefix in accordance with the following rules. (The forms in which a rule applies are listed in parentheses.)

(a) s + l → lhu (1s)

(b) n + l → l (2s)

(c) d + l → l (1d)

(d) h + l → lhu (2dp)

The subject markers combine with the lh-valence prefix in accordance with the following rules. (The forms in which a rule applies are listed in parentheses.)

(a) s + lh → lh (1s)

(b) n + lh → lh (2s)

(c) h + lh → lh (2dp)

(d) d + lh → l (1d)

The /d/ valence prefix combines with the stem of the verb in a complex fashion, following the same rules as does the /d/ of the first person dual subject prefix /id/, namely the D-Effect rules, discussed below. The situation in which valence /d/ interacts with the subject prefixes is when the verb stem begins with /n/. In this case the valence /d/ is realized as a /t/.

The subject markers combine with the d-valence prefix in accordance with the following rules.

(a) s + t → s (1s)

(b) d + t → t (1d)

(c) h + t → t (2dp)

The valence prefixes listed are abstract underlying forms and may be realized differently in particular forms. Thus, *hanaja* "he came back" is represented as d-ya, the underlying form that is converted into the surface *ja* by the D-Effect rule.

The D-Effect is a set of phonological interactions in Athabaskan languages between two /d/-final prefixes, the /d/-valence prefix and the first person dual subject marker, and the following consonant, which is either the initial consonant of the verb stem or a valence prefix.

The set of interactions between either the final /d/ of the first person dual subject marker or the /d/ valence prefix and the following consonant, is referred to as the D-Effect. These interactions take place only with certain following consonants, as illustrated below.

The D-Effect

C_2	Result	1d	3s	Gloss
'	t'	idut'alh	u'alh	chew
l (__V)	dl	idudlez	ulez	stew
lh	l	idulgi	ulhgui	dry
n	tn	hoodutnih	whunih	be awake
y	j	idujez	nyez	be tall
z	dz	nidudzoot	nuzoot	skate around

The condition in the case of /l/ is that /l/ be followed immediately by a vowel, that is, that it be the initial consonant of the verb stem. Where the /l/ is the valence prefix it will necessarily be followed by the initial consonant of the stem. In this case, the /d/ is invariably deleted without a trace. For example, the 3s of "run around" is *nulgaih,* with *l*-valence. The 1d is *nidulgaih.* Here the /d/ of the 1d subject prefix /id/ disappears.

Where there is no D-Effect from the 1d subject prefix /idud/, the final /d/ disappears without a trace, as illustrated below:

Disappearance of Final /d/ of /idud/

C_2	1d	3s	Gloss
b	nidube	nube	swim around
t	kuniduta	kunuta	look for
d	nidudaih	nudaih	dance
t'	hoonidut'i	hoont'i	be happy
tl	didutle	dutle	be soft
dl	idudlat	udlat	lap
tl'	idutl'oo	utl'oo	weave
ts	idutsah	utsah	cry
ts'	hooduts'it	whuts'it	fib
ch	diduchut	duchut	be soft
j	naidujooh	najooh	re-rack
k	niduke	nuke	go around by boat
g	toonaidugus	toonagus	wash
gh	didughut	dughut	saw wood

Whereas the final /d/ of the first person dual subject prefix is deleted if there is no D-Effect, this is not the case with valence /d/. When valence /d/ triggers no D-Effect, an epenthetic /u/ is inserted following the /d/ and it is retained.

Note that Carrier permits the valence prefix sequences *d-lh* (realized as *l*) and *lh-d* . Listings of non-zero valence prefixes are believed to be accurate; however, some stems listed as having the 0-valence prefix probably have non-overt d-valence prefixes. These are gradually being identified. Similarly, the verb stem shown may not be precisely as it surfaces in the actual form due to the D-Effect in first person dual subject forms. Thus, the verb stem shown for *idujez̲* "we (2) are tall" is *-yez̲,* which becomes *-jez̲* in the first person dual under the influence of the underlying *d* of the subject prefix *idud-.*

Another phonological rule that may obscure the form of the stem is the rule of */ts/-Deaffrication.* This rule deletes the /t/ portion of /ts/ when it is immediately preceded by /s/, leaving a geminate (double) /s/. For example, the stem of "be dirty" is /tsun'/, as in *dutsun* "he is dirty", but in *dussun* "I am dirty" the /s/ of the first person singular subject marker triggers the deletion of the initial /t/ of the stem.

The valence prefixes listed are abstract underlying forms and may be realized differently in particular forms. Thus, *nadust'ai* "I found (for myself)" is represented as *d-'ai,* with the surface *t'ai* the result of the D-Effect. Note that Carrier permits the valence prefix sequences *d-lh* (realized as *l*) and *l-hd.* Listings of non-zero valence prefixes are believed to be accurate; however, some stems listed as having the 0-valence prefix probably have non-overt d-valence prefixes. These are gradually being identified.

Similarly, the verb stem shown may not be precisely as it surfaces in the actual form due to the D-Effect in first person dual subject forms. Thus, the verb stem shown for *'oonaduje* "we (2) picked berries" is *-yeh,* which becomes *-je* in the first person dual under the influence of the underlying *d* of the subject prefix *idud-.*

The terms *d-class, n-class,* and *wh-class* refer to categories in the system of noun classification. These refer roughly to things that are stick-like, round, and saliently two- or three-dimensional respectively. For example, the *generic* form *lhuyul* is the word meaning "it is white" that is appropriate for most objects, such as a book or a cup. The *d-class* form *dulyul* is appropriate for describing a cane or a tree. The *n-class* form *nulyul* is appropriate for describing someone's face or a ball. The *wh-class* form *whulyul* is appropriate for describing a house. In a number of constructions *wh-class* forms are also used in reference to clauses. For example, "I like beaver tail" is *Tsache oonust'i.* Here the object of the verb "like" is the noun phrase *tsache* "beaver tail". "I like eating beaver tail" is *Tsache usyi hoonust'i.* Here the verb "I like" takes the prefix *wh* (realized as *h* before *oo*) because its object is the clause *Tsache usyi* "I eat beaver tail".

An important part of the Carrier lexicon is the set of *classificatory verbs*, used to describe the handling and location of objects of different types. The stems of classificatory verbs convey no information other than the type of object; what action is performed is determined by the choice of derivational prefixes. The following list illustrates the range of verbs that may be derived from the stem for handling two dimensional flexible objects (e.g. shirts).[1]

behanaitelhchoos	he/she is going to take it out
didutelhchoos	he/she is going to hold it up
dughaitelhchoos	he/she is going to hang it up
k'itelhchoos	he/she is going to put it on (the table)
k'unaitelhchoos	he/she is going to put it back on (the table)
sanaitelhchoos	he/she is going to bring it back
yughatelhchoos	he/she is going to give it to her
yughutelhchooz	he/she is going to lend it to her
nutelhchooz	he/she is going to carry it around
'atelhchoos	he/she is going to bury it
tatelhchoos	he/she is going to submerge it
natelhchoos	he/she is going to put it on the ground
yaiyutelhchoos	he/she is going to bring it ashore

The categories into which objects are divided are illustrated below where forms of "he will give me" appropriate to a variety of objects are given. In the first column are listed the abbreviations used for these categories in dictionary entries. Notice that some but not all of these bases permit cross-classification by means of the absolutive prefixes. Where a classificatory verb is also classified via the absolutive prefix system, this information is added to the abbreviation for the classificatory verb category. For example, the abbreviation *lro-d* means that the verb class is "long, rigid object" and that it is further marked by the *d* absolutive prefix.

[1] The two stem variants observed in these examples, *chus* and *chuz*, reflect different aspects. The former is momantaneous, the latter durative.

he will give me

sdo-gen	non-plural default object (chair)	**sghate'alh**
sdo-n	non-plural n-class object (ball)	**sghante'alh**
sdo-d	non-plural d-class object (name)	**sghadute'alh**
sdo-wh	non-plural wh-class object (house)	**sghaoote'alh**
mdo-gen	plural default objects (chairs)	**sghatelelh**
mdo-n	plural n-class objects (balls)	**sghantelelh**
mdo-d	plural d-class objects (names)	**sghadutelelh**
mdo-wh	plural wh-class objects (houses)	**sghaootelelh**
euo-gen	effectively uncountable generic objects (sugar)	**sghatedzih**
euo-n	effectively uncountable n-class objects (berries)	**sghantedzih**
euo-d	effectively uncountable d-class objects (toothpicks)	**sghadutedzih**
lro-gen	long rigid object (canoe)	**sghatetilh**
lro-d	long rigid d-class object (stick)	**sghadutetilh**
body	body (dog)	**sghatelhtelh**
coc	contents of open container (cup of tea)	**sghatekalh**
2df	2-dimensional flexible object (shirt)	**sghatelhchoos**
mushy	mushy stuff (mud)	**sghatetlah**
liquid	liquid (water)	**sghatelhdzah**
hay	hay-like (hay)	**sghadutelhdzah**
fluffy	fluffy stuff (down)	**sghantelhdo**

The choice between the plural and non-plural default verbs is not entirely straightforward. A single object calls for the non-plural verb, three or more for the plural verb. Two objects usually, but not always, take the nonplural verb. The plural default verb is also used for certain single items, such as ropes and fishnets, perhaps because these are considered to consist of multiple coils and meshes.

controlled	npdo	coc	body	2df	fluid	lro	mush	hay	fluff	pdo	euo
locative	npdo	coc	body	2df	fluid	lro	mush	hay	fluff	pdo	euo
uncontrolled	npdo	coc	body	2df	fluid	lro	mush	hay	fluff	pdo	euo
inherent	npdo	coc	body	2df	fluid	lro	mush	hay	fluff	pdo	euo

In the dictionary, classificatory verbs are indicated by the notation "classv" followed by the abbreviation for the category. In a few cases, the meaning of the verb is sufficiently specialized that the category information adds nothing to the definition. In most cases, however, the category information is important. For example, the forms *sghatelhchoos* and *sghatetilh* are both glossed "he or she is going to give me". The category abbreviations add the information that the first is restricted to two-dimensional flexible obects, the second to long, rigid objects.

Glosses of verbs use certain conventions in order to specify information that is not straightforwardly translated into English. English does not distinguish the dual (referring to two people) from the plural (referring to more than two people), and does not distinguish singular, dual, and plural in the second person, glosses use numbers to specify this information. Thus, "you (1)" means that the form refers to a single person, "we (2)", means that the form refers to the two of us, and "we (3+)" means that the form refers to the three or more of us. A gloss that uses the term "non-plural" means that the form may refer to either one or two but not to three or more.

The abbreviation *uo* stands for "unspecified object" and is used to mark transitive verbs in the form used without an overt noun phrase object. For example, *'usyi* is an "unspecified object" form meaning "I am eating" and is used without a noun phrase indicating what is being eaten. It can form a sentence all by itself. The corresponding specified object form is *usyi.* This cannot form a sentence by itself, but must be used together with a noun phrase indicating what is being eaten, as in the sentence *lhes usyi* "I am eating bread."

The basic paradigm of a Carrier verb consists of eight nine-member subject paradigms. These consist of the imperfective, perfective, future, and optative paradigms, each in the affirmative and negative. Of these, only the future is a true tense. The imperfective is often translatable as a present tense form but strictly speaking it refers to an action that is not yet complete and may be used in reference to the past. Similarly, the perfective is often translatable as a past tense form, but strictly speaking refers to completed action and so may not refer to on-going past action. The optative is roughly comparable to the subjunctive in languages such as French and German. Optative forms only occasionally appear as main verbs (e.g. *nuts'ooke* can mean "let's go around by boat" or *Bil bulh nondaih* "you should dance with Bill") but are frequently found in a variety of more complex constructions (e.g. *noske sulhni* "he told me to go around by boat" *Nghoo doolhjut whuch'a toonaingus.* "Brush your teeth lest they decay".). The optative negative is rarely used: its primary use is to say "let's not do such-and-such".

The following paradigm of the verb "to go in a loop by boat" exemplifies the basic paradigm of a Carrier verb.

IA	**singular**	**dual**	**plural**
1	nuske	niduke	nuts'uke
2	ninke	nahke	nahke
3	nuke	nuhuke	nuhuke

IN	**singular**	**dual**	**plural**
1	nuchasuske	nuchasiduke	nuchats'uske
2	nuchasinke	nuchasahke	nuchasahke
3	nuchaske	nuchahuske	nuchahuske

FA	**singular**	**dual**	**plural**
1	nuteskelh	nutadukelh	nuztekelh
2	nutankelh	nutehkelh	nutehkelh
3	nutekelh	na:tekelh	na:tekelh

FN	singular	dual	plural
1	nuchatuzeske	nuchatuzaduke	nuchazteske
2	nuchatuzanke	nuchatuzehke	nuchatuzehke
3	nuchateske	nucha:teske	nucha:teske

PA	singular	dual	plural
1	nususki	nusiduki	nuts'uzki
2	nusinki	nusahke	nusahke
3	nusuki	nuhuzki	nuhuzki

PN	singular	dual	plural
1	nuchaskel	nuchaidukel	nuchats'ikel
2	nuchainkel	nuchahkel	nuchahkel
3	nuchaikel	nuchahikel	nuchahikel

OA	singular	dual	plural
1	noske	noduke	nuts'ooke
2	nonke	noohke	noohke
3	noke	nuhooke	nuhooke

ON	singular	dual	plural
1	nuchasooske	nuchasoduke	nuchats'ooske
2	nuchasonke	nuchasoohke	nuchasoohke
3	nuchaooske	nuchahooske	nuchahooske

BIOGRAPHIES OF MEMBERS OF THE LANGUAGE COMMITTEE

Emma Baker

I was born in Stella, raised on a farm with four brothers and sisters. I began my education attending the Lejac residential school at age six. I loved knowledge. School was not easy; it was made more difficult to learn by strict teachers. I was determined to get educated, I attended the Vancouver Normal School to become a teacher, and I succeeded in becoming a qualified teacher. In my travels of collecting knowledge, I met Willie Baker, who I married in his home town in New Mexico, there I worked as a nurses aid, and had six children. In my travels, I experience much racism. My husband was killed in a horrific car accident. I made up my mind to come home with my children to attend a Carrier Language Teacher Training and became a qualified language teacher. My greatest desire was to teach our own language, history, culture to my Stellat'enne and whoever else would listen. My language and culture has always been important to me. My greatest desire was to preserve and deliver the dialect of our people.

Josephine Carter

Born in Stella, second daughter of the late Angeline Patrick. Has three children with one stepchild. Worked as teacher's aide and assisted in preserving our language with the late Teddy George. Josephine has kept her language and culture active. Her wealth of knowledge is immeasurable. She has participated in drafting this latest version of our dictionary.

Andrew Casimel

Hiu, I am Andrew Casimel. Lhts'ehyoo Frog Clan, born September 27, 1940, Stughachola Cheslatta, son of Margaret and Charlie Casimel. We lived in Grassy Plains until my mother's territory was flooded out by Alcan in 1952. We were left with no option but to move back to Stella in 1956. I attended Lejac Indian Residential School from 1956 to 1958. I graduated from Domano College in 1962. In 1962/63 took an automotive course in Prince George Vocational School. Josie and I got married July 4, 1964. We moved back from Vancouver and lived in Fort St. James from 1969 to 1979. It was where I had the rare privilege of living among the most gracious and wonderful Yinka Dune of Yekooche, Tlazt'en and Nakazdli whut'enne. The most profound and lasting experience for me was their knowledge of their Yinka dune culture and language. Tube snachailya.

Mary George

I am a Stella'whut'ene elder. In my earlier years I married Albert George, and made my home in Cheslatta. While living there, I served 1 term as the Cheslatta Chief. I am a fluent speaker of my community dialect. I have 7 children and many grandchildren I am proud of. My goal is help preserve our language and culture, even in having many negative experiences in the residential school; I have never surrendered my mother tongue.

George George, Sr.

George George Sr. was born in the village of Nadleh in 1933 to Maxime George from the Owl/Grouse clan and Agnes George (nee Leon) from the Frog clan Son of the late Chief Maxime George who was a Hereditary Chief of Nadleh for thirty-six years. Yatunayeh is Mr. George's traditional name in the balhats system. George attended Lejac Residential school from September 1943 to March 1949. After school, Mr. George worked in the logging industry from 1950 to 1964 and the mining industry from 1965 to 1993. George George Sr. has one son and many grandchildren. Throughout the years George George Sr. served as an elected Nadleh Whut'enne Chief for eight years and an elected Nadleh Whut'enne Council member for 20 years. Over the years, Mr. George volunteered on various committees, the latest being for the Rose Prince Pilgrimage.

Geraldine Gunanoot

I am a Yinka Dene born at Endako BC. I have worked on revitalization for over 25 years. I have worked on many tribal territories being instrumental in building and restoring their language and culture. I have extensive experience in developing books, collaborating with elders, interviewing, recording, translating, compiling information, curriculum development, book layout and publishing. While employed as an Education Coordinator for the Kitselas Band I published a book called *"Sprinkle in Time"* as well as revised a Role Model Handbook used by the School District 82. I also completed a Language Handbook "*Ancient Voice of the Kitselas People*" working with the Kitselas Language Professors. I also revised a Head Start Book for the Kitselas Band. Upon given an opportunity to work in my own community of Stella, I am delighted to be part of revitalization Stellat'enne language dialect through our Nadleh/Stellat'en Dictionary. I commend the Stellat'en fluent speakers, elders, Stellat'en language center team for their hard work and dedication to leave a legacy.

Lillian Louis

My name is Lillian Louis, I am a Stellat'en elder. I was born in a log cabin in Talhghazchek across from John's Island on Francois Lake, my father was working for a sawmill in the Grassy Plains area around the time I was born. We moved to Stella, where we lived in the old hall which my grandfather James Zachary worked to build. I was taken to Lejac residential school at 5 years old by a tall man dressed in all black who now I assume was a priest. The priest told my mom that I would be taken care of good and my mom forced herself to tell me, in Dakelh, that everything would be okay. When they sent me to residential school, I never spoke English, I only spoke Dakelh. I was at Lejac for four years. After my four years in Lejac, I went to school in Grassy Plains where my parents moved. For high school, I was sent to Kamloops to go to residential school there. I trained as a practical nurse for a year. Later, I was elected chief of Stellaquo in from 1976-80 for four years. My parents would always speak Dakelh to us when we were with them. English is only my second language, Dakelh is my first language. I have never forgotten Dakelh, it is important to remember my traditional language. As a fluent speaker, it was important to me to participate in working on this project to pass on our language to future generations.

Eleanor Lowe

I belong to the Grand Trunk Clan/Luksilyoo. My parents are my father Donald Edward inle and have one daughter and two sons, four granddaughters, three grandsons, one great-grandson and two great-granddaughters. With my partner Dennis Patrick I have four stepchildren. I have worked with Nadleh Whut March 2024. Through different language grants, we have made a huge impact in revitalizing and preserving our Dakelh dialect. This would not be possible without the help of our fluent speakers who are our precious knowledge keepers. I love our elders. Nenachalhuya!

Robert Luggi Sr.

Robert Luggi Sr. is of Lhedli T'en and Stellat'en decent. Robert was born in Prince George, B.C. and lived on the Shelley Reserve and at the age of two moved to Stellaquo. When he arrived at Stellaquo Robert only knew how to speak the English language. As a child he had attended the Lejac Residential School for 6 years. Robert has been married since 1963 and he and his wife, Theresa raised their 5 children in Stellaquo. Robert spent many years from the 60's to the 80's involved in Stellaquo recreation with many hours of fundraising for the fast ball and hockey teams. Robert is a member of Tsayoo Beaver clan and inherited his name, Dayoos from his uncle. Robert takes pride in sharing his knowledge, wisdom and fluency of our language and culture.

Adeline Nadon

Adeline (Addie) was born on her dad's farm near the Nadleh Whut'en Reserve close to Fort Fraser, BC on July 23, 1934 to Joachim (Wassa) and Virginia Sutherland. She was one of 7 children; 5 boys and 2 girls. Addie's mother passed away when Addie was only 5 years old and it was left to her and her sister, Terry, to take care of her brothers and father. Thus, began a lifetime of being a caregiver of which Addie became an expert. Although she never attended school, Addie learned how to read, write and do arithmetic from her dad. She was also fluent in her Dakelh language. She became a very avid reader and enjoyed keeping up on current events. Addie was born into the Bear clan, but when her father passed away, she proudly received his name,

Wiluknan, and being adopted into the Frog clan of which her father was a part.

Mom made it her business to make sure her children knew their heritage, so she would tell them stories of her father's Scottish origins and her mother's Dakelh origins. She made sure her children knew they were part of the Bear Clan and would bring Adeline to as many potlatches as possible. She made an effort to speak her language with her children; Antoine and Joseph who also spent many years growing up on the farm Addie was born. Addie enjoyed telling us stories of her hunting and fishing exploits around Nadleh and Fort Fraser.

At the age of 24 she met and married Louis Nadon. Addie was proud of her 4 children; 2 sons Antoine (Tony) and Joseph (Rodney) and 2 daughters, Adeline and Elizabeth. They spent the first 5 years of their marriage in Burns Lake, BC and moved to Terrace, BC in 1964 where she would spend the next 54 years of her life. While Louis worked as a logger, Addie worked as a housekeeper which entailed walking about 4 miles a day with baby Elizabeth on her back. Several years later, she worked as a cashier at the Westend store and then as a short-order cook at the Co-op in Terrace, BC. It was through these positions, that Addie became friends with so many. She made friends easily and found it very interesting to meet new people and find out all she could about them.

She found interest in life and was noted for her phenomenal memory even in her final years. Addie was passionate about her family (near and far) and she also loved to sew, knit, play the guitar (self-taught) and fishing. Addie also enjoyed playing cards, cribbage and Skip-Bo with her grandsons and friends. She enjoyed driving across the prairies to visit friends and family and always said how much she loved the big sky of the prairies. Addie dreamed of visiting the Bay of Fundy to see the highest tides, but sadly that dream was never realized.

Another passion was gardening; as such Spring was her favourite time of year because that's when she could start planning her gardens. She enjoyed discussing with her neighbours what she was going to plant and found even more joy in choosing the seeds to buy.

Another love she had was feeding the birds, so she made sure she could see the birdfeeders from her couch. Another joy of Addie's was watching the Canucks and comparing notes with her grandson, Jules, over the phone. It was this love that won her the opportunity to meet Ron McLean and Don Cherry the year Terrace won the Kraft Hockey challenge.

In her last few years, Addie had the privilege of participating in the creation of a Central Carrier/English dictionary along with several other elders from the Stellaquo and Nadleh reserves. She really enjoyed those days and took great pride in being able to share her knowledge and stories with them. Unfortunately, that had to stop because she had a stroke

in May 2018 and although only her speech was partially affected by the stroke, Addie was not able to drive the long distances any longer. She was 85 when she passed away.

Addie was pre-deceased by her parents Joachim and Virginia; her siblings Arthur, Alfred, Robert, Richard, Duncan and Therese (Terry); her husband, Louis; her son, Antoine; and son-in-law, Kim.

Addie will be lovingly remembered by her son, Joseph; daughters, Adeline (Steven) Ignas and Elizabeth Duerr; granddaughters Veronica and Geraldine ; grandsons Christopher and Andy; Jules (Sharon), Frazier and Shaun; 13 great-grandchildren; 2 great-great-grandchildren and numerous friends across the country.

Roy Nooski

My name is Roy Augusta Williams Nooski, I am 76 years old, from the Luksilyoo Clan. I was raised by my Grandfather Edward William. My aba inle is Donald Edward Nooski and sloo inle is Florence (née Augusta) and my grandmother is Elizabeth (née Sutherland). I went to Lejac Residential School in 1952, then we were bussed to St Joseph's in Vanderhoof. In 1959 when I reached the age of 16, I started working at a Sawmill for 90 cents an hour. For many years I worked in the lumber industry, worked as a truck driver, and for CNR for many years.

I finished my Grade 12 at Malaspina College, went on to receive my Alcohol & Drug Counsellor Certificate at Lower Fraser Valley University and volunteered at D/A Worker at Duncan, BC. I started working as Nadleh Band A & D workers, then for 5 years, I worked for Carrier Sekani Family Services as Drug & Alcohol Counsellor. Now, I am on-call Elder Cultural Teacher with CNC and UNBC.

I married Terry Thomas from Duncan, BC. We have 6 children together and we have 13 grandchildren. I served 2 terms for Nadleh Whut'en Band Council from 2013 to 2015, from 2018 to 2020, and 2022 – 2024, and I volunteer cutting grass for the Rose Prince Pilgrimage.

For my language, I help teach Nadleh Language Nest, a mentor for the Nadleh Silent Speaker Program, and I am also part of the Nadleh Stella Dictionary Working Group Committee. I am willing to teach anyone who is interested in learning our Nadleh Dakelh Dialect.

My philosophy is to help as many people as I can, giving out moose meat, bear meat, salmon, and trout. I make herbal medicine for people who are sick from cancer, kidney, TB, and other medicine for people who call from Prince Rupert, Terrace, Alberta, Nanaimo, Saskatchewan, Manitoba, and from the local area.

Dennis Patrick

Dennis Patrick is of Stellat'en descent. Son of Angeline Patrick and belong to the Caribou Clan. Attended Lejac Residential school. Went on to FLESS then NVIT after which attended UBC. Several certificates later were acknowledged for outstanding achievement in Education. Participated in creating Language app. Assisted in mapping out language dictionary for Nadleh and Stella t'enne. Many people helped in my road in healing and managing good decisions. Part of that includes Eleanor Nooski strong cultural guidance.

Bill Poser

Bill Poser is a linguist. He received his Ph.D. in Linguistics from the Massachusetts Institute of Technology in 1985. He has worked on Dakelh since 1992 and has produced documentation and teaching materials for several dialects. He designed the database and software that underlies this dictionary, wrote the grammar sketch, and collected part of the vocabulary included.

Virginia Reynolds
Luksilyoo Clan/Caribou

Virginia was born on the Stellaquo Reserve, 70 years ago to Angeline Patrick and did not learn English until after attending Lejac Residential School from the age of seven to the age of fifteen. After upgrading and completing several forms of training through work Virginia received her Chef Level One training in Burns Lake, BC and focused on catering. Having a close relationship with her mother helped Virginia keep her language by helping her mother with traditional activities and making herbal medicine. Virginia has one son, four daughters and eleven grandchildren. In her work as a fluent language speaker Virginia hopes that everyone will learn the language once again.

Virginia Reynolds
Language Work History

- Fluent Speaker – worked as language teacher with Nadleh Language Nest Program, visiting homes of the children, and telling Dakelh Stories.
- Nadleh Stella Dictionary Working Group Committee
- Audio and translation of Dakelh Words, Phrases & Stories for Nadleh Stella First Voices.
- Translation documents for Stella Language Center

REFERENCES

Borror, Donald J. & Richard E. White (1970) *A Field Guide to Insects: America North of Mexico.* Boston: Houghton Mifflin.

Brayshaw, T. Christopher (1996) *Trees and Shrubs of British Columbia.* Vancouver: UBC Press. Royal British Columbia Museum Handbook series.

Brough, Sherman G. (1990) *Wild Trees of British Columbia.* Vancouver: Pacific Educational Press.

Corkran, Charlotte C. & Chris Thoms (1996) *Amphibians of Oregon, Washington and British Columbia.* Vancouver: Lone Pine Publishing.

Cowan, Ian McTaggart & Charles J. Guiguet (1956) *The Mammals of British Columbia.* Victoria: British Columbia Provincial Museum.

Dunn, John Asher (1976) *A Practical Dictionary of the Coast Tsimshian Language.* Ottawa: National Museums of Canada. Canadian Ethnology Service paper no. 42.

Mackinnon, Andy, Pojar, Jim, and Ray Coupé (eds.) (1992) *Plants of Northern British Columbia.* Vancouver: Lone Pine Publishing.

Morice, Adrien-Gabriel (1888-1889) "The Western Denes Their Manners and Customs" *Proceedings of the Canadian Institute.* Toronto.

Morice, Adrien-Gabriel (1892) "Les Dénés occidentaux" *Missions de la Congrégation des Missionaires Oblats de Marie-Immaculée* 22:249-327.

Morice, Adrien-Gabriel (1893) "Notes Archaeological, Industrial, and Socio-logical on the Western Denes, with an Ethnographic Sketch of the Same" Toronto: Copp, Clark.

Morice, Adrien-Gabriel (1932) *The Carrier Language.* Mödling bei Wien, St. Gabriel, Austria: Verlag der Internationalen Zeitschrift "Anthropos".

Morice, Adrien-Gabriel (1933) "Carrier Onomatology," *American Anthropologist* n.s.35.632-658.

Describes and analyzes personal and place names.

Prunet, Jean-François (1990) "The Origin and Interpretation of French Loans in Carrier," *International Journal of American Linguistics* 56.4.484-502.

Schalkwijk-Barendsen, Helene M. E. (1991) *Mushrooms of Western Canada.* Vancouver: Lone Pine Publishing.

Schofield, W. B. (1992) *Some Common Mosses of British Columbia.* Victo-ria: Royal British Columbia Museum. Second edition.

Scott, Shirley L. (ed.) (1987) *Field Guide to the Birds of North America.* Washington, D.C.: National Geographic Society. Second edition

Tarpent, Marie-Lucie (ed.) (1986) *Hańiimagoońisgum Algaxhl Nisga'a [Nis-gha Phrase Dictionary].* New Aiyansh, British Columbia: School District 92.

Manufactured by Amazon.ca
Acheson, AB